Integrated Pest Management

Integrated Pest Management

David Dent

School of Pure and Applied Biology, University of Wales, Cardiff, UK

with contributions by

N.C. Elliott
USDA, Agricultural Research Service, USA

J.A. Farrell
New Zealand Institute for Crop and Food Research, New Zealand

A.P. Gutierrez
University of California, USA

J.C. van Lenteren
Wageningen Agricultural University, The Netherlands

M.P. Walton
University of Wales, UK

S.D. Wratten
Lincoln University, New Zealand

CHAPMAN & HALL

London · Glasgow · Weinheim · New York · Tokyo · Melbourne · Madras

Published by Chapman & Hall, 2–6 Boundary Row, London SE1 8HN, UK

Chapman & Hall, 2–6 Boundary Row, London SE1 8HN, UK

Blackie Academic & Professional, Wester Cleddens Road, Bishopbriggs, Glasgow G64 2NZ, UK

Chapman & Hall GmbH, Pappelallee 3, 69469 Weinheim, Germany

Chapman & Hall USA, 115 Fifth Avenue, New York NY 10003, USA

Chapman & Hall Japan, ITP-Japan, Kyowa Building, 3F, 2–2–1 Hirakawacho, Chiyoda-ku, Tokyo 102, Japan

Chapman & Hall Australia, Thomas Nelson Australia, 102 Dodds Street, South Melbourne, Victoria 3205, Australia

Chapman & Hall India, R. Seshadri, 32 Second Main Road, CIT East, Madras 600 035, India

First edition 1995

Typeset in 10/12 Paladium by Colset Pte. Ltd., Singapore
Printed in England by Clays Ltd, St Ives PLC.

ISBN 0 412 57370 9

A catalogue record for this book is available from the British Library

Library of Congress Catalog Card Number: 95-68342

∞ Printed on permanent acid-free text paper, manufactured in accordance with ANSI/NISO Z39.48-1992 and ANSI/NISO Z39.48-1984 (Permanence of Paper).

Contents

Contributors

D.R. Dent
School of Pure and Applied Biology, University of Wales, PO Box 915, Cardiff, CF1 3TL, UK

N.C. Elliott
USDA, Agricultural Research Service, Southern Plains Area, 1301 North Western Street, Stillwater, Oklahoma 74075, USA

J.A. Farrell
New Zealand Institute for Crop and Food Research Ltd, Canterbury Agriculture and Science Centre, Private Bag 4704, Gerald Street, Lincoln, Canterbury, New Zealand

A.P. Gutierrez
Department of Environmental Science, Policy and Management – Biological Control, University of California, Berkeley, CA 94720, USA

J.C. van Lenteren
Laboratory of Entomology, Wageningen Agricultural University, PO Box 8031, 6700 EH Wageningen, The Netherlands

M.P. Walton
School of Pure and Applied Biology, University of Wales, PO Box 915, Cardiff, CF1 3TL, UK

S.D. Wratten
Department of Entomology and Animal Ecology, PO Box 84, Lincoln University, Canterbury, New Zealand

Preface

Integrated pest management (IPM) has passed through a period of rapid expansion with the development of a whole host of agents and measures available for pest control. Over more recent years there seems to be a move towards consolidation of principles, approaches and practices in IPM. Part of this process of consolidation has, in my mind, focused attention, not on the detail of individual control measures, but on the more practical needs of moving the ideas and the techniques to the field and to dealing with the problems of IPM implementation. For me this has involved asking the simple question 'How do you develop an IPM system?'. From this simple question arise a number of others: 'What principles are involved in the design of an IPM programme; How do you integrate the work of a multidisciplinary research team'; 'How do you organize and run a programme of research and development'; and 'Once you have developed an IPM system how do you implement it in the field'?. These are but a few of the questions which come to mind. If you look through the literature there are few publications which even begin to address these questions, yet it seems obvious to me that being able to answer them is crucial to the successful development of IPM systems. The purpose of this book is to provide a practical guide to the principles, approaches and techniques involved in developing an IPM programme. It emphasizes the need for and the means of good research management and the integration of control measures to produce complete IPM programmes which fit the farmers' needs and farming systems.

The book has been written using the term pest in its broadest sense (including insects, weeds, pathogens, rodents, etc.) for researchers from all relevant scientific and economic disciplines and the extension workers involved in the development of IPM systems. The examples used in the text cover a whole gamut of cropping systems and pest problems from a wide chronological range.

The philosophy of IPM has been around now for over 20 years, but there are too few working examples of IPM systems in the field. There are many reasons for this, but one of the most striking failures has been the inability to produce integrated pest management programmes and systems. A great deal of very good research is carried out in individual specialist subjects to evaluate particular control measures, but rarely are these measures assessed in tandem or as part of an integrated research programme. This is probably one of the greatest areas for improvement in IPM, but it requires a greater emphasis on research management than is currently the case. I hope that this book will help fulfil this need by looking at the problem of IPM from a management perspective.

David Dent
August 1994

CHAPTER 1

Introduction

D.R. Dent

Agriculture continues to change in response to the needs of society. These needs have contributed to a gradual intensification of agriculture over a number of centuries. During the 1900s this process of intensification, at least in the Western world, has gathered momentum through the enlargement of fields, genetic uniformity of crops (species, cultivar and genotype levels), increases in plant density, harvest index, specialization and mechanization, and an increase in the international exchange of infected material (seed, plant and soil) (Zadoks, 1993). Agricultural intensification has been possible and has evolved from the contributions of all branches of agricultural science, including crop protection. The developments in crop protection have been driven by the changing pest problems faced by the farmers, the options available to them and their changing cash and labour requirements (Norton, 1993). These developments have though, since the 1950s, created some of their own problems, including those of pesticide resistance, secondary pest outbreaks and hostplant resistance breakdown. Also, the environmental pollution and hazards posed by some aspects of high input agriculture have led scientists to seek more sustainable alternatives (Chadwick and Marsh, 1993). The solution which modern crop protection scientists have devised to address these problems is that of integrated pest management (Gutierrez, 1987; Zadoks, 1993).

Integrated pest management (IPM) is considered here to mean: a pest management system that in the socioeconomic context of farming systems, the associated environment and the population dynamics of the pest species, utilizes all suitable techniques in as compatible a manner as possible and maintains the pest population levels below those causing economic injury (Smith and Reynolds, 1966; Dent, 1991). There are two components of this definition which require further elaboration; a definition of the term pest and a qualification of the use of 'system' in relation to pest management. A pest is considered here in its general sense, to mean birds, rodents,

Integrated Pest Management Edited by David Dent. Published in 1995 by Chapman & Hall, London. ISBN 0 412 57370 9

mites, insects, nematodes, fungi, bacteria, viruses and vectors. All so-called pest organisms have their own natural place in the world's ecosystem and any organism may potentially develop into a pest (Zadoks, 1993).

Spedding (1988) defined a system as a group of interacting components, operating together for a common purpose, capable of reacting as a whole to external stimuli. A system is unaffected by its own outputs and has a specified boundary based on the inclusion of all significant feedbacks. Four types of system are generally acknowledged in agriculture. They are: agroecosystems, farming systems, cropping systems and pathosystems (Rabbinge and de Wit, 1989) (Figure 1.1). In this hierarchy of systems (Ruthenberg, 1980) a system at one level is a sub-system of the system level above it. A system may consist of a number of sub-systems. A sub-system features a degree of independence of the whole system such that they can be studied separately and results of such studies incorporated into a model of the whole (Spedding, 1988). The pathosystem, which refers to the interaction of host and the pest parasite populations, is part of a cropping system that includes crop protection among its agronomic and other activities (Rabbinge and de Wit, 1989). IPM is a sub-system of the cropping system.

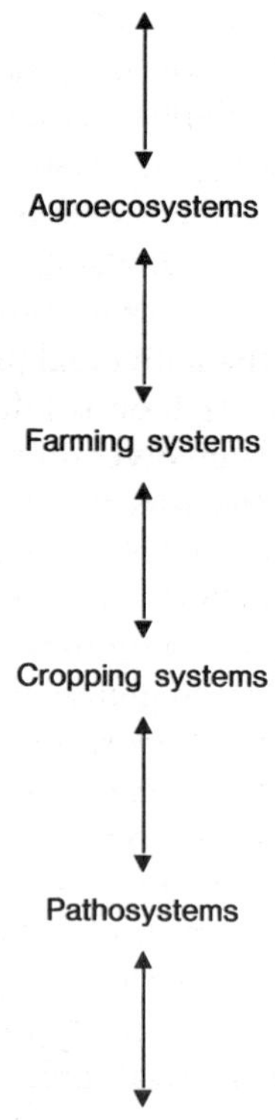

Figure 1.1 A hierarchy of systems in agriculture.

It is the operating system used by farmers in order to manage the control of crop pests. It has a degree of independence, it can be studied in isolation of the cropping system, has its own inputs and has the same output as the main system (i.e. yield) but relates to only some of the components and thus, to only some of the inputs (Spedding, 1988).

An IPM system is considered here to be distinct from an IPM programme which is considered to mean the research and development phase that culminates in the development of an IPM system. In addition it is worth noting that the overall goal of an IPM R & D programme will be the development of an IPM system and that this goal is achieved through the implementation of a programme strategy. An IPM system has a goal of providing the farmer with an economic and appropriate means of controlling crop pests. An IPM system will be achieved through implementation of the system strategy, often referred to as a delivery system. These terms are defined to avoid confusion and to distinguish more effectively between two distinct phases in IPM development and its utilization, i.e. an IPM programme and an IPM system.

Systems and systems analysis have figured prominently in the approaches adopted to the development of IPM. This has largely been due to the problem-based approaches which are so useful in facilitating work between disciplines (Heong, 1985; Norton and Mumford, 1993). The development of IPM presents problems which are dependent for their solution on the ability of participants to bring together work from a number of disciplines. Swanson (1979) considered problem solving from a systems perspective by using the four element loop depicted in Figure 1.2. A particular research activity may begin at any of the four elements, and equally may finish at any of the four. Each of the pathways between the elements is bidirectional reflecting the wide diversity in the individual processes involved in problem solving. The ideal problem-solving loop is I–II–III–IV, i.e. there is concern for the whole range of activities involved in modelling and model solving. Another loop I–II–IV–I omits the scientific model. In the use of this loop there is some confusion between conceptualization and modelling, and an attempt may be made to substitute a conceptual model for a scientific model. However, no matter how rich in detail the conceptual model might be, it does not substitute for a validated scientific model. Omission of the scientific model implies the loss of opportunity to develop the logical structure of the conceptual model in a systematic way and perform the important validation process. Thus, the approach weakens any potential implementation (Swanson, 1979). The problem and decision-making-based approaches of Norton and colleagues (e.g. Norton and Mumford, 1993) go a long way in IPM to bridging the gaps between the loops I–II–IV–I and II–III–IV–II and provide the decision tools required in problem specification, analysis and delivery. Such approaches are invaluable and figure largely throughout this book. However, the decision-based approaches often assume and require the availability of extensive information, the premise being that the

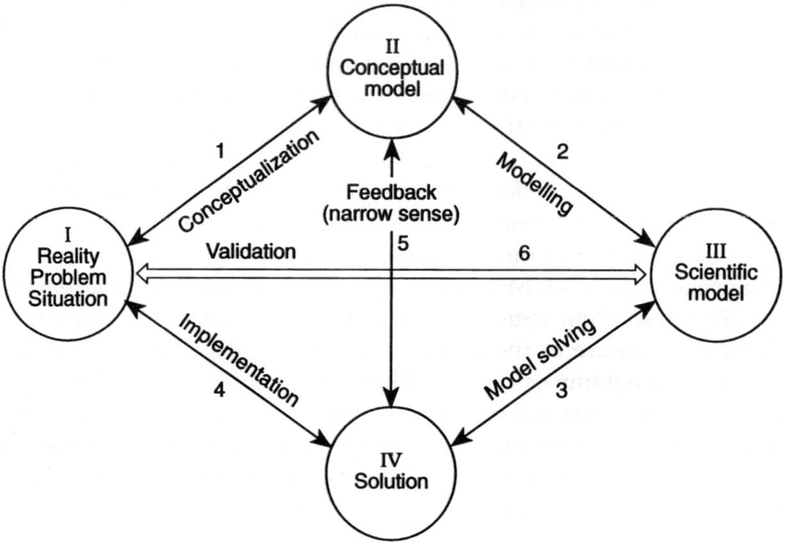

Figure 1.2 A systems view of problem solving. (After Swanson, 1979; see text for explanation).

technology and relevant data exist to enable the appropriate decision tools to be used to identify the most appropriate means for utilization of the technologies. In this context R & D are often seen as a means of meeting a perceived future demand or an information gap (e.g. Norton and Mumford, 1982). There are now a number of cropping and pest management systems where this approach is realistic and wholly appropriate; data have been generated over many years for the whole pest complex, their biology and ecology, and it is possible through appropriate analysis to identify key gaps in our knowledge and understanding, thus providing a direction for R & D. The emphasis in such situations will be on the need to develop IPM systems. For other patho- and cropping systems however, where there has not been a long history of R & D, the information and research gaps will be enormous. In these situations the emphasis will need to be placed on the development of IPM R & D programmes. The direction they take will be influenced by the availability of basic data. Their success will be as much dependent on proper research management as on the use of appropriate decision tools. This combined use of decision tools and research management in the development of R & D programmes leading to the development of appropriate IPM systems, has been too long overlooked.

IPM research tends to be either pest or control centred in its approach. Each aspect of a pest's taxonomy, biology and ecology is considered in depth by specialists from different disciplines, whether entomologists, pathologists, nematologists or weed scientists. Similarly, research on various control measures is usually carried out in isolation by attendant specialists, whether they are involved in pesticide, cultural control, host resistance research or whatever. One of IPM's failings in the past has been the inability to integrate the work of these different disciplines in order to develop IPM programmes relevant to the farmers' systems needs. Research programmes may go under the guise of 'multidisciplinary' but rarely are they 'interdisciplinary', despite the fact that interdisciplinary research is the most effective way of bringing about integration of pest control measures at the programme level. Multidisciplinary projects simply require everyone to do their 'own thing' together, with little or no necessity for any one participant to be aware of the other participants' work (Petrie, 1976). Interdisciplinary programmes on the other hand, may require modification of the disciplinary specialists' contribution while the enquiry is proceeding and the development of an integrated solution to a problem. Techniques exist which can help promote the management of interdisciplinary activity in pest management programmes, but they have rarely been mentioned in previous texts on IPM.

The first five chapters of the book are ordered according to the sequence in which they would be addressed in the development of an IPM programme where the aim is to develop an IPM system for farmers. The two chapters following this one deal with the basic principles of ecology, genetics and economics and the control measures which underlie the development of any IPM programme and/or system. These are the basic building blocks of IPM and a good knowledge of these individual components and an understanding of how they interact are essential for anyone interested in IPM research and systems development. Chapter 4 considers the different components involved in the initial stages of programme development. If you cannot define a problem adequately then you will not find appropriate solutions. This chapter considers the trigger events which lead to the initiation of a programme, the information needs during this preliminary stage, in terms of the status of research, socioeconomic, historical, ecological and appropriate yield loss data, and the goals which are set on the basis of all of this information.

Chapter 5 expands and develops the ideas of Chapter 2 on programme planning and management. This chapter sets out how to go about the process of devising and running an IPM programme, starting with the crucial first step of defining the type of system that the programme will produce, something often neglected in IPM R & D (other than perhaps in the most general and ambiguous terms). This leads into a consideration of organizational structures, planning and monitoring, management, leadership and team communication; subjects normally left unconsidered, but ones that will inevitably determine the success or otherwise of any R & D programme. Many of these approaches are utilized within existing programmes, often in a subconscious or implicit way. This chapter however, aims to make the ideas and techniques more explicit and hence, accessible to those wishing to improve the ways in which they plan and manage IPM programmes. It places particular emphasis on the need for integration of different disciplines and techniques for achieving this.

Chapters 6 and 7 consider two approaches that will tend to gain significance as the programme or system development gets under way. The chapter on systems analysis considers the ecological and economic models used in the analysis of pest management systems, emphasizing the role of decision-making tools. Chapter 7 deals with the issue of research paradigms; science is tending to become more and more specialized and research carried out by very specialized groups who share certain assumptions, beliefs and approaches. This chapter seeks to highlight this reductionist tendency and the effect it can have on the development of IPM programmes. The situation is illustrated with a review of some of the paradigms which exist within a few of the specialist disciplines within IPM.

Chapter 8 considers the practicalities of delivering or implementing an IPM system. Everything in developing an IPM programme and the system should be geared to this end. This chapter considers how this can be achieved and the problems that may be faced.

The final four chapters provide an appraisal of IPM in olives, cotton, glasshouses and cereals. The chapter on IPM in olives considers the process and problems of developing an IPM R & D programme in that crop, while the next chapter more specifically considers the role of modelling in the development of a cotton IPM programme. The chapters on IPM in glasshouses and cereals provide excellent examples of the incremental way in which IPM tends to develop in practice. These last four chapters provide practical illustrations of IPM programmes and systems development that seek to complement the more theoretical and guide-line oriented chapters in the first half of the book.

IPM is often regarded as just a combination of control measures that can be achieved through uncoordinated effort from within each specialist discipline. However, to achieve the integration so often identified as a goal of IPM research a different perspective is really required, one which focuses on the subject as a whole and acknowledges the dynamics of programme development and the need to adopt the processes and techniques required to achieve integration. This concept of integration has now been around for many years, and after an initial expansionist phase of research and development which saw a great increase in the availability of different control measures, IPM research is passing into a consolidation phase where consideration can be given to these more holistic needs of proper research management. It is hoped that the following will provide a basis for this next great advance in IPM development.

REFERENCES

Chadwick, D.J. and Marsh (1993) *Crop Protection and Sustainable Agriculture*, John Wiley & Sons, Chichester.

Dent, D.R. (1991) *Insect Pest Management*, CAB International, Wallingford.

Gutierrez, A.P. (1987) Systems Analysis in Integrated Pest Management, in *Protection intégrée: quo vadis? – "PARASITIS 86"*, (ed. V. Delucchi), PARASITIS, Geneva, pp. 71–82.

Heong, K.L. (1985) Systems Analysis in Solving Pest Management Problems, in *Integrated Pest Management in Malaysia*, (eds B.S. Lee, W.H. Loke and K.L. Heong), Malaysian Plant Protection Society, Kuala Lumpur, pp. 133–49.

Norton, G.A. (1993) Agricultural development paths management: a pragmatic view of sustainability, in *Crop Protection and Sustainable Agriculture*, (eds D.J. Chadwick and J. Marsh), John Wiley & Sons, Chichester, pp. 100–15.

Norton, G. and Mumford, J.D. (1982) Information gaps in pest management, in *Proceedings of the International Conference on Plant Protection in the Tropics, Kuala Lumpur, Malaysia*, 1–4 March, 1982, pp. 589–97.

Norton, G.A. and Mumford, J.D. (1993) *Decision Tools for Pest Management*, CAB International, Oxford.

Petrie, H.G. (1976) Do You See What I See? The Epistemology of Interdisciplinary Inquiry. *Journal of Aesthetic Education*, **10**, 29–43.

Rabbinge, R. and de Wit, C.T. (1989) Systems, models and simulation, in *Simulation and System Management in Crop Protection*, (eds R. Rabbinge, S.A. Ward and H.H. van Laar), Pudoc, Wageningen, pp. 3–15.

Ruthenberg, H. (1980) *Farming Systems in the Tropics*, 3rd edn, Oxford University Press, Oxford.

Smith, R.F. and Reynolds, H.T. (1966) Principles, definitions and scope of integrated pest control, in *Proceedings of FAO (United Nations Food and Agriculture Organisation) Symposium on Integrated Pest Control*, vol. 1, pp. 11–17.

Spedding, C.R.W. (1988) *An Introduction to Agricultural Systems*, 2nd edn, Elsevier Applied Science, London.

Swanson, E.R. (1979) Working with Other Disciplines. *American Journal of Agricultural Economics*, **61**(5), 849–59.

Zadoks, J.C. (1993) Crop protection: why and how, in *Crop Protection and Sustainable Agriculture*, (eds D.J. Chadwick and J. Marsh), John Wiley & Sons, Chichester, pp. 48–60.

CHAPTER 2

Principles of integrated pest management

D.R. Dent

2.1 INTRODUCTION

IPM seeks to integrate multidisciplinary methodologies to develop pest management strategies that are practical, effective, economical and protective of both public health and the environment (Smith *et al.*, 1976). The different specialist methodologies used to develop IPM programmes are derived from the disciplines of applied entomology, plant pathology, weed science and nematology. They include the diverse range of control measures that have been developed and the principles advocated for their use, within each discipline. Under-pinning the work of each of these functional disciplines however, are the more fundamental scientific principles of ecology, population genetics, socioeconomics and crop husbandry (Figure 2.1). These four disciplines together provide the theoretical framework on which so much of IPM is based. An understanding of the principles contributing to this framework is an essential first step in any consideration of IPM, and is especially pertinent to the needs and perspectives of an IPM programme or systems manager.

2.2 PRINCIPLES OF CROP HUSBANDRY

Crop husbandry is the practice of growing and harvesting crops, the main objective being to produce good, healthy crops as economically as possible without impoverishing the land (Lockhart and Wiseman, 1978). Hence, crop husbandry includes the operations of land preparation, planting, maintenance of plant health, harvesting and storage. The maintenance of plant health involves the provision of the best possible conditions for plant

Integrated Pest Management Edited by David Dent. Published in 1995 by Chapman & Hall, London. ISBN 0 412 57370 9

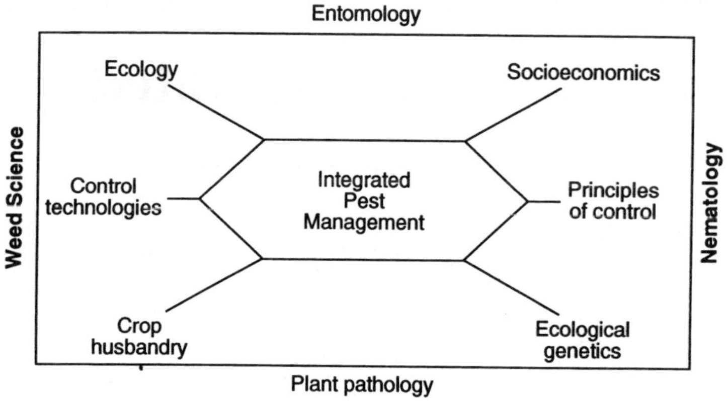

Figure 2.1 A schematic representation of the relationship between the more functional and fundamental disciplines underpinning IPM.

growth through appropriate use of fertilizers, irrigation and crop protection. In this context, pest management represents only one of many practices utilized by farmers in the successful production of healthy crops. While it is not possible to consider here the whole range of principles involved in good crop husbandry it is appropriate to briefly outline those principles and practices that directly interact with pest management. Of particular interest are the principles involved in the use of fertilizers, cultivation and crop rotations, since each of these practices can have a direct impact on pest management problems.

The growth of crops results in the continued removal of nutrients from the soil. If the soil is not to become impoverished then nutrients must be put back to replace those used in crop growth. When adding nutrients to the soil it is usual to apply more than that required by each crop because some nutrients are lost through drainage (e.g. nitrogen) while others will become fixed in the soil and unavailable to the plant (e.g. potash) (Lockhart and Wiseman, 1978). Nitrogen, a key requirement of plants for growth, can be supplied by application of fertilizers, organic matter or through use of leguminous crops which fix nitrogen in their root nodules. Phosphates and potash are supplied by soil minerals or application or organic manures and fertilizers. The farmer decides at the start of each season which of these amendments to apply to ensure that the necessary soil nutrients are available for the production of the next crop.

Fertilizers are primarily applied to maintain high yields of a crop (Cooke, 1982), but their use may have a direct effect on levels of pest attack. This effect may be positive or negative. A healthy crop may have both a greater tolerance to pest attack and the ability to compensate for damage. In addition a healthy crop will have rapid growth, thereby shortening the time the susceptible plant stage is available (Coaker, 1987). On the other hand, the use of fertilizers can also change the physiology of the plant making it more 'attractive' as a host for an insect pest, and thereby increasing its chances of reproduction and survival. If pest populations increase on these healthier crops then the benefits of the increased yields due to fertilizers will need to be offset by the increased pest population and the potential need for their control (van Emden *et al.*, 1969).

The cultivation and tillage of soil involves operations which attempt to alter the soil structure for the purpose of providing a suitable seed-bed in which the crop can be planted and grown. A good seed-bed should be prepared with a minimum amount of working and with little loss of soil moisture. A number of implements are available for this purpose including ploughs, cultivators, harrows and various hoes, their use being largely dependent on the type of crops to be grown, the soil type and weather (Lockhart and Wiseman, 1978; Wrigley, 1981). Cultivation has traditionally also had a secondary role in providing a method for weed control. This role has largely been replaced by the use of herbicides, but cultivation can, and still does, provide control of weeds in less intensive systems where herbicides are unavailable. Annual weeds may be tackled by: (i) working the stubble after harvest to encourage weeds to germinate, which are then destroyed by harrowing; (ii) similarly by preparing a 'false' seed-bed and then cultivating before sowing root crops; and (iii) through inter-row hoeing, particularly of root crops. A long-term fallow can be used in a rotation to kill perennial weeds. The frequent working of the fallow and the associated periods of drying-out will kill the perennial weeds. However, this technique is less common where labour costs are high, especially because it means a loss of profit through unused land.

A rotation is a cropping system in which two or more crops are grown in sequence. Although it is normally impractical to maintain soil fertility by crop rotations alone, a good rotation will usually give better average yields than continuous cultivation of the same crop or taking a succession of exhausting crops (Webster and Wilson, 1980). Crops may be classified as *exhaustive*, *cleaning*, *nitrogen-fixing* or *restorative*, according to their function in the rotation. Exhaustive crops such as cereals are removed from the field and are usually sold as cash crops. These crops tend to utilize soil nutrients, encourage weeds and if grown continuously on the same field, diseases can build up that seriously reduce yields. The 'cleaning crops', e.g. root crops, are so called because timely cultivation before sowing and during early growth can control most weeds. Although mainly high-value cash crops requiring deep soils with a heavy demand for plant nutrients, they permit the use of large amounts of fertilizer and thus, enable a farmer to build up soil fertility. Legumes (nitrogen fixing crops), e.g. clover, peas, beans, can also help restore soil nitrogen levels and hence, are potentially very useful for improving soil fertility. The restorative crops are those used for animal grazing, e.g. grasses and kale which return nutrients and organic

Table 2.1 Benefits of a good crop rotation

1	Reduce the financial risk by diversification of crops
2	Spread the labour requirement more evenly throughout the year
3	Reduce the risk of pests, particularly weeds
4	Build up and maintain soil fertility

matter to the soil. A good rotation will serve a number of different but important functions (Table 2.1). However, despite these benefits rotations are less commonly used today especially in intensive cropping systems.

The changes that have been made to crop husbandry practices as cropping systems have become more intensive, are often considered to have exacerbated and sometimes to have been the cause of pest problems (Pimentel, 1991). The emphasis now for practices designed to produce individual plants of greater yield (by increased use of irrigation, fertilizer and crop protection) and to produce a greater number of plants per unit area (by intensive monoculture) or per unit time (by multiple or continuous cropping), provides an unusually rich, dense continuously available substrate that favours pest populations (Corbet, 1970). In these circumstances pest management then becomes necessary in order to maintain the intensive nature of the crop production system. Hence, it is important that those involved in pest management realize and understand the nature of the root causes of pest problems and the role which crop husbandry practices may play in this process.

Arable land is a diminishing resource and there are increasing demands from a growing world population for increased yields per hectare. There is a need to understand the input of pest management in the context of the cropping system, i.e. in terms of the crop husbandry practices of which pest management is a part. Without this perspective, pest management research is reduced to a constant series of minor corrective modifications, i.e. tinkering with the system, which ultimately fail to address the real problems and hence, provide no sustainable solutions.

2.3 PRINCIPLES OF SOCIOECONOMICS

Economics is the study of the allocation of scarce resources for the satisfaction of human wants and the problems of choice that this involves (Harvey, 1977; Reichelderfer *et al.*, 1984; Stanlake, 1989). Economics is closely related to the other social sciences, ethics, politics, sociology, anthropology and psychology through its emphasis on the study of human behaviour. However, economics differs from these because it concentrates on the aspects of human behaviour involved with choice between alternatives in order to obtain the maximum satisfaction from limited resources (Harvey, 1977). The subject of economics is relevant to all aspects of human endeavour, including farming, crop protection and hence, IPM. The resources which are limited in this latter case may include capital for pest control inputs, labour or even pest management information. Other resources also may be scarce however, such as pesticide-susceptible pests or an uncontaminated/unpolluted environment (Mumford and Norton, 1984). The economist is however, not interested in the nature of the resources themselves; the subject for the economists' study is the mobilization of the resources to achieve the farmers' goals, and the efficiency of the methods they select for use (Harvey, 1977).

Economists use scientific method but because the subject matter is human behaviour (rather than the reactions of lower organisms or inanimate matter) they are denied the use of many of the experimental techniques available to natural scientists. For instance, they cannot test hypotheses by laboratory experiment because the laboratory is human society (Stanlake, 1989). This means that an economist can never be absolutely sure of their 'initial position' since 'facts' concerning people are difficult to ascertain. It is not possible to isolate a group of people to see how they react to change – surveys have to be used – nor is it possible to hold some conditions constant while evaluating the effect of one measure (Harvey, 1977). These limitations amount to substantial differences in approach from those assumed by biologists, and present the economists with difficulties in the precision of their findings; difficulties which must be appreciated by the natural scientists collaborating with economists in interdisciplinary IPM programmes.

The natural scientists involved in IPM, entomologists, pathologists, weed scientists, etc. have encouraged the involvement of economists in the field of pest management. This has largely been due to a clearly-identifiable role for economists in: (i) the consideration of yield losses; and (ii) for developing economic thresholds (Norgaard, 1976; Reichelderfer *et al.*, 1984). The economic threshold concept was first introduced in 1959 by Stern *et al.* (1959) where it was defined as 'the density of a pest population that will justify treatment'. What was meant by 'will justify' was not made entirely clear but it has subsequently been accepted to mean 'the density of the pest at which the loss through damage just exceeds the cost of control' (Mumford and Norton, 1984).

The basic components of the economic threshold concept have been well illustrated by Norton (1977) with an example involving the potato cyst eelworm and its control with the nematicide DD. It becomes profitable to apply the nematicide when

$$PD\theta K \geq C \tag{2.1}$$

where θ is the level of pest attack (eggs per gram of soil); P is price of the potatoes; D is loss in potato yield (tonnes per hectare) associated with 1 egg per gram of soil (the damage coefficient); K is the reduction in pest attack obtained with DD (a percentage); and C is the cost of application of DD per hectare (monetary units).

The economic threshold (θ^*) is thus;

$$\theta^* = \frac{C}{PDK} \tag{2.2}$$

With a damage function for eelworm that is approximately linear (Brown, 1969), a population of 1 egg/g of soil caused a loss in potato yield of 0.1 tonnes/hectare (ha). The nematicide reduced the nematode population by 80% when applied at time of planting (Jones, 1973), and given the estimated price of the crop at £40/tonne and cost of control at that time as £100/ha, then the threshold was calculated to be:

$$\theta^* = \frac{100}{40 \times 0.1 \times 0.8} = 31 \text{ eggs/g of soil.} \tag{2.3}$$

However, if any of the variables were to change then so would the economic threshold value. In practice such changes readily complicate the use of these thresholds, but more specifically the complexity of the interactions that can affect the damage function, e.g. altered planting dates, weather, crop variety, cultural practices (to name but a few) render the concept generally unmanageable in terms of data requirements.

The economists, while having made a positive contribution to the economic threshold concept have been more interested in a much wider and pervasive contribution of economics to the study of pest management (Norgaard, 1976; Reichelderfer *et al.*, 1984) (Figure 2.2), particularly in relation to utilization of models of optimization, decision theory and behavioural-based decision-making approaches (Mumford and Norton, 1984). The optimization and decision theory models outlined by Mumford and Norton (1984) make use of normative economics, providing a prescription for what may be done according to specific subjective criteria, while the behavioural decision model utilizes descriptive economics, concerned with economic facts and relations. Descriptive economics attempts to answer questions about how for instance, an increase in the cost of a control measure will affect a pest management strategy. Examples of the techniques used in normative economics, such as dynamic programming, decision tree analysis and pay-off matrices, are provided in later sections of the book when the problems of decision making are considered (see also Chapter 6). The use of prescriptive economics in IPM has tended to be more limited. The approach to date has been mainly confined to the consideration of the cost/benefits of insecticide use, particularly in relation to prophylactic versus responsive use of chemicals (e.g. Norton, 1985). Pesticide use lends itself well to such analyses because costs associated with use at various levels can be determined and the relative benefits in yield increase quantified. Such short-term economic analyses may also be complemented by a longer-term economic perspective (Ruthenberg and Jahnke, 1985). For

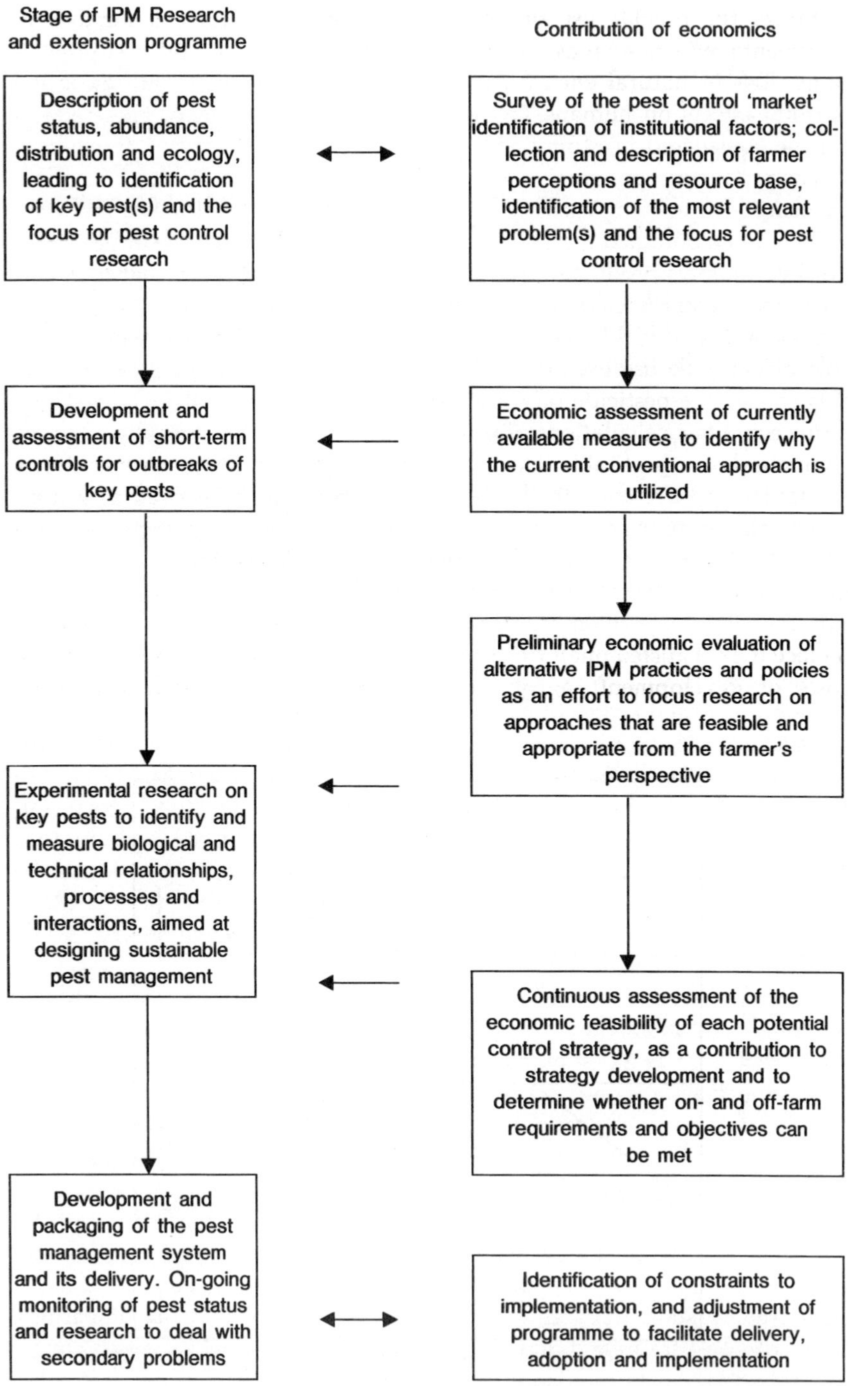

Figure 2.2 The contribution of economics to integrated pest management. (After Reichelderfer *et al.* 1984).

instance, the regular use of pesticides is known to have a number of detrimental effects such as, development of pesticide-resistant pest populations, loss of natural enemy populations, secondary pest outbreaks and harmful effects on humans, livestock and the environment. These effects can be considered in economic terms.

The production potential of a pesticide is comparable with that of a mine: by using it, it is being used up (Figure 2.3). Without pesticides a low crop yield may be obtained (point A, Figure 2.3). The intensive use of pesticides however, allows production to reach yield level C but maintenance of this yield level requires increasing pesticide inputs over time *t*. At the point *t*, the possibility of yield increases through pesticide use is exhausted, i.e. the cost of pesticide use exceeds its benefits. If, as an alternative, yield level B is chosen the pesticide remains effective and economical for longer. As a rule economic calculations of pesticide use show that technically effective compounds are profitable even if longer-term and indirect adverse effects are taken into account. In this way a pesticide can be viewed as a non-renewable resource and its use modified to maintain its long-term viability (Bartsch, 1978; Ruthenberg and Jahnke, 1985). Although 20 years ago the failure of pesticide use due to the development of resistance was not considered a serious problem (it was felt a new compound would be found to replace the exhausted one), this is no longer the case today. With increasing costs of development, agrochemical companies are placing increasing

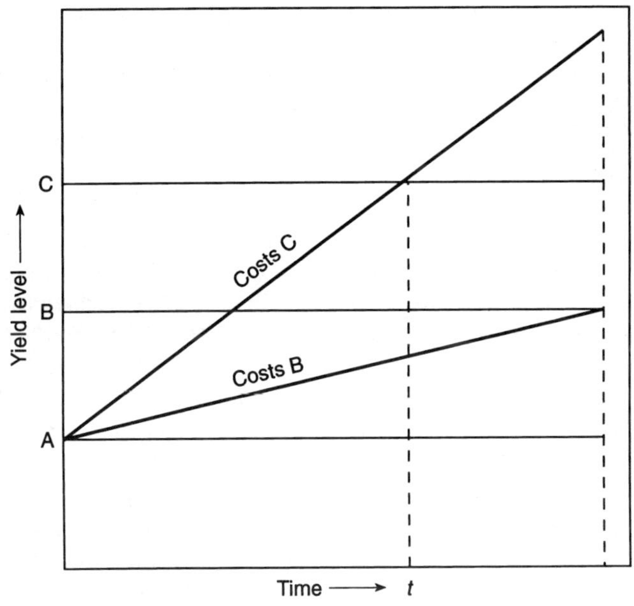

Figure 2.3 Costs and returns of varying levels of pesticide application. (After Bartsch, 1978; see text for explanation).

emphasis on maintenance of pesticide viability to ensure the product recoups its development costs and provides a viable return on the investment.

The development of any new technical innovation used for pest control should be viewed in economic terms. One way of classifying innovations, to permit economic analyses, is in terms of either yield-increasing innovations or labour-saving ones. Most commonly, research effort in IPM is directed towards yield-increasing (through pest reduction) innovations, but in economic terms it can be just as important for a farmer to reduce labour costs while maintaining the same level of pest control. For instance, the development of herbicides for weed control has been primarily a labour-saving innovation. Herbicides can be regarded as substitutes for hand labour, animal traction or tractor work, although practical agricultural labour capacities are usually insufficient for all weeding requirements. Hence, herbicide use may not only increase crop yield but also, and perhaps more importantly, provide a major labour saving. Such technological capabilities have an important influence on whether or not innovations will be adopted by farmers. How they complement and interact with existing farming practices will also play an important role.

The interactions of different farming and crop protection practices in economic terms is clearly important in IPM. Returning to the economics of pesticide use; crop response to pesticides is highly sensitive to the application of supplementary inputs. In a study of pesticide in cotton Bartsch (1978) specified the following interaction:

$$Y = 1406\ (1.14^{D_{50}})\ (1.53^{Dn})\ (1.41^{Ds}) \tag{2.4}$$

where Y is cotton yield in kg/ha; D_{50} is variable for planting time (early planting = 1.0); Dn is variable for fertilizer use (no fertilizer = 0); and Ds = variable for pesticide use (no pesticide = 1.0).

Yield of cotton was, thus, explained through the multiplicative interaction of the three variables, seeding time, fertilizer use, and pesticide application. Late planting and no fertilizer produced a yield of 1982 kg/ha while early planting and fertilizer application gave a return of 3458 kg/ha. If pesticides were not applied, then in the first case, yield was reduced by 576 kg, and in the latter case by almost 1000 kg; quite substantial reductions. This ability to use basic principles of economics to study the interaction between components of a system, to bring together diverse elements of a variety of disciplines within a single framework, makes economics a powerful tool for use in the development and integration of effort of IPM programmes and systems. The role of economics in the process of IPM programme development is illustrated in Figure 2.2. The principles of economics and their associated techniques figure prominently throughout the rest of this book.

2.4 PRINCIPLES OF ECOLOGY

Ecology is the scientific study of the interactions that determine the distribution and abundance of organisms (Krebs, 1978). Many of the early developments in ecology came from the applied fields of agriculture, fisheries and medicine. In agricultural pest management there is a need to manage both populations of plants for maximum yield and populations of pests that may reduce yield to unacceptable levels. Because pest management deals largely with populations, the research methods used are of necessity based on population ecology and population dynamic theory (Gutierrez, 1992).

Population ecology has tended to be dominated by studies of insects mainly because short generation times make experimentation productive and because insects are important as pests (Putman and Wratten, 1984). In fact, even much of the early epidemiological research conducted on plant pathogens was based on the population dynamics of insects (Carruthers, 1985). However, since the basic population characteristics representing the dynamics and stability of life-systems are common to all populations, regardless of their taxonomic classification, the theory may be equally applied to pathogens, insects or weeds (Carruthers, 1985).

The potential number of interactions involved in any pest life-system is enormous, involving a multitude of abiotic and biotic factors that combine to influence a population in any number of permutations. For this reason it is impractical, if not impossible, to observe all the potential interactions among all processes and species. Hence, ecological research tends to involve use of simplifying assumptions. These assume that much of the complexity of nature is either unimportant or can be subsumed within a few summary variables. Thus, the study of population dynamics and interactions is by necessity a process of ecological abstraction; identifying only those key factors required to describe and predict ecological patterns (Tilman, 1989).

The principles of ecology are considered here solely from the perspective of population ecology. This is dealt with in terms of population growth and regulation and population interactions. Population growth and regulation involve studies of birth, mortality, immigration and emigration rates and the factors influencing the fluctuations and changes in size of populations. The challenge for the pest management population ecologist is the quantification of the different growth rates in combination with an assessment of population interactions (competition and predator–prey interactions), in order to explain what is influencing the timing and magnitude of pest population fluctuations and the mean level around which the populations occur (Putman and Wratten, 1984).

The task of describing the role of population ecology in pest management would be a book in its own right (see Horn, 1988) and so the intention here is to provide only an indication of the different ways in which pest management is dependent on population ecology and the sorts of ways in which the techniques used by ecologists provide information and data relevant to the needs of pest management. It will become apparent just how inseparable the two subjects of ecology and pest management really are and hence, why IPM has often been referred to as an ecological approach to pest control.

2.4.1 Population growth and phenology

In essence, the population growth of any organism will depend on the action of five components: (i) the physical environment; (ii) a host or food source; (iii) space; (iv) the population itself; and (v) other species of organism (Southwood and Way, 1970). All five components act through the pathways of natality, mortality and dispersal where: (i) natality is the rate at which new individuals are added to the population by reproduction; (ii) mortality is the rate at which individuals are lost by death; and (iii) dispersal is the rate at which individuals immigrate into or emigrate out of the population (Figure 2.4).

The physical environment includes the climate, microclimate and weather which may influence temperature, humidity, wind speed, and the amount of ultraviolet (UV) radiation, but it also includes the soil environ-

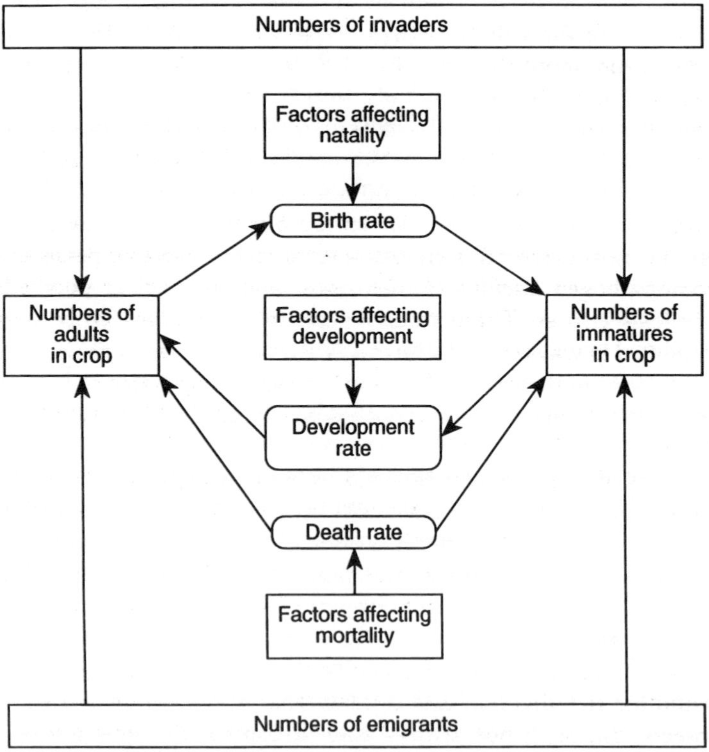

Figure 2.4 The population systems of the pest. (After Southwood and Way, 1970).

ment and the conditions experienced by a pest at its site of infection or infestation. The physical environment can have an enormous impact on pest populations. Where conditions are severe, pests may be unable to survive and mortality rates may be high, for instance, if temperatures are too high or low. The poor survival of insects outside their normal temperature range has been made use of as a means of control for stored product insects, e.g. storage silos maintained at −20°C for 8 hours, −16°C for 12 hours or −12°C for 24 hours provide conditions inhospitable enough to kill all stages of the grain weevil *Sitophilus granarius* (Benz, 1987). The physical environment of a pest may also be altered by habitat modification. Provided the optimum conditions for pest development and survival are known then modification of a habitat to produce a suboptimal environment may be undertaken, e.g. control of *Amblyomma americanum* involved creating a harsh environment for the tick through mechanical clearing of land and pasture establishment (Meyer *et al.*, 1982).

Low temperatures may also have a detrimental effect on pathogen survival. Outbreaks of the blue mould of tobacco (caused by *Peronospora tabacina*) in the USA is related to winter temperatures (Tarr, 1972) while in the case of the bacterial wilt of maize caused by *Erwinia stewartii*, the low winter temperature affects the vector rather than the disease itself. The primary vector of the wilt is the flea beetle which cannot survive winters where the mean monthly temperatures during December–February are below 32°C (Parry, 1990).

Among pathogens one of the most important environmental factors affecting mortality is UV radiation. Radiation in the UV band 210–330 nm kills many fungal spores (Zadoks and Schein, 1979) and can have a major impact on spore survival, especially during dispersal. UV radiation is also probably the most destructive environmental factor affecting persistence of the entomopathogen *Bacillus thuringiensis*, and the nuclear polyhedrosis and granulosis viruses. These valuable control agents have to be targeted and formulated in ways to improve their field persistence (Dent, 1991, 1993; Rhodes, 1993) or in the case of *B. thuringiensis* a programme of UV screening and selection may be undertaken to develop UV-resistant strains (Smyth, 1993).

The process of dispersal can be aided or impeded by environmental conditions. Organisms are often dependent on suitable environmental conditions for their dispersal, whether it is achieved by wind, water or animals. Temperature and wind speed are often limiting factors for activity in insects (temperature thresholds for flight, Taylor, 1963; Dent and Pawar, 1988) while spore release by fungi may be dependent on air humidity, radiation or wind conditions. Once airborne small organisms are usually at the mercy of the wind for the direction and distance that they may be carried. Most small insects, although not always able to control the direction of their flight, can often leave the airstream when they choose, by cessation of flight and dropping from the airstream, e.g. aphids, whiteflies. The movement of pests during their dispersal phase has often been taken as an opportunity to monitor population levels, most commonly through different forms of trap (Southwood, 1978; Dent, 1991). Since the organism and the trap performance can both be affected by the environmental conditions (e.g. wind speed affects the performance of pheromone traps; Campion *et al.*, 1974; Lewis and Macaulay, 1976) but often in different ways, the use of traps to provide estimates of population size has been fraught with many difficulties (Dent and Pawar, 1988; Dent, 1991). Migration of some insect pests has been linked to prevailing winds (Pedgley, 1982; Tucker *et al.*, 1982) providing characteristic seasonal patterns of infestation. Rainfall patterns have also been implicated in a number of situations (Tucker and Pedgley, 1983; Tucker, 1984).

The rate of reproduction of pathogens and insects is dependent on temperature. Usually, up to a critical maximum, the number of offspring produced increases with increasing temperature (e.g. *Metopolophium festucae*; Dent, 1983), hence, the value for natality will be dependent on temperature. Such relationships are however, influenced by the hostplant type and condition (Dent and Wratten, 1986). The host cultivar or species can have a significant effect on pest reproduction and survival. Where a pest has a range of host species or cultivars these will be evaluated to determine ones on which it is least fecund. If these hosts are crop cultivars such studies are often conducted with a view to identifying levels of antibiotic resistance that can be used in resistance breeding programmes. There are numerous examples of such work (Russell, 1978; Maxwell and Jennings, 1980).

The immigration and emigration of insects will also be influenced by the host species and cultivar (e.g. Müller, 1958). Insect species exhibit marked preferences for host cultivars on which they settle and oviposit (Lowe, 1978, 1982; Dent, 1986; Givovich *et al.*, 1988). Preferences may be caused by release of hostplant volatiles, hostplant colour, shape or morphology. Information on these preferences is used in the breeding of cultivars that exhibit antixenotic resistance. Chemical ecologists are also very interested in the chemicals involved in the attraction or repellancy of pests since these can be studied and developed as pest control agents (see section 3.6.1).

In contrast to insects, dispersing pathogens do not select their host, or even the substrate on which they end up – they are deposited, by rain splashings, leaf rubbing or turbulent transfer. The process may occur more than once, but the dispersal process effectively ensures the spread of the infection or infestation through space. The spatial characteristics of a pest infestation are of particular importance and provide one of the components used to assess the severity of a pest problem (the other is temporal distribution). The dispersion of the pest is also important from a population sampling perspective. Individuals may be randomly or uniformly distributed, but most commonly under natural conditions they tend to have clumped distributions. Knowledge of spatial distribution of pests is absolutely essential for the proper application of appropriate control measures. For instance, a uniform distribution of pest infestation over a wide area may allow a regional pest control programme to be undertaken (section 5.2), whereas, clumped local distributions of pests may require the application of control measures by individual farmers and only to specific areas of a crop (e.g. patches of weeds).

All populations have a temporal component, they all change with reference to time. Birth rates, death rates and migration rates are all examples of temporal characteristics of populations. These particular characteristics are classically aggregated into a common population parameter, the intrinsic rate of natural increase (r_m). This measure, derived from the birth rate minus the death rate under fixed conditions, has been used as a means of describing the effects of environmental factors on insect population dynamics (Siddiqui and Barlow, 1973; Wyatt and Brown, 1977), patterns of antibiotic resistance (Dent, 1983; Birch and Wratten, 1984; Holt and Wratten, 1986) and for assessing the potential threat of a virus vector (Puche and Funderburk, 1992). The r_m value is also extremely useful for a rapid comparison of different epidemics by the same pathogen (Zadoks and Schein, 1979) and can be used to provide quantitative assessment of the effects of various control measures and weather on pathogens.

The intrinsic rate of natural increase of a population has its derivation in the logarithmic equation:

$$\log N_t = \log N_0 + r_m t(\log e) \tag{2.5}$$

where the population size at time t is N_t, N_0 is the initial population size and log e the exponential constant. This equation however, describes unlimited expotential growth, whereas actual populations tend to reach a stable upper limit where population growth is limited by available resources. This limit is known as K – the carrying capacity. The equation which incorporates K is

$$\frac{dN}{dt} = r_m N\left(\frac{K - N}{K}\right) \tag{2.6}$$

As the population increases, the rate of increase r, where $r = r_m (K - N)/K$, is progressively reduced, and it approaches zero (birth rate = death rate) when the population size has reached the value of K (Varley *et al.*, 1973). K is the carrying capacity of the environment as determined by the availability of food, space and predators, while r is the intrinsic growth rate, free from environmental constraints.

The parameters r and K are important descriptors of population change and are much used in discussions of the dynamics of pest populations. A particularly powerful concept that utilizes these parameters as a metaphor is that of r- and K-selection (MacArthur and Wilson, 1967). As a concept it has specific relevance to pest management, providing useful insights into the life-history strategies of pests and pointers to appropriate strategies for control (section 2.6) (Conway, 1984). Depending on the form of the population size function ($F(N)$) (Figure 2.5) it is possible to distinguish between

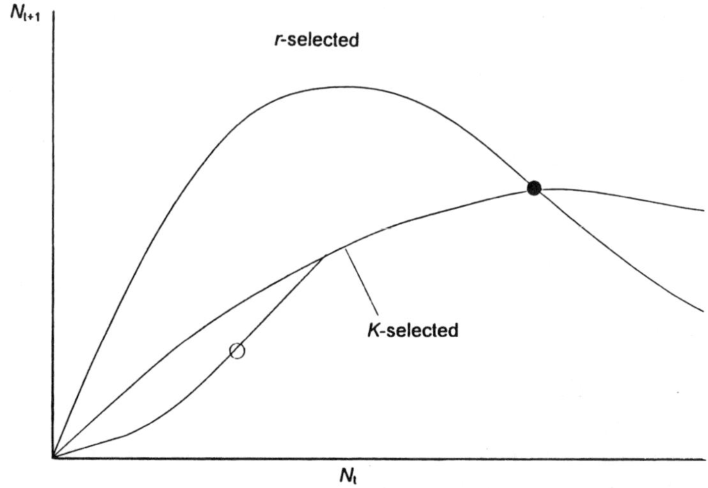

Figure 2.5 Reproduction curves for *r*- and *K*-selected pest and pathogen populations. •, Stable; ○, unstable equilibrium. (After Conway, 1984).

r-pests and *K*-pests. *r*-Pests are opportunists, exploiting temporary habitats having a high rate of population increase, short generation time, strong dispersal and host finding abilities, such as the African bollworm *Helicoverpa armigera*, the desert locust *Schistocerca gregaria*, the black bean aphid *Aphis fabae*, black stem rust *Puccinia graminis*, amd the brown stem rust *Puccinia recondita*. *K*-Pests, by contrast, have lower rates of increase, greater competitive abilities, more specialized food preferences and by comparison with other members of the same taxa, greater size. *K*-Pests tend to exploit more stable, permanent habitats such as forests, orchard and rangeland systems. Examples of *K*-pests include the codling moth *Cydia pomonella*, tsetse fly *Glossina* species and the fungi *Stromatinia gladioli* and *Sclerotium rolfsii*. In between these two extremes, Southwood (1977) has proposed an intermediary category of species (at least for insects) that are normally regulated at various levels by natural enemies.

Populations can only be regulated where a negative feedback process operates, for instance, a density-dependent mortality factor. The ability to identify the density-dependent factors which naturally operate to regulate pest populations, can provide information useful in the design of pest control programmes. A technique used by ecologists to identify key mortality factors in the life-cycle of an organism is the age-specific life-table. The construction of life-tables is dependent on a census based on counts of the pest in samples of known size and should ideally count the numbers which die in each time interval and identify the causes of death. Data need to be collected over a number of complete life-cycles of the pest and hence, the process can take a number of years (8–15) to accumulate sufficient data for analyses (see Southwood, 1978; Dent, 1991 for description of life-table analyses). Age-specific life-tables are however, only applicable to pests having discrete generations, which largely limits their use to temperate regions. The key mortality factor is determined by identifying the contribution of each mortality (calculated as the differences between two life-stage densities) to the total mortality. This is done either graphically (Varley and Gradwell, 1960) or quantitatively (Podoler and Rogers, 1975). Where life-tables and key-factor analysis are carried out to determine the role of natural enemies in pest population changes, the relationship between mortality and pest density will need to be known. Only when the mortality is density-dependent will the natural enemies have a potential role as control agents (Dent, 1991).

Life-tables, although most often used for insect pests (e.g. Burn, 1984; Room *et al.*, 1991), can also be constructed for natural enemies, particularly parasitoids. Data from such tables can be used to calculate the area of discovery '*a*' of the parasitoid for use in analytical population models (e.g. Hassell, 1978; section 6.3.1). Such models have figured largely in the development of ecological population theory, especially for the study of the dynamics of predator–prey interactions.

2.4.2 Population interactions

Progress in the study of population interactions has been modest, limited mostly to simplified representations of two species interactions (Gutierrez, 1992). The majority of the theoretical work carried out has been based on relatively simple parasitoid–prey models (see section 7.5 for a more detailed description). These have been considered useful in biological control for the identification of important aspects of interactions and potentially useful attributes of potential control agents (Dent, 1991). However, their emphasis on searching for factors that reduce the equilibrium level of the pest population in a stable pest–parasitoid interaction has contributed little to increasing the success rate of classical biological control introductions or for understanding the reasons for failures (Gutierrez *et al.*, 1993).

Among the arguments put forward for improving the success of classical biological control introductions has been that of multiple species introductions. There has been a long-standing debate over whether or not interspecific competition between two parasitoids may produce an increase in the pest equilibrium level relative to control by a single parasitoid system (Kakehashi *et al.*, 1984). Theoretical models of multiple species interactions have tended to predict in favour of multiple introductions (May and Hassell, 1981; May *et al.*, 1988; Waage and Hassell, 1982). Indeed the bulk of empirical evidence suggests that the overall impact of a complex of natural enemies exceeds that of any one natural enemy, regardless of its efficiency in the absence of the other species (Huffaker *et al.*, 1976; Horn,

1988). Although interspecific competition is assumed by most insect ecologists, evidence for its routine occurrence is far from equivocal (Arthur, 1982; Schoener, 1982). In fact, among phytophagous insects there is little evidence to demonstrate the widespread occurrence of interspecific competition (Strong *et al.*, 1984). Among other organisms however, the situation is entirely different, interspecific competition having important implications for both population growth and survival.

The theory of interspecific competition has tended to focus on what is known as the 'competitive exclusion principle'. Originally formulated by Gause (1934) the principle of competitive exclusion argues that when two species compete with one another intensely enough over a limited resource, then over a long enough time one or other will become extinct. But if two species have similar but distinct resource requirements (they compete with each other weakly) then they may both persist indefinitely in the same environment. The utility of the theory and its very existence in some cases has been seriously questioned on both theoretical and empirical grounds. Nevertheless, much of modern ecology takes as its point of departure the veracity of the general proposition that if competition is sufficiently weak, then two species may co-exist, but if it is sufficiently strong one or another will perish (Vandermeer, 1989).

Interspecific competition has implications for the utilization of intercrops, weed science and for biological control using pathogen antagonists, hence it is an area of ecology which has important implications for pest management. Weeds directly compete with crop plants for light, nutrients and water. Weed competition is a serious problem, particularly in the tropics where crop losses are two to three times greater than in the temperate zone (Wrigley, 1981). Shade produced by weeds will depress crop growth due to deprivation of light, while root systems compete for available nutrients and by drawing on restricted water supply may desiccate crop root systems. When soil fertility is low, both weed and crop growth may be limited. The addition of fertilizer may stimulate the weeds more than the crop and may even reduce crop yield because of this increased competition (Dawson and Holstun, 1971).

Adding fertilizer and manure to the soil can be used as a means of promoting biological control. The principle used involves attempts to deprive a pathogen of essential nutrients by encouraging the growth of nutrient-consuming antagonists. This may be achieved by rapid incorporation of large amounts of 'food material' into the soil. The incorporation of cellulose or barley straw into soil contaminated with *Fusarium solani* f.sp. *phaseoli*, which causes bean rot, stimulates population growth of saprophytes which deprive the pathogen of the nitrogen essential for germination and the penetration of the bean host. Similarly 'green manuring', the incorporation of green plant material (e.g. grass cuttings) into soil, is useful for control of pests such as the common scab (*Streptomyces scabies*) of potatoes, although in this case no specific organism has been identified as the primary competitor (Parry, 1990). Generalized build-up of soil saprophytes as competitors are responsible for the phenomenon known as suppressive soils, thought to be responsible for suppression of *Fusarium*, *Phytophthora* and *Pythium* sp., and 'take all decline'.

A number of biological control agents depend on competition for available resources for their effectiveness, e.g. fluorescent pseudomonads which compete for iron and control fungal pathogens (Whipps and McQuilker, 1993), *Erwinia* spp. isolated from rose petals reduces the number of rose blight lesions caused by *Botrytis cinerea* (Redmond *et al.*, 1987), mycorrhizae that compete for root space with plant parasitic nematodes (Kerry, 1993) and the commercial use of *Peniophora gigantea* to prevent growth of the fungus *Heterobasidion annosum* on pine tree stumps (Parry, 1990). These examples clearly illustrate the potential value of pathogen competitor–antagonists as biological control agents.

The development of appropriate control strategies is usually highly dependent on a sound knowledge of the ecology of the pest and its host. The theoretical approaches, particularly that of ecological systems analysis have been very useful in this context (Chapter 5). However, without experimental population data and an understanding of the types of species and environmental interactions that can occur, even simulation models are limited in their application. Pest management is totally dependent on the principles and practice of ecology. Pest managers require a good understanding of ecology if they are to develop sustainable pest management systems.

2.5 POPULATION GENETICS

Agroecosystems are important arenas for evolutionary change. This is because many of the control measures used are designed to reduce pest population levels by increasing mortality or decreasing reproductive capability. Differences among pest individuals in survival and fecundity by these control agents result in natural selection (Via, 1990). If pest populations harbour genetic variability for physiological or behavioural characteristics that circumvent control measures then the evolution of better adapted pests, that are more difficult to control, is inevitable. In such circumstances an understanding of the mechanisms involved in pest evolution becomes an essential ingredient in the development of sustainable pest management systems.

2.5.1 Key concepts in population genetics

The evolution of pests and other organisms basically consists of changes in the genetic composition of populations. The study of changes in the gene frequency of populations falls within the discipline of population genetics.

In a genetic sense, a population is conceived of in terms of a Mendelian population, which is 'a community of interbreeding individuals' (Sinnott *et al.*, 1958). In the simplest of genetic models, a population is considered to be infinitely large, with random mating between individuals and an equal sex ratio. All adults have an equal probability of producing offspring and the generations are non-overlapping (Daly, 1991). Of course real populations deviate from this ideal model and it is through the study of population genetics that these differences and particularly their impact on changes in gene frequency, are assessed.

In the study of the role of population genetics in pest management it is essential that a number of key concepts be remembered. The first of these are the Mendelian laws of inheritance which include the two major principles of segregation and independent assortment of genes, together with some less fundamental generalizations such as the dominance and recessiveness of traits in hybrids. In addition, there is the important principle associated with the Hardy–Weinberg law. This law is concerned with frequencies of genes and of homozygous and heterozygous genotypes in Mendelian populations. Hardy and Weinberg showed that in the absence of forces which change gene frequencies, populations may have any proportions of dominant and recessive traits, and the relative frequencies of each gene allele tend to remain constant from one generation to the next (Sinnott *et al.*, 1958). This implies that there are no changes in gene frequency from one generation to the next, which is of course only true if the population is very large and there is: (i) no genetic drift; (ii) no migration; (iii) no gene mutations; and (iv) when there are no selective forces acting on the population. In the real world all of these factors are operative and act to change the gene frequency of a population.

Genetic drift is a random process caused by the fact that real populations are limited rather than infinite in size, in which case gene frequency changes occur because of sampling errors. In a situation where the number of reproductive individuals in a population is consistently large for each generation there is always a high probability of obtaining a good sample of the genes of the previous generation. However, where only a few reproductive individuals start a new generation, such a small sample of genes may deviate widely from the gene frequency of the previous generation (Strickberger, 1976). Furthermore, after a change in the gene frequency has taken place in this way, no matter how small it may be, there may not be any restoring force to change the frequency back (Parkin, 1979). Sampling of this kind will inevitably occur and will result in a small change in gene frequency each generation until an allele is lost. The other allele is then the sole representative in the population and is said to be 'fixed'. When the number of reproductive individuals is small the population reaches fixation more rapidly.

Another type of event which can easily occur because of restrictions in population size is that of mating between relatives, or in-breeding. As the size of a population decreases, the probability of mating among relatives increases, with distant relatives less likely to do so than close ones. Each time relatives mate there is a chance they pass on alleles that are identical by descent, which will tend to produce identical homozygotes. These will accumulate over time until all members carry the same gene at a locus and genetic variability is lost. This is a phenomenon that has been well exploited by plant breeders, who through procedures which control the mating between inbred individuals, derive homogeneous populations exhibiting desirable crop characteristics, e.g. uniform rates of growth, pathogen resistance.

The three remaining forces, migration, mutation and selection which can have an impact on population gene frequencies differ from genetic drift in that they usually act in a direction, changing gene frequencies progressively from one value to another. Migration, referred to as 'gene flow' by geneticists, occurs when individuals disperse between populations and successfully reproduce in their new locality. The effect gene flow will have on the recipient population will be dependent on the difference in frequencies between the two populations and the proportion of 'exotic' genes which are incorporated in each generation. Adequate information about gene flow can be made only from detailed observations of both dispersal and the genetic structure of populations, and hence requires lengthy, detailed study. Knowledge of gene flow is however, important in pest management, especially with regard to pesticide resistance management, since it is the general consensus that the extent of immigration of pesticide-susceptible individuals into treated areas is one of the more critical influences in the development of pesticide resistance.

A factor which affects gene frequency which is of a less critical influence in pest management terms, is that of gene mutations. Although mutations are the ultimate source of new genes, mutation alone is not a prime force in maintaining gene frequencies in natural populations. Of more importance is the intensity of selection which acts on populations by discriminating between genotypes, through either an influence on reproductive success or mortality. The term 'fitness' is used to describe relative reproductive success, such that, when a genotype can produce more offspring than another in the same environment, then it has superior fitness. Fitness is often difficult to evaluate since in natural situations the factors influencing fitness may be numerous and subtle in their effect. In situations manipulated by man the factors responsible and the extent of their impact on fitness may be more obvious. In pest management the use of control measures that cause pest mortality, if widely used in space and time, can in certain circumstances act as highly intensive selective agents, e.g. crop cultivars, pesticides. Genotypes able to overcome the effect of the control measure in time proliferate and are irresponsive to continued and increasing levels of use of the same control measure (Roush and McKenzie, 1987). Such situations have occurred commonly in both the use of chemical insecticides and pathogen-resistant crop cultivars.

2.5.2 Modes of inheritance

In most higher animals and plants, selection takes place primarily in the diploid stage. When, for instance, chemical insecticide resistance in the pest is under monogenic control (and there is only one resistant and one susceptible allele) there are three possible genotypes for a single gene difference (e.g. RR, RS, SS; R = resistant, S = susceptible) so that the effectiveness of selection depends upon the degree of dominance. Dominance is a term used to describe the relative phenotypic resemblances between the heterozygotes and the homozygous parents. If the heterozygote (RS) more closely resembles the resistant parent then resistance is described as dominant while if the heterozygotes more closely resemble the susceptible homozygote then the resistance is said to be recessive. Unlike traits such as human eye colour (blue eyes recessive to brown) often used as examples in genetics textbooks, resistance to insecticides is rarely if ever completely recessive or dominant (Roush and Daly, 1990).

Identifying the mode of inheritance of traits has generally involved the genetic crossing between strains exhibiting the desired traits with those individuals that don't. The intention is to try to determine the number of genes involved and the dominance relationships between the genotypes. This has been a particularly productive area for plant breeders seeking to enhance plant resistance to pathogens. Often under monogenic control and a dominant trait, crop plant resistance to pathogens has been identified and transferred through conventional breeding techniques to agronomically useful crop cultivars (e.g. Allard, 1960; Simmonds, 1979). Such work has been carried out since the beginning of this century.

In addition to the expanding knowledge of the inheritance of resistance to diseases at the beginning of the century, parallel advances were also being made towards an understanding of genetic variation in fungal pathogens. Early studies on the inheritance of virulence were carried out in the 1930s on the fungus *Ustilago* by Nicolaisen (1934). This work was closely followed by that of Flor (1942) who from his work with *Melampsora lini* on flax postulated that for each gene conditioning a resistance mechanism in the host, there is a corresponding gene in the parasite that conditions pathogenicity; the gene-for-gene relationship (section 6.4). This gene-for-gene paradigm plus the long history of discovering useful resistance genes in wild crop progenitors, led to the examination of virulence and the resistance structure of natural plant and pathogen populations and a theory of co-evolution emphasizing reciprocal selection through gene-for-gene interactions (Marquis and Alexander, 1992). This contrasts markedly with the theory on plant–insect co-evolution based on: (i) whether or not there is a prolonged 'arms race'; (ii) the determinants of the number of species associated with a plant species; or (iii) factors accounting for the host specificity of insects (Futuyama and May, 1991). Emphasis in plant–herbivore systems has been placed on the mechanisms of resistance utilized by the host plant, but to the relative neglect of virulence traits in the herbivore. The opposite has been true in plant–pathogen systems where the examination of the mechanisms of resistance has been second to understanding the mode of inheritance and genetics of the interaction (Marquis and Alexander, 1992). These differences in approach are reflected in the way entomologists and pathologists have addressed the problems of host-plant resistance, with the entomologists taking a mechanistic approach based on the work of Painter (1951), while the pathologists have developed the pathosystem concept developing the idea of the role genetic interactions in resistance studies (Van der Plank, 1963, 1975; Robinson, 1980a,b) (section 6.4).

2.5.3 Vertical and horizontal resistance

Van der Plank (1963) postulated that resistance to diseases in plants could be placed in one of two categories, vertical or horizontal resistance. The absolute definition of vertical resistance is that it involves a gene-for-gene relationship. For a pathogen to colonize a plant successfully the gene or genes for resistance in the plant must be matched by a corresponding gene or genes in the pathogen. The plant is susceptible to all individuals of a pathogen having the appropriate corresponding genes or more than all the genes. Pathogen individuals characterized by having the same virulence genes are known as a race, and exhibit a recognizable differential interaction with complementary host races or cultivars. The concept of a race, defined by a specific gene-for-gene interaction has been adopted by some entomologists (although referred to as a biotype) in their attempts of breeding plants resistant to insects. However, the evidence in support of such gene-for-gene interactions in host–insect relationships has been hotly contested (e.g. Claridge and den Hollander, 1982; Diehl and Bush, 1984; Claridge, 1991). In one of the best studied pest–host interactions (from a plant breeding point of view) *Nilaparvata lugens* (brown planthopper) on rice, claims of gene-for-gene relationships (e.g. Saxena and Barrion, 1987) have been disputed, with authors arguing the case for virulence being under polygenic (den Hollander and Pathak, 1981) or environmental control (Claridge *et al.*, 1984). If the resistance is ultimately shown to be polygenic rather than monogenic then it will fall into the horizontal resistance category of Van der Plank (1963).

Horizontal resistance is defined simply by the fact that it does not involve a gene-for-gene relationship (Robinson, 1980a). Horizontal resistance may be under monogenic control but not involve a gene-for-gene interaction, or as is more commonly acknowledged horizontal resistance is usually under polygenic control. When dealing with polygenic effects we move into the realms of quantitative genetics.

The term polygenes was coined by Mather (1955) to describe genes, which individually have only a small effect, but cumulatively control the magnitude of the quantitative character (Parkin, 1979). These genes are no different from the major genes of larger effects in their inheritance or behaviour but because a particular gene cannot be associated with a particular characteristic or response it becomes necessary to measure the phenotype in a quantitative manner. Hence, quantitative genetic theory is based on expected phenotypic resemblances of relatives and the variances of characters among related groups are then used to estimate the genetic variation in phenotypic characters (Via, 1990). Many morphological and life-history characters show continuous distributions expected for polygenic traits. Even some physiological characters that are controlled by major genes (e.g. in insecticide resistance) are also frequently affected by polygenic modifiers. A small number of plant breeders have also made use of polygenically controlled horizontal resistance to insects and pathogens through mass selection techniques to provide more durable levels of resistance (e.g. Hanson *et al.*, 1972, Jones *et al.*, 1976). However, it is likely, through the techniques of gene manipulation that qualitative genetics will continue to dominate quantitative genetics in the future of pest management.

In the rush to develop pest management products through gene manipulation techniques it must be remembered that control measures that have previously exploited or have been dependent on monogenic characteristics for their effectiveness, can be readily over-ridden by the pests they are meant to control. It may be hoped that the growing collaboration and interdisciplinary approaches between ecologists and geneticists that are becoming evident, may take this issue on board and through modelling and experiment provide a framework for the rational use of these new major gene-based products.

2.6 PRINCIPLES OF CONTROL

The principles of control transcend the normal classifications of control measures in terms of their chemical, physical and/or biological characteristics (e.g. chemical pesticides, biological control, hostplant resistance) in favour of a more functional classification based on various criteria affecting the selection and the use of the control measures in pest management systems. Such a functional approach is necessary, in part because of the general move away from the exclusive use of a single control measure, with the aim of eradicating a pest population, to the more sustainable approach, based on management of populations through the integration of measures that maintain pest levels below those causing economic injury. Under these conditions it is important that appropriate control measures are selected, that they are compatible and their combined use is practicable and effective. This section considers the utilization of control measures on the basis of pest characteristics, the type of impact required on the pest population and the operational factors that affect the way in which control measures are deployed. In addition, there is a brief appraisal of the strategies employed to prolong the useful life of some control measures or to limit the impact of their failure.

2.6.1 Selection of control measures: pest types

Each discipline has tended to devise different classifications of control measures and means for deciding on which are the most appropriate for a given situation. Foremost among the frameworks identified for insect pests (but also considered relevant to pathogens) has been that based on the principle of *r–K* selection and the control measures appropriate for *r*- and *K*-pests (Conway, 1984, 1991; Table 2.2). The high fecundity, short generation time and good mobility of *r*-pests combined with frequent invasions and damaging outbreaks necessitates control measures which can be used when required and which can have a fast, immediate impact on pest population levels. Natural enemies can rarely respond quickly enough and all too often pesticides provide the only viable means of preventing excessive pest damage. The rapid rates of increase and short generation times can mean rapid selection for resistance to pesticides and/or virulent pest strains in situations where control measures exert intense selection

Table 2.2 The principal control measures appropriate for different pest strategies. (From Conway, 1984)

Control measure	r-*Pests*	*Intermediate pests*	K-*Pests*
Insecticides	Early wide-scale applications based on forecasting	Selective insecticides	Precisely targeted applications based on monitoring
Biological control		Introduction or enhancement of natural enemies	
Cultural control	Timing, cultivation, sanitation, and rotation →		← Changes in agronomic practice, destruction of alternative hosts
Resistance	General, polygenic resistance →		← Specific, monogenic resistance
Genetic control			Sterile mating technique

pressure. *r*-Pests are particularly adept at breaking resistance in crop cultivars where resistance is under monogenic control. Polygenic resistance is more likely to be durable against *r*-pests. Cultural control will be useful where it serves to reduce the likelihood and size of the pest attack. Early or late planting, for example, may prevent crop damage, while the production of weed-free seed and mechanical cultivation may prevent the build-up of *r*-selected weeds (Conway, 1984).

K-Pests have characteristically lower fecundities, longer generation times, are less mobile and make a greater investment in host specialization, occupying relatively narrow niches. Hence, they are particularly vulnerable to control measures that affect the niche, such as hostplant resistance (monogenic resistance is appropriate here) or cultural control measures that render a crop unattractive, e.g. changes in planting density, or cropping patterns such as intercrops. Pesticides are appropriate, particularly for pests that cause direct damage to the harvestable product such as fruit and root crops. Applications need to be precisely timed however.

Intermediate between *r*- and *K*-pests are those pests which are particularly susceptible to control through conservation, augmentation or introduction of natural enemies (Southwood and Comins, 1976). Where pesticides are considered necessary they should be applied selectively to ensure they do not decimate natural enemy populations. Conway (1984) argues that the intermediate pests should be seen as the initial target in an IPM system, within which the strategies for control of the remaining *r*- and *K*-pests are then integrated.

As a general model, the *r*–*K* framework for control has gained widespread acceptance among entomologists. However, alternative models are available (Stenseth, 1981, 1987; Berryman, 1987) although they share with the *r*–*K* profile a lack of practical utility. Despite this, they represent some of the few frameworks available and are important for generating awareness of the need to approach the problem of control from a more functional perspective.

2.6.2 Selection of control measures: effects on pest populations

There are two fundamental ways in which control may be used to manage damaging pest population levels. Measures are either used to reduce the initial numbers infesting the crop (X_0 or N) or they are used to reduce the rate of population increase (r) after colonization. An IPM programme would aim to reduce both X_0 and r.

Maintaining low initial population sizes may be achieved by a whole range of methods. Quarantine procedures that prevent or reduce the likelihood of the introduction into a country or region of exotic pests would come under this category. The use of certified seed and fungicide seed treatments will reduce the likelihood of pathogen infection. Pathogen populations may also be reduced by removal and destruction of susceptible plants or diseased plant parts. Phytosanitary practices, roguing and elimination of alternative hosts all act to reduce initial levels of inoculum. The growing plant, if necessary, may then be protected from inoculum by use of protectant fungicides. Crop rotations have been used as a means of control for centuries, for reducing initial population size of many types of pests. Rotations tend to be most effective against pest species that have a narrow host range and a limited range of dispersal (Wright, 1984). In a rotation of potato and wheat the oviposition and first appearance of the Colorado potato beetle (*Leptinotarsa decemlineata*) were delayed (Lashomb and Ng, 1984). This delay was attributed to the physical and environmental barriers that slow emigration from the wheat by the overwintering beetles. Vertical resistance will reduce the size of initial infestation. Individual parasites not able to match gene-for-gene with the crop cultivar will either die or, if an insect, disperse to other hosts. Antixenotic resistance that lowers host attractiveness or repels the insect also serves to lower X_0.

The intrinsic rate of increase r may be reduced by control measures that have an effect on mortality, reproductive rate, development or generation time and the time over which a pest is capable of reproduction. Pesticides are commonly used to kill various developmental stages of the pest, and are particularly effective, in many cases in killing a substantial proportion of a pest population in a very short time. Natural enemies will also reduce r through increasing mortality and, with chronic infestations, through increasing development times and decreasing reproduction. Insect growth regulators will increase development times and mortality while the sterile insect technique will reduce r through decreasing mating success and hence, reproduction. Control measures which can increase development or generation times are particularly useful for pests normally having high r values. Where explosive epidemics occur for instance, one of the first aims should be to increase the latency period (Van der Plank, 1972). This can often be done by plant breeding through horizontal resistance, or antibiotic resistance. Such measures will often also reduce reproductive rate and increase pest mortality.

The interaction of measures to control X_0 and r can either be assessed through use of simple logic or through the use of analytical or simulation models (Chapter 6). Common sense informs us that a pest population with a high rate of increase will be more difficult to control than a population with low initial numbers and a low rate of increase. In situations where initial population size is high but subsequent rates of increase are low, then the application of measures that will reduce initial population size will usually substantially reduce the pest population size. Depending on the state of the host and/or the weather that might favour a higher rate of increase, a marked reduction in initial infestation is unlikely to be offset by improved conditions. When the initial level of infestation is low but the subsequent rate of increase is high the effect of reducing the size of the initial population is to delay the onset of a damaging pest outbreak (Van der

Plank, 1972). However, control measures aimed at reducing initial population size may readily be offset by even a small increase in rate of increase due to environmental conditions favouring the pest.

Zadoks and Schein (1979) provide a graphical representation of a hypothetical situation in which disease progress over time is simplified to a straight line and the population levels are manipulated by various control measures (Figure 2.6). The initial population size X_0 is manipulated by the actions of **a** (sanitation), **b** (change of planting time), the resulting population size (disease severity X_t) is manipulated by **d**, treatment with an eradicant fungicide and **e**, treatment with a protectant fungicide whereas '*r*' is affected by the actions **c**, partial hostplant resistance and **f**, residual resistance in the adult stage, or regular treatments with fungicides. Numer-

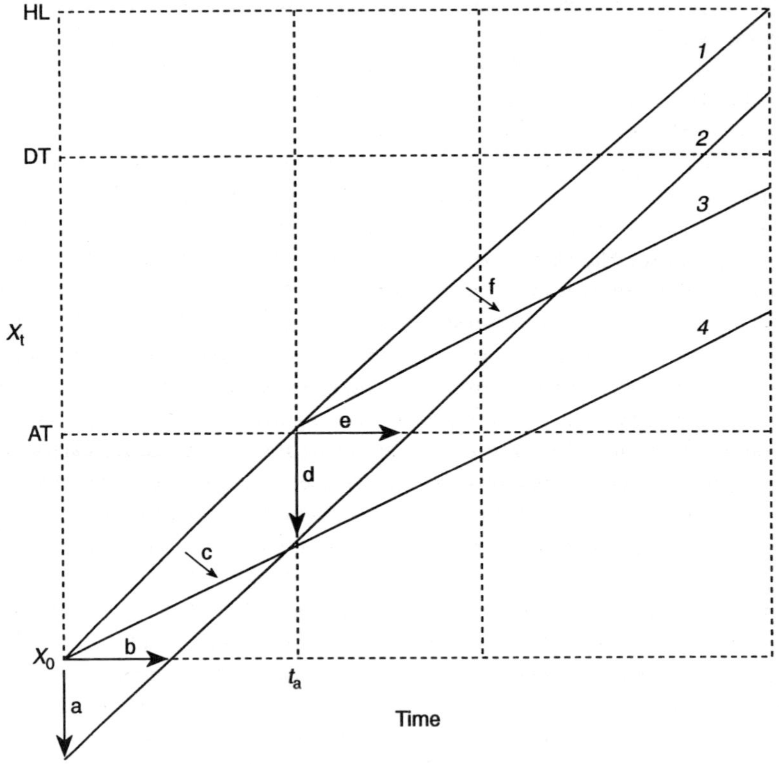

Figure 2.6 A model demonstrating the effects of various disease control actions in terms of the reduction of *r* (rate of increase), X_o (initial inoculum) or X_t (disease severity). 1, Original disease progress curve; 2, same after reduction of X_o or X_t (actions **a** and **d**) or delay of the epidemic (actions **b** and **e**); lines 1 and 2 have the same *r* value; 3, *r* changed after action **f** taken at action time t_a; 4, *r* changed from the beginning of the season by action **c**. HL, harvest level; DT, damage threshold; AT, action threshold. (After Zadoks and Schein, 1979).

ous illustrations of different permutations of this type of hypothetical situation could be devised, but in practice it is necessary to consider the way in which each control measure acts on the pest population and then their effect in combination. This is best done through the use of simulation models. Such models can also determine the effect of various operational factors on the level of pest control achieved for different deployment strategies.

2.6.3 Operational factors

The deployment and efficacy of a control measure will depend on a whole range of operational factors which in turn will be dependent on the characteristics of the pest, the cropping system, the associated environment and the nature of the control agent itself. It would be an impossible task to review the various combinations and permutations of control measures, pest attack scenarios and environmental constraints which exist. However, there are a few general principles concerning the way operational aspects of control are approached, and these are considered here.

A single control measure may be utilized either in a prophylactic or a responsive way. A control measure is used prophylactically when it is applied without any evaluation of whether or not, it will produce an economic gain for the farmer. In contrast, a control measure is considered responsive when an evaluation of the potential gain from its application has been determined (Vandermeer and Andow, 1986). In practice, use of a control measure in a responsive way is dependent on the availability of an appropriate pest monitoring and forecasting system, and the control measures normally involved are pesticides. With pesticide use, it is generally considered preferable for farmers to switch from prophylactic, calender-based applications to responsive, need-based applications, simply to improve targeting and reduce the environmental impact of unnecessary pesticide applications. However, the extent to which farmers are willing to make this change is dependent on a number of factors. The difficulties associated with a choice between the two strategies, prophylactic (calender) spraying and responsive, monitoring and spraying programmes are illustrated by the hypothetical net revenue lines depicted in Figure 2.7 (Norton, 1985). In years when the level of pest attack is very low the farmer may benefit from not monitoring or controlling the pest, but with monitoring still providing a more profitable option than prophylactic applications. As the level of pest attack increases the benefits of monitoring and spraying outweigh those of both no control and prophylactic control. However, in situations where pest attack is more frequent the need for the monitoring programme is reduced, and the costs exceed the benefits, making prophylactic control the most financially attractive option (Figure 2.7). The choice between the use of either prophylactic or responsive measures is rarely simple, largely because of the range of different conditions that can affect the decision (Table 2.3). It is evident from Table 2.3 that there are

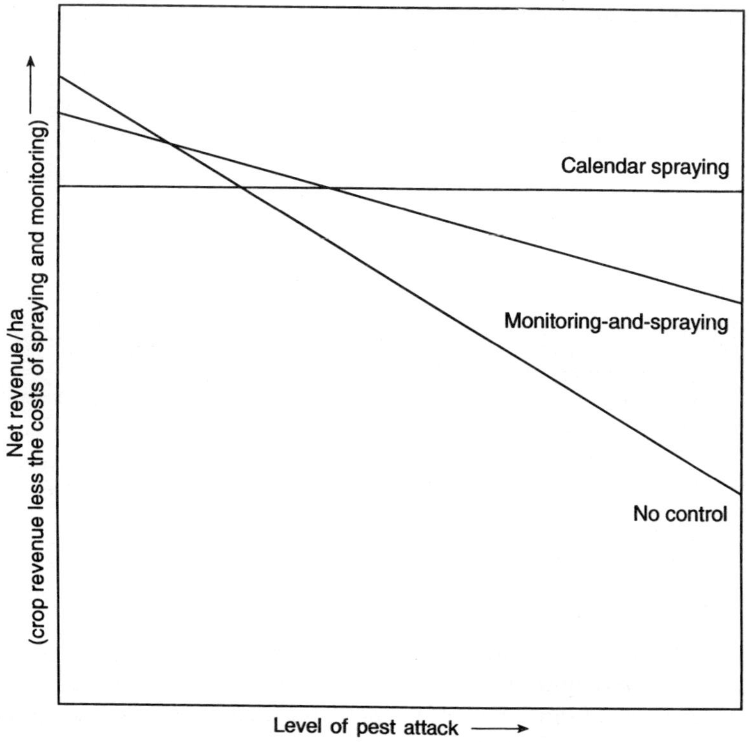

Figure 2.7 Hypothetical net revenue curves for a prophylactic and a responsive-spray strategy at different levels of pest attack. (After Norton, 1985).

many situations where prophylactic control measures are more appropriate than responsive ones; hence, the widespread value of measures such as host-plant resistance, and the cultural control methods.

One of the drawbacks often associated with responsive measures is the need for reapplication due to a lack of persistence. Persistent chemical pesticides are now considered environmentally undesirable which can mean repeated applications of less persistent chemicals or other agents such as microbial pesticides. This may cause operational problems for a farmer. The microbial pesticides are particularly poor at persisting at the target site, with sunlight probably being the most destructive environmental factor affecting the persistence of entomopathogens in general and particularly commercial microbial pesticides (Ignoffo, 1992). This lack of persistence has important implications for the efficacy of the control agent, since if an agent does not persist at the target site then the duration for which it can act against the pest may be severely limited. Timing of application then becomes critical. The application must be made at the time at which the pest is present and at a susceptible stage. For instance, it is important to

Table 2.3 Relative conditions which favour use of prophylactic and responsive control measures. (From Mumford and Norton, 1987)

Prophylactic	*Responsive*
Regular pest attack (in space and time)	Irregular pest attack
Frequent pest attack	Infrequent pest attack
Endogenous pests	Exogenous pests
Multiple pest complexes	Single major pest
High rate of pest increase	Slow rate of increase
High damage cost	Low damage cost
Less effective control	More effective control
Long lasting control	Short duration control
Few bad secondary control effects (pollution, resistance, etc.)	Serious bad secondary effects
Low control cost	High control cost
Limited choice of controls	Wide choice of controls
Poor natural control	Good natural control
Monitoring technique difficult	Monitoring technique easy
High managerial cost	Low managerial cost
Poor access to advice	Good access to advice
More risk aversion	Less risk aversion

apply insecticides against the early instars of podborers such as *Helicoverpa armigera* before they bore into the pod or fruit where they are difficult to control. The size of insects may also affect the dosage required to kill them; hence, lower doses can be applied to control early instars. Timing may also be important in relation to augmentative releases of parasitoids. Parasitoids should not be released before a pest population is large enough to allow natural enemy establishment but not too late that they cannot exert sufficient levels of control before damage is caused. Where a combination of measures is used, for instance, pesticides and natural enemies, then the timing of pesticide application needs to be undertaken to minimize the harm to natural enemy populations and other beneficial organisms. The interaction between pesticides and natural enemies is largely viewed as a negative one, but insecticides can be used to selectively improve the effectiveness of biological control through improving the natural enemy : prey ratio (van Emden, 1983). Of all the potential interactions between control measures most emphasis has been placed on those between pesticides and biological control; however, there are others that are equally important and which can potentially have important implications for effective pest control, i.e. the interaction between insecticides and plant resistance and also between plant resistance and biological control.

Insects that have developed on resistant plant cultivars are usually of smaller size than those from susceptible cultivars. Since the toxicity of an insecticide is related to the body weight of the insect, then the same percentage mortality of the pest can be achieved on the resistant plant as the susceptible one through a lower dose of the insecticide (Selander *et al.*, 1972;

van Emden, 1983, 1991). The effect is thought to be mainly concerned with a physiological sensitivity of the insect to the insecticide in relation to the stress the insect experiences from feeding on the resistant plant (van Emden, 1991).

The smaller size of insects on resistant plants may have a negative interaction with biological control simply because the smaller prey will result in the emergence of smaller parasitoids having a reduced fecundity. Also, the secondary plant substances such as alkaloids involved in the plants' resistance can be toxic to parasitoids within their hosts (Herzog and Funderburk, 1985). These negative effects may be balanced however, by a higher mortality of the pest on resistant than on susceptible plants due to the enhanced effect of natural enemies. This may come about through the effect of resistance slowing the rate of pest population increase and providing more time for natural enemies to have an impact (van Emden, 1983) or because smaller pest insects fail to adequately defend themselves, are more prone to dislodgement caused by restlessness and natural enemy activity or in addition because predators may consume a greater number of smaller prey (van Emden, 1991). All such factors may have a significant impact on reduction in pest numbers.

These two-way interactions between control measures are relatively simple to consider. However, there are potentially more complex interactions (either positive or negative) that could occur in practice, and to evaluate these it is necessary to consider a simulation modelling approach.

2.6.4 Mismanagement strategies

The concept of IPM is readily justified on the basis that it provides an appropriate alternative to the problems which have been caused through the widespread introduction of monogenic resistant cultivars, the misuse of pesticides, the destruction of natural populations of biological control agents and the abandonment of cultural controls. van Emden (1983) refers to this process as pest mismanagement. The subject of mismanagement is dealt with here on the basis of the principles that have been adopted in the development of mismanagement strategies; strategies aimed at either prolonging the useful life of successful control agents or to limit the impact of their failure. In doing so the section draws on the similarities between the strategies used in pesticide resistance management (PRM) and those involved in the management of vertical hostplant resistance genes.

The cost associated with the loss of an important pest control agent through its mismanagement will be dependent on the cost of research, development, registration, and the cost associated with production. In the agrochemical industry during the 1950s and 1960s the loss of a chemical insecticide through the development of pest resistance was not considered too serious because new technical products were continually coming on-line and R & D and registration costs were relatively low. Provided the product life-time (i.e. the length of time before resistance developed) was not too short the industry could make a 'healthy' profit. However, the overall costs of developing chemical pesticides gradually increased as new compounds became more complex and as registration and production costs soared. With an increasing number of organisms, particularly insects and mites, developing resistance to chemical pesticides (447 species by 1984; Roush and McKenzie, 1987) and the increasing difficulties of recouping development costs, it was imperative that strategies be developed to reduce the likelihood of resistance developing. In the event that resistance did occur it was also necessary to identify the means by which continued use of the pesticide could be achieved. A number of operational tactics were introduced to reduce the likelihood of resistance development. These included varying the dose and frequency of applications, using local rather than area-wide applications, use of monitoring systems and less persistent chemicals, and better targeting of specific pest life-stages to ensure a more rational use of pesticides (Croft, 1990). Where a number of insecticides provide control of a pest, resistance may be delayed through use of insecticide rotations or cocktails. Insecticide rotations ensure that pests are exposed to a range of chemicals and hence, selection against any one becomes less likely (Sawicki and Denholm, 1987). With insecticide cocktails a number of different insecticides are combined in a single application. This method is only effective if the total mortality from the cocktail is the same as with a higher dose of a single insecticide and if the proportional mortality rates (from each of the insecticides) in various sections of the population are correlated (Comins, 1984). Sequences of pesticides could also provide a delaying tactic provided there are a number of pesticides available for the control of a particular pest.

The strategies of pesticide rotations, cocktails and sequence are similar to those utilized by plant breeders to prolong the useful life of cultivars with pest resistance based on monogenic characters. Plant breeders talk in terms of use of cultivar sequences, of rotations, of pyramiding genes and of multiline varieties (Gallun and Khush, 1980). The sequential release of resistant genes involves 'holding genes in reserve' and as soon as resistance breakdown appears inevitable a new resistance gene is incorporated into the cultivar. Genes may also be used in rotation; cultivars are withdrawn after breakdown of resistance but are reintroduced at a later period when the original virulent pest races are known to have become scarce. Alternatively, just the resistant genes could be recycled. The pyramiding of genes for resistance involves the introduction of a number of resistance genes into a single crop cultivar. Breakdown of resistance in the cultivar would then only occur if a pest had the same combination of virulence genes. The disadvantage of this approach is that should such a virulent pest occur then a number of valuable resistance genes would be lost at once; a scenario similar to the pattern of cross resistance in an insecticide cocktail. The closest analogy to an insecticide cocktail however, is that of the multiline, which is a crop composed of phenotypically similar component lines differing only in their genes for resistance, derived from a common breeding programme. This approach has been found useful with some pathogens but the approach has not been widely attempted with insect hostplant resistance.

The importance of discussing these mismanagement strategies relates as much to their potential use in the future as to their past use. Given the monogenic nature of many gene manipulation technologies that are likely to find their way into general use in pest management in the next 10 years, their long-term success will be dependent on the use of appropriate deployment strategies. The lessons learnt from PRM and the mismanagement of hostplant resistance breakdown must be applied to the strategies adopted for the use of these new technologies. Without such adherence to the basic principles of gene deployment valuable pest control agents will be quickly squandered.

REFERENCES

Allard, R.W. (1960) *Principles of Plant Breeding*, John Wiley and Sons, New York, London.

Arthur, W. (1982) The evolutionary consequences of interspecific competition. *Advances in Ecological Research*, **12**, 127–87.

Bartsch, R. (1978) *Economic Problems of Pest Control*, Afrika-Studien 99 edn, Ifo-Institut, Munich.

Benz, G. (1987) Integrated Pest Management in Material Protection, Storage and Food Industry, in *Protection intégrée: quo vadis? – PARASITIS 86*, Geneva, pp. 31–69.

Berryman, A.A. (1987) The theory and classification of outbreaks, in *Insect Outbreaks*, (eds P. Barbosa and J.C. Schultz), Academic Press, San Diego, London, pp. 3–30.

Birch, N. and Wratten, S.D. (1984) Patterns of aphid resistance in the genus *Vicia*. *Annals of Applied Biology*, **104**, 327–38.

Brown, E.B. (1969) Assessment of damage caused to potatoes by potato cyst eelworm, *Heterodera rostochiensis* Woll. *Annals of Applied Biology*, **63**, 493–502.

Burn, A.J. (1984) Life-tables for the carrot fly *Psila rosae*. *Journal of Applied Ecology*, **21**, 891–902.

Campion, D.G., Bettany, B.W. and Steedman, R.A. (1974) The arrival of male moths of the cotton leaf-worm *Spodoptera littoralis* (Boisd.) (Lepidoptera: Noctuidae) at a new continuously recording pheromone trap. *Bulletin of Entomological Research*, **64**, 379–86.

Carruthers, R.I. (1985) The use of simulation modelling in insect–fungal disease research, in *Proceedings of the 1985 Summer Computer Simulation Conference*, Chicago, 22–24 July. The Society for Computer Simulation, San Diego, CA, pp. 503–8.

Claridge M.F. (1991) Genetic and biological diversity of insect pests and their natural enemies, in *The Biodiversity of Microorganisms and Invertebrates: Its Role in Sustainable Agriculture*, (ed. D.L. Hawksworth), CAB International, Wallingford, pp. 183–94.

Claridge, M.D. and den Hollander, J. (1982) Virulence to rice cultivars and selection for virulence in population of the brown planthopper *Nilaparvata lugens. Entomologia Experimentalis et Applicata*, **32**, 213–21.

Claridge, M.D., den Hollander, J. and Haslam, D. (1984) The significance of morphometric and fecundity differences between the 'biotypes' of the brown planthopper *Nilaparvata lugens. Entomologia Experimentalis et Applicata*, **36**, 107–14.

Coaker, T.H. (1987) Cultural methods: the crop, in *Integrated Pest Management*, (eds A.J. Burn, T.H. Coaker and P.C. Jepson), Academic Press, London, pp. 69–88.

Comins, H.N. (1984) The mathematical evaluation of options for managing pesticide resistance, in *Pest and Pathogen Control: Strategic, Tactical and Policy Models*, (ed. G.R. Conway), John Wiley & Sons, Chichester, pp. 454–69.

Conway, G.R. (1984) *Pest and Pathogen Control: Strategic, Tactical and Policy Models*, John Wiley & Sons, Chichester.

Conway, G.R. (1991) Man versus pests, in *Theoretical Ecology: Principles and Applications*, 2nd edn, (ed. R.M. May), Blackwell, Oxford, pp. 356–86.

Cooke, G.W. (1982) *Fertilizing for Maximum Yield*, 3rd edn, Longman Scientific & Technical, London.

Corbet, P.S. (1970) Pest management: objectives and prospects on a global scale, in *Concepts of Pest Management*, (eds R.L. Rabb and F.E. Guthrie), North Carolina State University, Raleigh, pp. 191–208.

Croft, B.A. (1990) Developing a philosophy and program of pesticide resistance management, in *Pesticide Resistance in Arthropods*, (eds R.T. Roush and B.E. Tabashnik), Chapman & Hall, London, pp. 277–96.

Daly, J.C. (1991) Methods for studying the genetics of populations of *Heliothis*, in *Heliothis: Research Methods and Prospects*, (ed. M.P. Zalucki), Springer-Verlag, Berlin, pp. 157–70.

Dawson, J.H. and Holstun, J.T. (1971) Estimating losses from weeds in crops, in *Crop Loss Assessment Methods*, FAO Manual on the evaluation and prevention of losses by pests, diseases and weeds, CAB, Farnham/Rome, pp. 1–4.

den Hollander, J. and Pathak, P.K. (1981) The genetics of the 'biotypes' of rice brown planthopper, *Nilaparvata lugens. Entomologia Experimentalis et Applicata*, **29**, 76–86.

Dent, D.R. (1983). The biology and hostplant relationships of the grass aphid *Metopolophium festucae cerealium*. PhD Dissertation, University of Southampton, 184 pp.

Dent, D.R. (1986) Resistance to the aphid *Metopolophium festucae cerealium*: effects of the host plant on flight and reproduction. *Annals of Applied Biology*, **108**, 577–83.

Dent, D.R. (1991) *Insect Pest Management*, CAB International, Wallingford.

Dent, D.R. (1993) Products versus techniques in community-based pest management, in *Community-Based and Environmentally Safe Pest Management*, (eds R.K. Saini and P.T. Haskell), ICIPE Science Press, Nairobi, pp. 169–80.

Dent, D.R. and Pawar, C.S. (1988) The influence of moonlight and weather on catches of *Helicoverpa armigera* (Hübner) (Lepidoptera: Noctuidae) in light and pheromone traps. *Bulletin of Entomological Research*, **78**, 365–77.

Dent, D.R. and Wratten, S.D. (1986) The host-plant relationships of apterous virginoparae of the grass aphid *Metopolophium festucae cerealium. Annals of Applied Biology*, **108**, 567–76.

Diehl, S.R. and Bush, G.L. (1984) An evolutionary and applied perspective of insect biotypes. *Annual Review of Entomology*, **29**, 471–504.

Flor, H.H. (1942) Inheritance of pathogenicity in *Melampsora lini. Phytopathology*, **32**, 653–69.

Futuyama, D.J. and May, R.M. (1991) The coevolution of plant-insect and host-parasite relationships, in *Genes in Ecology*, (eds R.J. Berry, T.J. Crawford and G.M. Hewitt), Blackwell Scientific Publications, Oxford, pp. 139–66.

Gallun, R.L. and Khush, G.S. (1980) Genetic factors affecting expression and stability of resistance, in *Breeding Plants Resistant to Insects*, (eds F.G. Maxwell and P.R. Jennings), John Wiley & Sons, New York, Chichester, pp. 63–85.

Gause, G.F. (1934) *The Struggle for Existence*, Williams and Wilkins, Baltimore.

Givovich, A., Weibull, J. and Pettersson, J. (1988) Cowpea aphid performance and behaviour on two resistant cowpea lines. *Entomologia Experimentalis et Applicata*, **49**, 259–64.

Gutierrez, A.P. (1992) The physiological basis of ratio-dependent predator-prey theory: the metabolic pool model as a paradigm. *Ecology*, **73**, 1552–63.

Gutierrez, A.P., Neuenschwander, P. and van Alphen, J.J.M. (1993) Factors affecting biological control of cassava mealybug by exotic parasitoids: a ratio-dependent supply-demand driven model. *Journal of Applied Ecology*, **30**, 706–21.

Hanson, C.H., Busbice, T.H., Hill, R.R., Hunt, O.J. and Oakes, A.J. (1972) Directed mass selection for developing multiple pest resistance and conserving germplasm in alfalfa. *Journal of Environmental Quality*, **1**(1), 106–11.

Harvey, J. (1977) *Modern Economics. An Introduction for Business and Professional Students*, 3rd edn, The Macmillan Press Ltd, London.

Hassell, M.P. (1978) *The Dynamics of Arthropod Predatory-Prey Systems*, 1st edn., Princeton University Press, Princeton, New Jersey.

Herzog, D.C. and Funderburk, J.E. (1985) Plant resistance with cultural practice interactions with biological control, in *Biological Control* in *Agricultural IPM Systems*, (eds M.A. Hoy and D.C. Herzog), Academic Press, Orlando, pp. 67–88.

Holt, J. and Wratten, S.D. (1986) Components of resistance to *Aphis fabae* in faba bean cultivars. *Entomologia Experimentalis et Applicata*, **40**, 35–40.

Horn, D.J. (1988) Integrated insect pest management, in *Ecological Approach to Pest Management*, (ed. D.J. Horn), Elsevier Applied Science Publishers, London, pp. 133–49.

Huffaker, C.B., Simmonds, F.J. and Laing, J.E. (1976) The theoretical and empirical basis of biological control, in *Theory and Practice of Biological Control*, (eds C.B. Huffaker and P.S. Messenger), Academic Press, Orlando, pp. 42–78.

Ignoffo, C.M. (1992) Environmental factors affecting persistence of entomopathogens. *Florida Entomologist* **75**(4), 516–25.

Jones, A., Dukes, P.D. and Cuthbert, F.P. (1976) Mass selection in sweet potato: breeding for resistance to insects and diseases and for horticultural characteristics. *Journal of American Society Horticultural Science*, **101**(6), 701–4.

Jones, F.G.W. (1973) Management of nematode populations in Great Britain. *Proceedings Tall Timbers Conference Ecology Animal Control Habitat Management*, **4**, 81–107.

Kakehashi, N., Suzuki, Y. and Iwasha, Y. (1984) Niche overlap of parasitoids in host-parasitoid systems: its consequence to single versus multiple introduction controversy in biological control. *Journal of Applied Ecology*, **21**, 115–31.

Kerry, B. (1993) The use of microbial agents for the biological control of plant parasitic nematodes, in *Exploitation of Microorganisms*, (ed. D.G. Jones), Chapman & Hall, London, pp. 81–104.

Krebs, C.J. (1978) *Ecology. The Experimental Analysis of Distribution and Abundance*, 2nd edn, Harper & Row, New York.

Lashomb, J.H. and Ng, Yuen-S. (1984) Colonization by Colorado potato beetles, *Leptinotarsa decemlineata* (Say) (Coleoptera: Chrysomelidae), in rotated and nonrotated potato fields. *Environmental Entomology*, **13**, 1352–6.

Lewis, T. and Macaulay, E.D.M. (1976) Design and elevation of sex-attractant traps for pea moth, *Cydia nigricana* (Steph.) and the effect of plume shape on catches. *Ecological Entomology*, **1**, 175–87.

Lockhart, J.A.R. and Wiseman, A.J.L. (1978) *Introduction to Crop Husbandry*, 4th edn, Pergamon Press, London.

Lowe, H.J.B. (1978) Detection resistance to aphids in cereals. *Annals of Applied Biology*, **88**, 401–6.

Lowe, H.J.B. (1982) Some observations on susceptibility and resistance of winter wheat to the aphid *Sitobion avenae* (F.) in Britain. *Crop Protection*, **1**(4), 431–40.

MacArthur, R.H. and Wilson, E.O. (1967) *The Theory of Island Biogeography*, Princeton University Press.

Marquis, R.J. and Alexander, H.M. (1992) Evolution of resistance and virulence in plant-herbivore and plant–pathogen interactions. *TREE*, **7**(4), 126–9.

Mather, K. (1955) Polymorphism as an outcome of disruptive selection. *International Journal of Organic Evolution*, **9**, 52–61.

Maxwell, F.G. and Jennings, P.R. (1980) *Breeding Plants Resistant to Insects*, John Wiley & Sons Inc., New York.

May, R.M. and Hassell, M.P. (1981) The dynamics of multiparasitoid–host interactions. *The American Naturalist*, **117**, 234–61.

May, R.M., Hassell, F.R.S. and Hassell, M.P. (1988) Population dynamics and biological control. *Phil. Transactions of the Royal Society of London*, **318**, 129–69.

Meyer, J.A., Lancaster, J.L. and Simco, J.S. (1982) Comparison of habitat modification, animal control, and standard spraying for control of the lone star tick. *Journal of Economic Entomology*, **75**, 524–9.

Müller, H.J. (1958) The behaviour of *Aphis fabae* in selecting its host plants, especially different varieties of *Vicia faba*. *Entomologia Experimentalis et Applicata*, **1**, 66–72.

Mumford J.D. and Norton, G.A. (1984) Economics of decision making in pest management. *Annual Review of Entomology*, **29**, 157–74.

Mumford, J.D. and Norton, G.A. (1987) Economic aspects of integrated pest management, in *Integrated Pest Management, Protection Integrée: Quo Vadis? An International Perspective*, (ed. V. Delucchi), Parasitis **86**, Geneva, pp. 397–408.

Nicolaisen, W. (1934) Die Grundlagen der Immunitätszüchtung gegen *Ustilago avenae* (Pers.). *Jens Z Zeucht A Aflzeucht*, **19**, 1.

Norgaard, R.B. (1976) The economics of improving pesticide use. *Annual Review of Entomology*, **21**, 45–60.

Norton, G.A. (1977) Background to agricultural pest management modelling, in *Proceedings of Conference on Pest Management*, (eds G.A. Norton and C.S. Holling), IIASA, Laxenburg, Austria, pp. 161–76.

Norton, G.A. (1985) Economics of pest control, in *Pesticide Application: Principles and Practice*, (ed. P.T. Haskell), Clarendon Press, Oxford, pp. 175–89.

Painter, R.H. (1951) *Insect Resistance in Crop Plants*, The MacMillan Co., New York.

Parkin, D.T. (1979) *An Introduction to Evolutionary Genetics*, Edward Arnold (Publishers) Ltd, Whitstable, Kent.

Parry, D.W. (1990) How do we control disease?, in *Plant Pathology in Agriculture*, (ed. D.W. Parry), Cambridge University Press, Cambridge, pp. 86–158.

Pedgley, D.E. (1982) *Windborne Pests and Diseases: Meteorology of Airborne Organisms*, Ellis Horwood Ltd, Chichester.

Pimentel, D. (1991) Diversification of biological control strategies in agriculture. *Crop Protection*, **10**(1), 243–53.

Podoler, H. and Rogers, D. (1975) A new method for the identification of key factors from life-table data. *Journal of Animal Ecology*, **44**, 85–114.

Puche, H. and Funderburk, J. (1992) Intrinsic rate of increase of *Frankliniella fusca* (Thysanoptera: Thripidae) on peanuts. *Florida Entomologist*, **75**(2), 185–9.

Putman, R.J. and Wratten, S.D. (1984) *Principles of Ecology*, Croom Helm, London.

Redmond, J.C., Marois, J.J. and MacDonald, J.J. (1987) Biological control of *Botrytis cinerea* on roses with epiphytic microorganisms. *Plant Disease*, **71**, 799–802.

Reichelderfer, K.H., Carlson, G.A. and Norton, G.A. (1984) *Economic guidelines for crop pest control*. FAO Plant Production and Protection, Paper 58.

Rhodes, D.J. (1993) Formulation of biological control agents, in *Exploitation of Microorganisms*, (ed. D.G. Jones), Chapman & Hall, London, pp. 411–39.

Robinson, R.A. (1980a) New concepts in breeding for disease resistance. *Annual Review of Phytopathology*, **18**, 189–210.

Robinson, R.A. (1980b) The pathosystem concept, in *Breeding Plants Resistant to Insects*, (eds F.G. Maxwell and P.R. Jennings), John Wiley & Sons, New York, pp. 157–82.

Room, P.M., Titmarsh, I.J. and Zalucki, M.P. (1991) Life tables, in *Heliothis: Research Methods and Prospects*, (ed. M.P. Zalucki), Springer-Verlag, Berlin, pp. 69–80.

Roush, R.T. and Daly, J.C. (1990) The role of population genetics in resistance research and management, in *Pesticide Resistance in Arthropods*, (eds R.T. Roush and B.E. Tabashnik), Chapman & Hall, London, pp. 97–152.

Roush, R.T. and McKenzie, J.A. (1987) Ecological genetics of insecticide and acaricide resistance. *Annual Review of Entomology*, **32**, 361–80.

Russell, G.E. (1978) *Plant Breeding for Pest and Disease Resistance*, Butterworth, Oxford.

Ruthenberg, H. and Jahnke, H.E. (1985) *Innovation Policy for Small Farmers in the Tropics*, Oxford University Press, Oxford.

Sawicki, R.M. and Denholm, I. (1987) Management of resistance to pesticides in cotton pests. *Tropical Pest Management*, **33**(4), 262–72.

Saxena, R.C. and Barrion, A.A. (1987) Biotypes of insect pests of agricultural crops. *Insect Science and its Applications*, **8**, 453–8.

Schoener, T.W. (1982) The controversy over interspecific competition. *American Science*, **70**, 586–95.

Selander, J.M., Markkula, M. and Tiittanen, K. (1972) Resistance of the aphids *Myzus persicae* (Sulz.), *Aulocorthum solani* (Kalt.) and *Aphis gossypii* Glov. to insecticides and the influence of host plant on this resistance. *Annales agriculturae Fenniae*, **11**, 141–5.

Siddiqui, W.H. and Barlow, C.A. (1973) Effects of some constant and alternating temperatures on population growth of the pea aphid *Acyrthosiphon pisum* (Homoptera: Aphididae). *The Canadian Entomologist*, **105**, 145–56.

Simmonds, N.W. (1979) *Principles of Crop Improvement*, Longman, Harlow, Essex.

Sinnott, E.W., Dunn, L.C. and Dobzhansky, T. (1958) *Principles of Genetics*, 5th edn, Tata McGraw Hill Publishing Co Ltd, London, New York.

Smith, R.F., Apple, J.L. and Bottrell, D.G. (1976) The origins of integrated pest management concepts for agricultural crops, in *Integrated Pest Management*, (eds J.L. Apple and R.F. Smith), Plenum Press, New York, London, pp. 1–16.

Smyth, K. (1993) Isolation and characterisation of *Bacillus thuringiensis* strains for the control of *Helicoverpa armigera*. PhD Dissertation, University of Wales, Cardiff.

Southwood, T.R.E. (1977) Habitat, the templet for ecological strategies? *Journal of Animal Ecology*, **46**, 337–65.

Southwood, T.R.E. (1978) Systems analysis and modelling in ecology, in *Ecological Methods with Particular Reference to the Study of Insect Populations*, 2nd edn., Chapman & Hall, London, pp. 407–19.

Southwood, T.R.E. and Comins, H.N. (1976) A synoptic population model. *Journal of Animal Ecology*, **45**, 949–65.
Southwood, T.R.E. and Way, M.J. (1970) Ecological background to pest management, in *Concepts of Pest Management*, (eds R.L. Rabb and F.E. Guthrie), North Carolina State University Press, Raleigh, pp. 6–29.
Stanlake, G.F. (1989) *Introductory Economics*, 5th edn, Longman, Harlow, Essex.
Stenseth, N.C. (1981) How to control pest species: application of models from the theory of island biogeography in formulating pest control strategies. *Journal of Applied Ecology*, **18**, 773–94.
Stenseth, N.C. (1987) Evolutionary processes and insect outbreaks, in *Insect Outbreaks*, (eds P. Barbosa and J.C. Schultz), Academic Press, Orlando, pp. 533–64.
Stern, V.M., Smith, R.F., van den Bosch, R. and Hagen, K.S. (1959) The integrated control concept. *Hilgardia*, **29**, 81–101.
Strickberger, M.W. (1976) *Genetics*, 2nd edn, Macmillan Publishing Co. Inc., New York.
Strong, D.R., Lawton, J.H. and Southwood, R. (1984) *Insects on Plants Community Patterns and Mechanisms*, Harvard University Press, Cambridge, MA.
Tarr, S.A.J. (1972) *Principles of Plant Pathology*, Macmillan, London.
Taylor, L.R. (1963) Analysis of the effect of temperature on insects in flight. *Journal of Animal Ecology*, **32**, 99–117.
Tilman, D. (1989) Population dynamics and species interactions, in *Perspectives in Ecological Theory*, (eds J. Roughgarden, R.M. May and S.A. Levin), Princeton University Press, New York, pp. 89–100.
Tucker, M.R. (1984) Forecasting the severity of armyworm seasons in East Africa from early season rainfall. *Insect Science Application*, **5**(1), 51–5.
Tucker, M.R. and Pedgley, D.E. (1983) Rainfall and outbreaks of the African armyworm, *Spodoptera exempta* (Walker) (Lepidoptera: Noctuidae). *Bulletin of Entomological Research*, **73**, 196–9.
Tucker, M.R., Mwandoto, S. and Pedgley, D.E. (1982) Further evidence for windborne movement of armyworm moths, *Spodoptera exempta*, in East Africa. *Ecological Entomology*, **7**, 463–73.
Van der Plank, J.E. (1963) *Plant Diseases: Epidemics and Control*, Academic Press, New York.
Van der Plank, J.E. (1972) Basic principles of ecosystems analysis, in *Pest Control: Strategies for the Future*, National Academy of Sciences, Washington DC, pp. 109–18.
Van der Plank, J.E. (1975) *Principles of Plant Infection*, Academic Press, Orlando.
van Emden, H.F. (1983) Pest management – routes and destinations. *Antenna*, **7**, 163–8.
van Emden, H.F. (1991) The role of host plant resistance in insect pest mismanagement. *Bulletin of Entomological Research*, **81**, 123–6.
van Emden, H.F., Hughes, R.D., Eastop, V.R. and Way, M.R. (1969) The ecology of *Myzus persicae*. *Annual Review of Entomology*, **14**, 197–270.
Vandermeer, J. (1989) *The Ecology of Intercropping*, University Press, Cambridge.
Vandermeer, J. and Andow, D.A. (1986) Prophylactic and responsive components of an integrated pest management program. *Journal of Economic Entomology*, **79**, 299–302.
Varley, G.C. and Gradwell, G.R. (1960) Key factors in population ecology. *Journal of Animal Ecology*, **29**, 399–401.
Varley, G.C., Gradwell, G.R. and Hassell, M.P. (1973) *Insect Population Ecology an Analytical Approach*, Blackwell Scientific Publications, Oxford.
Via, S. (1990) Ecological genetics and host adaptation in herbivorous insects: the experimental study of evolution in natural and agricultural systems. *Annual Review of Entomology*, **35**, 421–46.

Waage, J.K. and Hassell, M.P. (1982) Parasitoids as biological control agents – a fundamental approach. *Parasitology*, **84**, 241–68.

Webster, C.C. and Wilson, P.N. (1980) *Agriculture in the Tropics*, 2nd edn, Longman Scientific & Technical, London.

Whipps, J.M. and McQuilker, M.P. (1993) Aspects of biocontrol of fungal plant pathogens, in *Exploitation of Microorganisms*, (ed. D.G. Jones), Chapman & Hall, London, pp. 45–79.

Wright, R.J. (1984) Evaluation of crop rotaton for control of Colorado potato beetles (Coleoptera: Chrysomelidae) in commercial potato fields on Long Island. *Journal of Economic Entomology*, **77**, 1254–9.

Wrigley, G. (1981) *Tropical Agriculture The Development of Production*, 4th edn., Longman Scientific & Technical, London.

Wyatt, I.J. and Brown, S.J. (1977) The influence of light intensity, daylength and temperature on increase rates of glasshouse aphids. *Journal of Applied Ecology*, **14**, 391–9.

Zadoks, J.C. and Schein, R.D. (1979) *Epidemiology and Plant Disease Management*, Oxford University Press, Oxford.

CHAPTER 3

Control measures

D.R. Dent

3.1 INTRODUCTION

The range of available control measures for use in an integrated pest management (IPM) programme is gradually increasing, providing an ever wider choice, often of more environmentally friendly methods. The different control measures are the essential tools with which an IPM programme/system is constructed. A good knowledge of each of the measures is required if they are to be integrated into a common system. Since they each represent highly specialized subjects in their own right they can only be referred to in brief here. For more detailed information readers should consult Haskell (1985; pesticides), Jervis and Kidd (1995; natural enemies), Singh (1986; hostplant resistance), Vandermeer (1989; inter-cropping) and for insect pest management techniques in general Dent (1991).

3.2 PESTICIDES

The use of chemical pesticides for the control of all manner of pests has dominated pest management since the 1950s. It is not difficult to understand why this is so, when one considers their effectiveness (visibly so in many cases), their relatively low cost, ease of use and their versatility. Chemical pesticides have suited the needs of farmers, industry and policy makers alike, as an efficient means of pest control which assists in maintaining the productivity of high input, intensive cropping systems. However, with food shortages now much less likely in the developed world combined with an increasing concern by the general public for the environment, the political landscape is changing in favour of alternative, more environmentally friendly means of control. However, the various pesticides: insecticides, fungicides, rodenticides, herbicides, etc., are likely to remain important components of the pest manager's armoury for the foreseeable future.

Integrated Pest Management Edited by David Dent. Published in 1995 by Chapman & Hall, London. ISBN 0 412 57370 9

3.2.1 Insecticides

Chemical insecticides are usually classified in one of two ways, either in terms of their chemical characteristics or alternatively according to their type of activity or mode of action. There are four chemical types of insecticide; organochlorines, organophosphates, carbamates and pyrethroids (Table 3.1). The organochlorines, or the chlorinated hydrocarbons, are broad-spectrum and highly persistent insecticides that are most effective against biting and chewing insects. They were discovered and largely developed between 1942 and 1956 and played an important part in the early success of synthetic insecticides (Barlow, 1985), the most celebrated among them being DDT (see section 7.2).

The organophosphate insecticides are less persistent than the organochlorines, but this means that the timing of application has to be more accurate. Originating from nerve gases, the organophosphates combine with the enzyme acetylcholinesterase at nerve junctions and prevent transmission of nervous impulses. Some are highly selective while others have an extremely high mammalian toxicity. Hence, they need to be handled with care since doses may be cumulative (Hill and Waller, 1982).

The carbamates are general insecticides with a mode of action similar to the organophosphates although their effect is more easily reversed, resulting in recovery of the insects if sufficiently high doses are not taken up. Often used where pesticide resistance has developed in other chemical classes, carbamates are generally of short to medium persistence.

Relatively low persistence is also a characteristic of the synthetic pyrethroids. Their natural counterparts, derived from pyrethrum flowers, are unstable in light and air, as were the earlier synthetic pyrethroids, allerthrin and resmethrin. The later examples of the group however, e.g. permethrin, cypermethrin and delta-methrin, are much more stable and have proved very effective against most groups of insects. Most pyrethroids, but not all, have very low mammalian toxicity and hence, are relatively safe for the spray operators during mixing and application.

Many of the organophosphates have systemic activity, where the insecticide enters the vascular tissues of the plant, either through the foliage or the roots and is translocated throughout the plant to the sites of pest attack. This mode of action is particularly useful for the control of sucking pests such as aphids and planthoppers. Leaf miners are readily controlled by

Table 3.1 The different chemical groups of insecticides

Chemical group	*Examples*
Organochlorines	DDT, aldrin, dieldrin, endosulfan, lindane
Organophosphates	Chlorpyrifos, dimethoate, malathion, phorate
Carbamates	Aldicarb, carboryl, pirimicarb
Pyrethroids	Permethrin, cypermethrin, delta-methrin

pesticides with translaminar activity. These pesticides are capable of penetrating the leaf cuticle and passing across the leaf lamina; however, they tend to have only limited or no systemic activity. Fumigants are used for the control of stored product pests. They are effective in the vapour phase and can be extremely hazardous to humans.

Insecticides which act primarily as a stomach poison have to be consumed by the insects; hence, they are only effective against chewing insects such as lepidopterous larvae, coleoptera and orthopteran species. The organochlorine, organophosphate and the carbamate classes of insecticides all act as stomach poisons. The limitation of stomach poisons and contact insecticides alike is that they must be targeted effectively in order to ensure transfer to the insect. Insecticide deposits applied to surfaces not frequented by the insects will not be effective.

3.2.2 Herbicides

Herbicides may be classified according to their selectivity, their application characteristics, the nature of their action or to their chemistry. Herbicides that are described as non-selective are those which kill all vegetation, e.g. sodium chlorate, whereas selective herbicides will control weeds without causing harm to the crop plants. One of the most widely known non-selective herbicides is paraquat or 1,1′-dimethyl-4,4′-bipyridylium dichloride. First synthesized in 1954, paraquat is a herbicide which rapidly destroys green plant tissue but is absorbed and rendered inactive almost instantaneously by the soil. Paraquat was first marketed in 1962 as PP910 for use in industrial weed control but later renamed Gramoxone.

2,4-D or 2,4-dichlorophenoxyacetic acid is a selective herbicide which was discovered around 1942, and has been used widely since then. The nature of the selectivity of 2,4-D and some other herbicides is dependent on the crop type and its growth stage. 2,4-D damages wheat if applied before the four-leaf growth stage or after jointing (Caseley, 1994).

In general, selectivity of herbicides may be due to differential retention, uptake, movement, metabolism or biochemical action of the herbicide in the crop and the weed (Kirkwood, 1987). The widest margin of selectivity is found with herbicides which are unable to interact at the target binding site in the crop. For instance, fluazifop-butyl inhibits acetyl co-enzyme A in grasses but in broad-leaved plants binding is prevented by the topography of the target niche and there is no toxic effect (Caseley, 1994).

Herbicide action may be either through contact with plant surfaces to which they are applied, or systemic action, where the chemical moves within the plant tissue to areas remote from the site of application. Paraquat is a contact herbicide and acts on the membrane systems of the tissues with which it comes in contact. By contrast 2,4-D is a systemic herbicide which tends to be slower in action than contact herbicides because the compound has to be translocated to its sites of metabolic activity.

Herbicides classified according to application characteristics may be further divided on the basis of the region and timing of application. Herbicides are usually applied to either the foliage or the soil. The soil-applied herbicides need not be selective if no crop is present. Where a crop is present they are normally absorbed by the roots and transported to the shoot in the transpiration stream (Kirkwood, 1987). Foliage-applied herbicides penetrate the outer waxy cuticle and are absorbed into the leaf tissues. There may be some translocation in the phloem.

The classification on the basis of timing of application is organized with reference to the crop. Thus there are pre-sowing, pre-emergent or post-emergent herbicides. Pre-sowing (or pre-planting) treatments are applied before the crop is sown. Some may be incorporated into the soil by shallow cultivation. A pre-emergent contact herbicide can make use of the fact that the crop, although sown, is not yet exposed to the chemical since it is pro-

Table 3.2 The different types of herbicide groups and an example of each. (From Caseley, 1994)

Groups	*Example*
Bipyridiniums	Paraquat
Anilides	Propanil
Niriles	Bromoxynil
Triazines	Atrazine
Triazinones	Metribuzin
Ureas	Diuron
Uracils	Lenacil
Diphenyl ethers	Acifluorfen
Anilides	Diflufenican
Pyridazinones	Norflurazon
Aryloxyphenoxy-alkanoic acid esters	Dichlofop-methyl
Haloaliphatic acids	Dalapon
Oximes	Sethoxydim
Thiocarbamates	EPTC
Amides	Diphenamid
Carbamates	Asulam
Chloroacetanilides	Alachlor
Dinitroanilines	Pendimethalin
Arylcarboxylic acids	Dicamba
(Aryloxy) alkanoic acids	2,4-D, MCPA
Quinoline carboxylic acids	Quinclorac
Nitriles	Bromoxynil
Imidazolinones	Imazethapyr
Sulfonylureas	Metsulfuron
Organophosphorus compounds	Glyphosate Glufosinate

tected by the soil (Lacey, 1985). Alternatively, a germination inhibitor may be applied to the soil as a post-emergent treatment to control the weeds after the crop has emerged. The timing of application basically provides an additional means of selectivity; however, this may also be achieved purely through herbicide target sites which depends on the chemical group used for the active ingredient. Although most herbicide groups affect either photosynthesis or cell division and growth some affect a number of target sites.

The herbicide groups (Table 3.2), their structures, sites of action and selectivity have recently been reviewed by Caseley (1994). The subject is beyond the scope of this text but interested readers should consult the latter text or alternatively Kirkwood (1987).

3.2.3 Fungicides

The recent advances in the chemical control of plant diseases has greatly expanded the effectiveness of fungicide use (Waller, 1985). Within the overall framework of pesticide use however, there are a number of important pathogen characteristics which create specific problems for control with fungicides that are worthy of mention. First among these is the fact that pathogens and plants are physiologically very similar. This makes phytotoxicity a major problem in the development of fungicides. Another important point to note is that pathogens spend most of their life-cycle within the tissues of their hostplants. This makes the placement and timing of fungicide application a crucial factor in successful control. Lastly, the application of fungicides at the time which disease symptoms appear is often too late. Since monitoring of pathogen dispersal in many cases is not a feasible option, there is a need for use of prophylactic chemical control. Given all of these things, it is no surprise that there are a variety of ways in which fungicides can be classified. Püntner (1981) lists the following: (i) by chemical composition; (ii) by application site; (iii) by timing of application and mode of action; and (iv) by product behaviour.

Fungicides classified according to their chemical composition are usually referred to as inorganic or organic fungicides, or as antibiotics (Table 3.3). The earliest fungicides were inorganic materials such as sulphur, copper and mercury compounds. The most renowned of the early fungicides, the Bordeaux mixture, consists of copper sulphate and hydrated lime and was used for the control of vine downy mildew. The first organic fungicides were the dithiocarbamates discovered in 1934. Over the subsequent 10 years products such as thiram, zineb and maneb were developed (Parry, 1990). The first antibiotics examined for use against pathogenic fungi however, were those employed in human chemotherapy, namely penicillin. Despite the obvious success of penicillin in medicine it never achieved commercial significance as a systemic fungicide (Cremlyn, 1978). Antibiotics have however, been developed and now represent a very important and useful class of fungicides.

Table 3.3 Categories and examples of fungicides

Inorganic fungicides	*Organic fungicides*
Copper compounds	General protectants (dithiocarbamates)
Bordeaux mixture	Anilazine
Copper (II) hydroxide	Captan
Copper naphthenates	Cuframeb
Copper (I) oxide	Dichlorfluanid
	Dithianon
Sulphur compounds	Drazoxolon
Lime sulphur	Folpet
Sulphur	Halacrinate
	Mancozeb
Mercurial compounds	Maneb
Mercury (I) and mercury (II) chloride	Prochloraz
	Thiram
Methoxyethyl mercury salts	Zineb
Non-mercurial metallic compounds	Specific narrow-range protectants
Triphenyltin compounds	Binaparcyl
Methyl arsenic sulphide	Dichloran
	Dinocap
Antibiotics	Edifenphos
Cycloheximide	Iprodine
Streptomycin	
Tetracycline, terramycin	Protectants for seed and soil application
Validamycin	Bronopol
	Dichlone
	Guazatine
	Hexachlorobenzene
	Quintozene
	Systemic fungicides
	Benomyl
	Bupirimate
	Carboxin
	Dimethirimol
	Ethirimol
	Imizalil
	Metalaxyl
	Thiabendazole
	Triadimefon
	Tridemorph

Fungicides may be applied as seed dressings, or as foliar or soil treatments. The treatment of seed with fungicides was one of the first methods used for chemical disease control. It represents a simple and economical means of application which ensures accurate placement of the fungicide. Placement is also important for foliar fungicides. If the susceptible crop surfaces do not receive adequate coverage then pathogens which infect the aerial parts of plants may spread quickly and produce devastating epidemics. In a growing crop repeated fungicide applications may be necessary to ensure new plant structures are protected. The duration of effective fungicidal cover determines the interval between applications and is related to crop growth, weather conditions and the persistence of the fungicides (Waller, 1985).

The application of fungicides to soil may take a number of different forms. It may involve direct application to roots when plants are transplanted (e.g. in the control of club root *Plasmodiophora brassicae* on brassica crops), as a fungicidal drench (a method sometimes used to check root diseases in horticultural and nursery stocks) or through soil fumigation. The latter form of control however, tends to be expensive, is applied before the crop is sown and is often limited to intensive glasshouse systems.

Fungicides can also be classified according to the timing of their application, i.e. their timing in relation to the stage of infection. Protectant fungicides are applied before infection to the plant surface. Bacterial cells or fungal spores which come into contact with the fungicide are then killed. Some protectants also have some eradicant activity, particularly against diseases such as downy mildews which develop primarily on the leaf surfaces (Parry, 1990). Although protectant fungiicides play a major role in protection of many crops world-wide their mode of action largely remains unclear. They tend to be broad-spectrum in their effect and have remained free from the problems of development of pathogen resistance. The same cannot be said of curative fungicides. These are fungicides which are applied after initial infection, for instance, during the incubation period (Püntner, 1981). The fungal structures within the plant may be killed before they produce visible disease symptoms. When infections have already become visible eradicative fungicides are used for control; these are also used for preventing further sporulation after spread of the disease.

The curative fungicides have been prone to pathogen resistance because of their tendency to target specific sites of action. The curative fungicides are largely systemic; they are capable of being absorbed by the roots, seeds or leaves of the plant and are then translocated within the plant. The uptake and translocation of systemic fungicides is not well understood in many cases. The penetration of the plant cuticle and entry into the apoplast (nonliving parts of the plant including all walls and xylem) and symplast (living plant parts including the phloem and protoplast) depend on the physiochemical properties of the fungicide and the hostplant (Parry, 1990). Most systemic fungicides are translocated in the apoplast and since the most important flows in the apoplast are in the xylem, application to the roots is likely to result in more extensive distribution throughout the plant than if treatments are applied to the foliage (Manners, 1982). In contrast to the behaviour and movement of systemic fungicides within the plant, residual fungicides applied to the foliage act as protectants but repeated applications may be necessary to replenish deposits lost by runoff from rain or abrasion.

3.2.4 Nematicides

Nematicides in commercial use are categorized into four groups: (i) halogenated aliphatic hydrocarbons (e.g. methyl bromide, ethylene dibromide; (ii) methyl isothiocyanate precursor compounds (e.g. metham-sodium, dazonet); (iii) organophosphates (e.g. ethoprophos, fenamiphos, thiomazin); and (iv) carbamates (e.g. carbofuran, aldicarb, oxaryl (Whitehead and Bridge, 1985). The first two groups are soil fumigants while the latter act as nemostats. A nemostatic pesticide is one that reduces movement of the nematode as concentrations increase and may cause cessation of all movement as doses rise to lethal concentrations. Only small amounts are required for control, amounts which are usually non-toxic to plants. Hence, they can be applied before sowing or even around the roots of established plants. In contrast, soil fumigants need to be applied in large amounts in order to kill nematodes and their eggs in the soil. The nematicides methyl bromide, metham-sodium and dazonet have the advantage that they also kill or inhibit soil fungi and may kill weed seeds or delay their germination (Whitehead and Bridge, 1985).

Other pesticides also have nematicidal action. Herbicides such as dalapon may inhibit nematode reproduction (Way and Cammell, 1985) while several fungicides such as benomyl and thiabendazole inhibit attack of plants by *Heterodera tabacum* (Rodriguez-Kabana and Curl, 1981). However, some pesticides can also increase the severity of non-target pests; hence, care needs to be taken in their selection.

3.2.5 Rodenticides

Traditionally rodenticides have been characterized by two classes, (i) acute (single dose, quick acting) or (ii) chronic (multiple dose, slow acting) rodenticides. Acute rodenticides cause death within 24 hours but tend to be more hazardous for humans and hence, there are severe restrictions on their use. Their high toxicity means that only small quantities are required, which reduces labour and product costs (Greaves and Jones, 1985). They have the disadvantage that surviving rodents are shy of poisoned bait, making it difficult to control them on subsequent occasions. This situation does not occur with chronic poisons however, because a lethal dose is usually consumed before the onset of any symptoms; hence, the rodent makes no association between the poison and subsequent symptoms. Chronic poisons which are mostly blood anti-coagulants (causing death through haemorrhage) have another advantage in that the concentration of the rodenticide in the bait can be kept low, sufficient to cause cumulative toxicity but low enough to reduce the hazard to non-target animals. Because of this combination of high efficacy and relative safety, chronic poisons are usually preferred to acute ones.

The distinction between chronic and acute poisons is now less marked because of the introduction of rodenticides with intermediate effect, e.g. calciferol. With such increasing diversity of approaches there is now a general consensus which tends towards the idea that each rodenticide, its formulation and method of application should be assessed individually in relation to the pest situation in which its use is proposed (Greaves, 1989). Rigid classifications have less value in such situations.

For the vast majority of applications, rodenticides are used in the form of poisoned baits. Almost every kind of edible material has been used as a bait. They are used either in a preventative way in places continually subject to infestation or as a remedial treatment after unacceptable levels of damage have occurred. The use of rodenticides in these ways has effectively displaced other forms of rodent control.

3.2.6 Formulation and application of pesticides

Pesticides usually consist of an active ingredient (a.i.; the toxic element of the pesticide) and a number of non-pesticidal materials which permit ease of application, handling, storage and may also enhance toxicity. Hence, pesticides are formulated to meet the particular needs of the pest, crop, and environment in which they are to be used and for their means of application. There are seven generally recognized categories of formulation: (i) solutions; (ii) wettable powders; (iii) emulsion concentrates; (iv) suspension concentrates; (v) dusts; (vi) granules; and (vii) baits. If the active ingredient is soluble in water or an organic solvent then pesticides can be prepared as concentrated solutions. Concentrated solutions are commonly used in ultra-low volume applications (ULV).

Most fungicides are formulated as wettable powders. Active ingredients which will not dissolve in water are finely ground and mixed with an inert material such as kaolin and a wetting agent. Dispersants may be added to delay sedimentation in the spray tank.

Technical products which are insoluble in water can be dissolved in an organic solvent combined with emulsifying agents. This emulsion concentrate can then be diluted with water in the spray tank before application. The emulsifying agent ensures that the technical product and organic solvent solution disperse evenly within the water. Once on the plant the water evaporates and the dispersed technical product solution exerts a toxic effect.

Suspension concentrates are characterized by the active ingredient formulated as a finely ground solid which is held in a suspension of water that is then diluted further with water in the spray tank. The solids are fixed in suspension by the addition of polymers (Barlow, 1985).

The application of pesticides as a seed dressing is usually achieved through use of dusts containing high concentrations of active ingredients and 'stickers' to improve contact with the seed. Dusts may be also applied to foliage through use of a dust blower machine; however, the pesticide content must be low since high application rates are required.

Bait formulations combine an attractant with a pesticide. The attractant may be a food lure, or an aggregation or sex pheromone, but it serves to lure the pest to the pesticide which is either a contact or stomach poison. Bait formulations are commonly used to control rodents (section 3.2.5) and household pests such as ants and cockroaches.

Granular formulations are used where targeting of the pesticide is less important because the active ingredient is carried by an intermediary agent such as soil or water. Granules (small solid particles of 2–5 mm diameter) are widely used to apply systemic organophosphorous insecticides for seedling crops and for formulation of nematicides. The characteristics of release can be built into the design of the granule.

Controlled release formulations of pest control agents have been extensively reviewed by Wilkins (1990). They are characterized by their ability to deliver to the local environment of a pest a constant weight of chemical per unit time. Hence, formulations such as microencapsulation or laminated strips are very useful for increasing the efficiency of pesticides. They are also safe to handle and present fewer hazards to beneficial organisms.

The different pesticide formulations provide for a range of methods for their application. Liquid formulations are applied at different volumes, ultra-low, medium- or high-volume applications and they may be sprayed from aircraft or by various means on the ground. The techniques of pesticide application have been extensively reviewed by Matthews (1992).

3.3 HOSTPLANT RESISTANCE

Hostplant resistance may be defined as the collectable, heritable characteristics by which a plant species, race or clone, or individual may reduce the possibility of successful utilization of that plant as a host by a pest species, race, biotype or individual (Beck, 1965). From the point of view of the farmer, horticulturist and others, the use of resistant crop cultivars represents one of the simplest and most convenient methods of pest control (Dent, 1991). As a control method it has generally proved very successful. Resistant cultivars have been produced for pathogens (excluding viruses), insects, nematodes and parasitic weeds. The greatest successes have been achieved with pathogen hostplant resistance breeding through introduction of major genes for resistance; however, there have also been too many cases of breakdown of resistance with the selection for virulent pathotypes. This has, in a number of cases, led to what is known as the 'boom and bust' cycle of resistance cultivar production. A recent example of this cycle has been resistance breeding in spinach against the spinach downy mildew *Peronospora effusa* (Koike *et al.*, 1992). Up until the early 1950s spinach crops were attacked by what came to be designated as race 1 downy mildew. In response to this, researchers developed a new resistant spinach cultivar 'Califlay'. However, in 1958 race 2 downy mildew appeared in

California in the USA, rendering 'Califlay' and related cultivars susceptible once again. Plant breeders countered with cultivars such as 'Early Hybrid 424' which were resistant to both races 1 and 2. These resistant cultivars were effective for almost 20 years until race 3 was reported, again in California, in 1978. Breeders again managed to find another source of resistance which was introduced with cultivars 'Polka', 'Shasta' and 'St Helen's' and others, which were resistant to all three races. In 1989 a new race, race 4, was reported as severe epidemics of downy mildew spread throughout Californian spinach-growing areas and in Texas. Significant losses were recorded in 1989 and 1990 but new spinach cultivars are currently under development that are resistant to all four races. Presumably these cultivars will be relatively free of downy mildew until the next race of *P. effusa* develops (Koike *et al.*, 1992). This example is typical of the boom and bust cycle which has been experienced in many pathogen–crop interactions, particularly among rusts and cereals (Russell, 1978; Johnson and Gilmore, 1980). The length of time that a cultivar remains resistant varies considerably but in cereals a clear succession of resistance-breaking races has continually plagued resistance development. There is now a general acknowledgement of the need for breeding of horizontal resistance which should provide more durable resistance (Robinson, 1987) (section 7.4).

The situation with insect pests and breeding for hostplant resistance has been somewhat different. In general, the level of resistance to insect pests found in plants has been less dramatic than with pathogens and has been referred to as partial resistance (probably horizontal resistance). For this reason and others the emphasis on breeding has been somewhat different. There has been a marked tendency to search for resistance in plants on the basis of morphological or biochemical characteristics of plants (Norris and Kogan, 1980). Morphological bases of resistance include plant colour, architecture and anatomy. Commonly, morphological characteristics, such as the presence or absence of pubescence or surface waxes and tissue toughness, are sought in the breeding programmes.

The biochemical bases of resistance can be divided into two broad categories, those influencing behavioural responses and those influencing physiological responses of insects. The types of chemicals responsible for insect resistance are numerous but most are secondary metabolites (chemicals that are not required for the general growth and maintenance of the plant but which are produced for the purposes of plant defence). Biochemical means of providing resistance are highly regarded because identification of a useful chemical could potentially provide a fast, analytical means for screening and segregating plant populations. This would markedly increase the speed of the screening process and decrease development time for new cultivars.

The traditional lengthy plant breeding methods that have been utilized for incorporating resistance into crop cultivars are now being supplemented by the techniques of genetic manipulation. This, and a further more detailed consideration of hostplant resistance is given in section 7.4.

3.4 BIOLOGICAL CONTROL

The term biological control has in the past meant different things to different pest control practitioners. Plant pathologists have tended to use the term to denote control methods that include crop rotation, alteration of soil pH, use of organic soil amendments as well as the use of one organism against another (Baker, 1985; Schroth and Hancock, 1985). However, the definition which is now becoming more widely accepted among the disciplines involved in pest management is the one used by the entomologists, where biological control is seen as the use of living organisms, excluding hostplants, as pest control agents (Greathead and Waage, 1983). That the entomologists' definition should become more universally acceptable perhaps reflects the greater involvement and the long history of successes which the entomologists have had with biological control. The most notable of these include control of the cotton, cushiony scale *Icerya purchasi* on citrus in the USA by the Vedalia beetle *Rodalia cardinalis* during the 1880s, the control of prickly-pear in Australia and more recently the control of the cassava mealybug *Phenacoccus manihoti* in Africa through the introduction of the parasitoid *Epidinocarsis lopezi* (Neuenschwander and Herren, 1988). These well-known examples of successful biological control have involved the introduction of an exotic natural enemy for the control of an accidentally introduced exotic pest. This form of control is commonly known as 'classical biological control' utilizing a strategy of 'introductions' of exotic natural enemies. There are however, a number of other approaches to biological control.

3.4.1 Biological control strategies

There are five different types of biological control strategy; introduction, augmentation, inoculation, inundation and conservation. As stated above, biological control through introductions is most frequently used against introduced pests which arrive in a new area (where they become permanently established) without an associated natural enemy complex. The stages typically involved in the introduction of a natural enemy for the control of an exotic pest include: (i) identification of the pest and its area of origin; (ii) an exploratory expedition overseas; (iii) an inventory of natural enemies associated with the pest; (iv) field screening and evaluation; (v) importation, culture, possible pre-culture tests; (vi) release; and (vii) occasionally where funds permit, post-release evaluation (Dent, 1991). Where successful, this classical use of biological control offers permanent levels of control which has few risks associated with it and, above all, it provides a very cost-effective solution.

The costs and accruing benefits from successful biological control have been considered by a number of authors (DeBach, 1964; Simmonds, 1968; Mohyuddin and Shah, 1977; Greathead and Waage, 1983; Habeck *et al.*, 1993) and in all of these analyses it has been shown that the benefits accrued have paid for the cost of control many times over. Cate (1990) considered that the long-term benefit/cost ratios are very high ($30/$1) when averaged across all attempts, the highest of any pest control approach, and this when most of the benefit/cost ratios reported in the literature have been conservatively estimated (Habeck *et al.*, 1993). Thus, introductions represent an important and highly effective means of controlling exotic pests. In addition the approach can also be used against native pests which lack effective natural enemies or where the mortality exerted by natural enemies has been upset by intensive agriculture; although, at present there are few examples of use of introductions in this way.

The cost also has been an important issue in the development of augmentative measures of biological control, to the extent that DeBach (1974) considered that augmentation of natural enemies should be given the lowest priority in biological control research. The main reason for this was the cost of production and application of augmentative releases using insect parasitoids. Augmentative control tends to be used in situations where natural enemies are absent or population levels are too low to be effective, so numbers are augmented by the use of laboratory-cultured natural enemies. In the case of insect natural enemies the costs of rearing, handling, distributing and release of a large number of insects, usually parasitoids, at an appropriate time has in some cases proved uneconomical (Oatman *et al.*, 1976; King *et al.*, 1985). However, in other situations, for instance, control of the glasshouse whitefly *Trialeurodes vaporariorum* with *Encarsia formosa* and the use of the phytoseiid mite *Phytoseilus persimilis* to control the two-spotted spider mite *Tetranychus urticae*, the approach has proved very economical. This success has been effected in the establishment of commercial companies producing these and other natural enemies for augmentative control. The problems of costs of production are not so acute with a number of pathogen control agents, especially where basic commercial fermentation techniques may be used. Given suitable control agents the potential for augmentative control utilizing pathogen natural enemies must provide an area for future commercial exploitation.

Augmentative control is used where there is no interest in long-term stability but where there is a need to reduce pest numbers below economic thresholds (van Lenteren, 1986). Hence, a number of releases over a season may be required because the control is only temporary.

Seasonal control may be achieved through use of inoculations. These are used where a native natural enemy is absent from a particular area, or an introduced species is unable to survive permanently (Greathead and Waage,

1983). The inoculative releases are made at the beginning of the season to colonize the area for the duration of the season (or the crop) and so prevent pest build-up.

The releases made with biological control through inundation involve very large numbers of native or introduced natural enemies in a similar way to the application of chemical pesticides. The natural enemy is usually a pathogen and is often formulated so that it can be applied using conventional pesticide spray equipment. Sometimes used as substitutes for chemical pesticides, inundative control agents are applied for short-term control when pest populations reach damaging levels. The most successful agent in this category is the bacterium *Bacillus thuringiensis* (Dent, 1993) which is used to control mainly lepidopteran and dipteran pests, although a number of other entomopathogens based on fungi and viruses have also found a specific niche.

The least studied techniques in biological control are those associated with conservation of natural enemies (Dent, 1991; Cock, 1994). Conservation involves the use of techniques to encourage and conserve the natural populations of natural enemies in the crop environment so that sufficient numbers are present to control an immigrant or developing pest population. The techniques which have been successfully developed include sowing flowering plants in field margins to provide a food source for adult parasitoids or predators (e.g. Cowgill *et al.*, 1993) and the construction of beetle banks in cereal fields to encourage overwintering of predatory beetles (Thomas *et al.*, 1991). There are potentially many more techniques which could be developed for conserving natural enemies, but it is not an area that receives sufficient attention or financial support.

3.4.2 Types of biological control agent

Entomophagous and phytophagous insects are the major agents used in biological control. The entomophagous insects are classed as either predators or parasitoids, each with completely different characteristics which contribute to their effectiveness as biological control agents. Parasitoids are parasitic during their immature stages when the larvae develop within (endoparasite) or on (ectoparasite) their host. Individual parasitoids consume only one host during their development to produce adults which are free-living and usually feed on pollen, nectar, honeydew or sometimes on the body fluids of their host. As a group, parasitoids constitute 300 000 species of Hymenoptera and Diptera exhibiting a wide range of host specialities and habits. The host-specific parasitoids are considered most suitable for use in biological control.

Similarly with insect predators, the polyphagous species are considered less suitable than monophagous species, because the former are less likely to concentrate feeding on the pest species in the presence of an abundant alternative prey species. However, in general, predators have the advantage over parasitoids in that each individual consumes a number of prey during their lifetime, and unlike parasitoids the immature stages are also actively searching for and consuming prey pest species. Among the most common predators of insect pests are beetles, predatory bugs, crysopids and the syrphid larvae.

Phytophagous insects are widely used for the biological control of weeds. Once again host specificity is important to ensure the natural enemy feeds only on the target weed. However, it is plant pathogens which offer some of the best prospects for weed control, and in general the role of pathogens in biological control is increasing.

Pathogens – bacteria, viruses, fungi, protozoa and parasites (particularly nematodes) – exhibit attributes which are of great use in biological control. True parasites, such as parasitic nematodes, differ from parasitoids in that they do not kill their host, they only weaken and debilitate them. Despite this they have proved useful control agents, to the extent that there now exist a number of commercial companies involved in the culture and sale of nematodes for the control of garden and horticultural pests. Parasitic pathogens will often kill their host outright and then liberate millions of spores or 'resting stages' which are dispersed to infect other host individuals. Their relative pathogenicity, speed of action and ease with which some of them can be cultured have ensured their use both for augmentation and inundative releases. Pathogens may also act as control agents through competitive exclusion or the production of antibiotics. This group, known as antagonists, are particularly useful in the biological control of plant pathogens.

One final group that deserves some mention are vertebrate biological control agents. Although in general vertebrates are too polyphagous for use in biological control they have on a number of occasions been found useful as a means of pest control. Examples include the use of ducks in Chinese rice paddies to control insects and weeds (Zhang, 1992), shrews in the control of larch sawfly in USA (Buckner, 1966), goats in Australia for the control of blackberry (*Rubus* spp.) in pine forests (Mitchell, 1985) and fish for control of mosquitoes (Hoy *et al.*, 1972; Bence, 1988).

3.4.3 Successful biological control

The deliberate use of natural enemies as agents for biological control can be traced back to the medieval date growers in Arabia who seasonally transported cultures of predatory ants from nearby mountains to oases to control phytophagous ants which attacked date palm (van den Bosch and Messenger, 1973). There have been hundreds of cases of successful biological control since that time for a whole range of pest species. The following account provides a brief outline and a few examples for each category of pests: insects, plant pathogens, nematodes and weeds.

Pests have been controlled with the use of insect biological control agents through introductions, augmentation and conservation. A classical example of successful control through an introduction of an exotic phytophagous insect for the control of a weed species is that of control of the prickly pear *Opuntia* spp. in Australia with the pyralid moth *Cactoblastis cactorum* (Dodd, 1959). More commonly, insect natural enemies (usually parasitoids) are introduced to control insect pests, for instance, the parasitoid introduced to Togo from India for the control of *Rastrococcus invadens*, a serious pest of mango and citrus in West Africa (Agricola *et al.*, 1989; Moore, 1993). There have been numerous examples of successful biological control through introductions. Of the 506 attempts made there have been 208 successful introductions (Greathead, 1984), covering a range of cropping systems from cereal crops to fruit, forage and plantation systems. Attempts at augmentation have been less successful, either for economic reasons or because of variable results under field conditions e.g. use of *Trichogramma* for control of Lepidoptera (Wightman *et al.*, 1990). Notable exceptions are the use of natural enemies for control of some glasshouse crops (see Chapter 9). For example, 90% of the commercial tomato growers in the Netherlands use *Encarsia* to control whiteflies (Sütterlin and van Lenteren, 1994). The Chinese have made effective use of biological control agents for augmentative releases with some success (Cock, 1985). The egg parasitoid *Anastatus* spp. is cultured by farmers and released to control the *litichii* stink bug *Tessaratona papillosa*, and *Trichogramma* spp. produced by local communes have long been released for control of pests such as corn-borers, rice leaf rollers, sugarcane borers and cotton pests (Huffaker, 1977). In Europe, forest systems under threat from the European Spruce beetle *Dendroctonus micans* can be protected by innoculative releases of *Rhizophagus grandis* (Gregoire *et al.*, 1990).

The conservation of insect natural enemies in agricultural systems remains a neglected area of research, although greater awareness of the need to reduce pesticide use has concentrated efforts more on ways of conserving natural enemy populations. Reduction in pesticide use may in some cases be sufficient to encourage population growth of natural enemy populations to levels where they may exert sufficient levels of pest control (Sotherton *et al.*, 1989; Chiverton and Sotherton, 1991). In other situations it may be necessary to take more active steps to encourage natural enemies. The undersowing of orchards (Zandstra and Motooka, 1978) or field margins (Cowgill *et al.*, 1993) with flowering plants, the use of mustard meal mulches (Ahlström-Olsson and Jonasson, 1992) or construction of beetle banks (Thomas *et al.*, 1991) have proved to be particularly valuable in this context. However, such measures are yet to be widely adopted.

Pathogens used as pest control agents have generally been used as microbial pesticides. Microbial insecticides account for only a 1.6% share of the world insecticide market, but 95% of these sales involve products based on isolates of the bacteria *Bacillus thuringiensis* (Richards and Rogers, 1990). Serotypes of *B. thuringiensis* have been identified and selected and now pre-

Table 3.4 Examples of commercialized microbial insecticides based on *Bacillus thuringiensis*. (From Richards and Rogers, 1990)

Serotype	*Target pest*	*Production examples (manufacturer)*
Bacillus thuringiensis ser. *aizawai*	Wax moth larvae	Certain (Sandoz)
B. thuringiensis ser. *israelensis*	Mosquito Blackfly	Vectobac (Abbott) Skeetal (Novo) Teknar (Sandoz)
B. thuringiensis ser. *kurstaki*	Lepidopterous larvae	Dipel (Abbott) Biobit Foray (Novo) Javelin, Thuricide (Sandoz)
B. thuringiensis ser. *san diego*	Colorado potato beetle	M-One (Mycogen)
B. thuringiensis (conjugate)	Colorado potato beetle	Foil (Ecogen)

sent an opportunity for a number of commercial products for control of a range of pest types (Table 3.4). *B. thuringiensis* has proved highly successful as a microbial insecticide and is now used in a number of crops as a substitute for chemical pesticides. Other bacteria which have also proved useful as microbial pesticides include *Serratia entomophila* and *Agrobacterium radiobacter* for control of grass grubs and crown gall (caused by *Agrobacterium tumefaciens*) respectively (Richards and Rogers, 1990; Ryder and Jones, 1990).

Virus diseases of insects and their role in the natural regulation of insect populations have been recognized for many years (Winstanley and Rovesti, 1993). There are three types of virus that are entomopathogenic, are considered harmless to humans, and are sufficiently virulent for use as control agents; the nuclear polyhedrosis viruses (NPV), the granulosis viruses (GV) and the cytoplasmic polyhedrosis viruses (CPV). Of these only the NPVs and GVs are widely used.

Commercial preparations of *Spodoptera exigua* NPV have been registered for the control of *S. exigua*, a pest of potted plants and cut flowers, ornamentals and vegetables in the Netherlands (Cuijpers *et al.*, 1994) and NPVs of *Neodiprion sertifer, Lymantria dispar, Panolis flannea* and *Euproctis chrysorrhea* for forest pest control are now widely used (Evans, 1990). Control of *Oryctes rhinocerus* (a pest of coconut and oil palms) and *Anticarsia gemmatalis* (a pest of soybean in Latin America) with their respective baculovirsues have also proved effective (Bedford, 1981; Jackai *et al.*, 1990), and practical formulations are being sought for the NPV of *Spodoptera littoralis*, a pest of cotton in Egypt (Jones, 1988).

The granulosis virus of *Cydia pomonella* was isolated in the early 1960s and, after extensive field tests and development, is now registered in

Germany (Winstanley and Crook, 1990). This virus is expected to have widespread application against *C. pomonella* in Europe and the USA.

Fungal biological control agents are an important group because of their wide-ranging application. Fungal agents may be used in introductions for the control of weed species, as mycoherbicides or augmentative releases to control indigenous weed pests (Table 3.5) or fungal entomopathogens can be used in inundative, inoculative or augmentative releases to control insect pests. A commercial preparation of *Beauvaria bassiana*, Bouverine®, has been used successfully in the former USSR for the control of the Colorado beetle (Ferron, 1981) while in Indonesia *Metarrhizium anisopliae* has been successfully used for the control of *Oryctes rhinocerus* (Munaan and Wikardi, 1986). Another commercial product is based on the fungal entompathogen *Verticillium lecanii* which is now used quite extensively for control of aphids, whitefly and thrips, particularly for glasshouse crops (Richards and Rogers, 1990).

Since the 1950s there has been increasingly active research effort to discover and develop plant pathogenic fungi as biological control agents against weeds (Greaves and MacQueen, 1990). This effort has concentrated

Table 3.5 Fungal plant pathogens used as biological weed control agents utilized in the augmentative classical or mycoherbicide strategies. (From TeBeest, 1993)

Pathogen	*Host*
Augmentative strategy	
Puccinia lagenophorae	*Senecio vulgaris*
Puccinia canaliculata	*Cyperus esculentus*
Classical strategy	
Entyloma compositarum	*Ageratina riparia*
Puccinia chondrillina	*Chondrilla juncea*
Mycoherbicide strategy	
Alternaria cassiae	*Cassia obtusifolia*
Alternaria crassa	*Datura stramonium*
Alternaria macrospora	*Anoda cristata*
Colletotrichum coccodes	*Abutilon theophrasti*
Colletotrichum gloeosporioides	*Aeschynomene virginica*
Colletotrichum gloeosporioides	*Hakea sericea*
Colletotrichum gloeosporioides	*Hypericum perforatum*
Colletotrichum gloeosporioides	*Jussiaea decurrens*
Colletotrichum gloeosporioides	*Malva pusilla*
Colletotrichum malvarum	*Sida spinosa*
Colletotrichum orbiculare	*Xanthium spinosum*
Fusarium solani	*Curcubita texana*
Phytophthora palmivora	*Morrenia odorata*
Sclerotinia sclerotiorum	*Cirsium arvense*
Sclerotinia sclerotiorum	*Taraxacum officinale*

on two approaches, the use of fungal agents for classical introduction to control exotic weed species and the use of indigenous pathogens to control indigenous weeds (the mycoherbicide approach).

There have been two notable successes involving classical biological control of a weed species with fungal pathogens. The first was the introduction of the rust *Puccinia chondrillina* into Australia in the early 1970s to control the skeleton weed *Chondrilla juncea* (Emge *et al.*, 1981). This introduction was successful because it was reliably pathogenic, spreading through the weed population and reducing weed growth to uneconomic levels, it maintained its effectiveness from year to year, and of course it is highly specific to the host (Bailey, 1990). Another notable success involved the *Cercosporella* fungus which was collected from Jamaica and introduced into Hawaii in 1974 for the control of the weed *Ageratina riparia* (Trujillo, 1976). Within 9 months there had been an 80% reduction in the weed population and over 50 000 ha of upland pasture had been rehabilitated (Trujillo, 1985).

Despite these examples however, there have been few other successes, even though there are many reasons to believe that fungi can be successfully managed for the biocontrol of weeds (Ayres and Paul, 1990). Progress has been slow because quarantine regulations and host screening tests have been lengthened, which increases the costs of introduction. This means there are now significant financial barriers to the use of fungi for classical biological control (Evans and Ellison, 1990).

In contrast, the situation with the use of fungal agents as mycoherbicides looks more promising. Two commercial products have been largely responsible for a rapid increase in research and development activity in mycoherbicides, DEVINE® and COLLEGO® (Greaves and MacQueen, 1990). The former is a formulation of *Phytophthora palmivora* chlamydospores which is used for the control of the strangler vine *Morrenia odurata* in citrus groves in Florida. Although a successful control agent, this product has not been a great commercial success because it persists in the soil, providing several seasons of control. The other product COLLEGO® also has been highly effective. This is a wettable powder formulation of *Colletotrichum gloeosporioides* f. sp. *aeschynomene* used for the control of the northern jointvetch *Aeschynomene virginica* in rice and soybean (Greaves and MacQueen, 1990).

Entomogenic nematodes have a wide spectrum of action, do not infest vertebrates or plants, persist at low levels in natural habitats and are effective for the control of insects in cryptic and subterranean habitats (Hominick, 1990; Kaaya, 1993). There are three genera of rhabditid nematodes which are considered to be ideal agents for biological control, *Neoaplectana*, *Steinernema* and *Heterorhabditis*. The *Heterorhabditis* and *Steinernema* nematodes have been used for the control of black vine weevils *Otiorhynchus sulcatus* in many field situations including strawberries, vineyards and ornaments. However, it is in the glasshouse crops where they are most effective for control of *O. sulcatus* and also for the control of the mushroom pests *Megaselia bovista* and *Lycoriella mali* (Richardson, 1990). It is likely that as new formulations and means of production are developed the use of nematodes as biological control agents will increase (Georgis, 1990; Popiel and Hominick, 1992).

3.5 CULTURAL CONTROL

The underlying principle of cultural control is the modification of management practices so that the environment is less favourable for pest invasion, reproduction, survival and dispersal and so achieve reductions in pest numbers (Takahashi, 1964). The measures involved are all indirect preventative methods and often the success or otherwise of such measures cannot be assessed exactly, as in the case of direct chemical control (El Amin and Ahmed, 1991). Many of the techniques used are variations of standard agronomic practices which have to be carried out to grow a successful crop. Others are highly specific to particular patho- and cropping systems. Only those measures which are more generally applicable are dealt with here, namely: crop rotations, cultivation practices, mulches and solarization, barriers and trap crops, planting characteristics, phytosanitation and irrigation. Intercropping or mixed cropping, usually considered under the heading of cultural control has been dealt with separately in section 7.3.

3.5.1 Rotations

Crop rotation is the systematic growing of different kinds of crops in recurrent succession on the same piece of land (Martin *et al.*, 1976). Essentially an agronomic practice used to improve or maintain soil fertility (section 2.2), it also provides a means of crop protection, particularly against pests which are relatively host specific and have poor powers of dispersal. Certain weeds are associated with specific crops and if the same crop is grown continuously for a number of seasons then the weed population builds up to unacceptable levels. Changing to a different crop by rotation helps to break the cycle, changing the selection pressure for a given species. Crop rotation also permits the use of different herbicides (Shenk, 1994).

The use of pesticides has reduced the significance of crop rotation as a means of pest control for most diseases and insect pests. For certain soil and trash-borne diseases however, and some soil-living stages of insects, rotation still remains one of the cheapest and most effective means of control. The take-all pathogen of cereals *Gaeumannomyces graminis* can be effectively reduced by just a 1- or 2-year break from susceptible crops (Parry, 1990). In general, the efficacy of the rotation depends largely on the ability of the soil-borne pathogens to survive in the soil. The insect pests *Agrotes* spp. (wireworms), *Melolontha melolontha*, *Amphimallon solsti-*

talis (chafers) and *Tipula* species (leather jackets) can be effectively controlled with a typical grass/legume/root crop rotation (Coaker, 1987), the success of a rotation being dependent on the preference for specific hosts. It is more difficult to devise a rotation for wireworms which are polyphagous and cause damage to most crops and cereals than to those monophagous species such as the wheat bulb fly *Delia coarctata*. It is advisable in devising a rotation to use crops with sharp contrast in biological characteristics, life-cycle, planting time and agronomic requirements, to maximize the change in conditions and thereby make the cropping environment unfavourable for the pest species. In this way rotations will ensure maximum impact on pest population levels.

Crop rotations have become less popular as agriculture has intensified and greater emphasis has been placed on continuous monocropping. However, rotations can provide, in some situations, a cheap and effective means of pest control. Hence, where appropriate, they should be employed as an alternative to perhaps less sustainable pest control techniques.

3.5.2 Tillage practices

Tillage practices directly influence the physical and chemical properties of the soil, soil moisture and temperature, root growth and nutrient uptake, populations of weeds, insect pests and vectors of plant pathogens. These factors may in turn influence the viability of plant pathogens and the susceptibility or resistance of the host (Sumner *et al.*, 1981). Little research has been done on the influence of tillage on nematodes, viruses or bacterial pathogens in either temperate or sub-tropical climates, most studies having been restricted to insects and fungal pathogens. Insect pests are affected either indirectly by the creation of inhospitable conditions and by exposing the insects to their natural enemies or directly by physical damage inflicted during the actual tillage process (Stinner and House, 1990). Deep ploughing for instance, can be applied as an indirect measure to bury or expose the diapausing or pupal stages of insect pests such as *Heliothis* and *Spodoptera* species (El Amin and Ahmed, 1991) and of the sunflower seed weevil (*Smicronyx fulvus*) (Gednalske and Walgenbach, 1984). The use of the disc plough caused mortality of overwintering larvae of another pest of sunflower *Dectes texanus* through destruction of the roots and exposure of the larvae to the soil environment (Rogers, 1985).

The habitat provided by ploughed soil is very different from the one produced by minimum cultivation or direct-drilling techniques. The lack of cultivation may favour or inhibit pest populations (Coaker, 1987). Several foliar pathogens are known to survive in maize residues in minimum tillage systems including *Helminthosporium turcium* and *H. maydis*, *Phyllostica maydis*, *Physoderma maydis*, *Colletotrichum graminicola* and *Cercospora zeae-maydis* (Sumner *et al.*, 1981). Most of these diseases are more severe in minimal tillage than when maize debris is buried by ploughing. Deep ploughing will reduce the incidence of sorghum downy mildew and increase the yield of a susceptible sorghum cultivar compared with conventional ploughing (Tuleen *et al.*, 1980). Cultivation can also be used to get rid of volunteer plants and crop residues that are important sources of inoculum. Where a pathogen's ability to survive in the soil is poor then inoculum production may be limited by ploughing in the material. For instance, the survival time of *Xanthomonas campestris pv campestris* in crop debris can be reduced if fields are disced or ploughed (Strange, 1993), and the recovery of *Pyrenophora tritici-repentis* from buried wheat stubble was lower than that from the soil surface (Summerell and Burgess, 1989).

Clearly the use of different tillage practices at appropriate times can provide a range of methods for the control of pests. To obtain the best from them however, a good knowledge of the pests' life-cycle is required, especially when there is a need for careful timing of the cultivation (Coaker, 1987).

3.5.3 Mulches and solarization

Mulches may be living or non-living (Akobundu and Poku, 1987). They have their origin in early agriculture where covering soil and plants with organic or inorganic matter was used to form a protective barrier against frost or warm soil to improve plant growth (DeVay, 1991). Mulching was also used to limit soil water evaporation to control weeds, to improve soil tilth and manage erosion, as well as a means of controlling pest insects, pathogens and nematodes.

Live mulch (also known as smother crops) involves the undersowing of a food or forage crop under an existing crop to reduce the resources available to weeds. In the case of legumes they also provide nitrogen to the crop (Shenk, 1994). Mulching normally refers however, to the use of plant material, usually crop residues, which are placed between and in the rows of a growing crop to suppress the germination of weed seeds and to retard the growth of any weeds that do grow.

More recently the development of mulches has taken a very significant step forward with the use of transparent polyethylene film for solar heating of the soil (solarization) (Katan *et al.*, 1976; Katan, 1981). This technique, reviewed by DeVay *et al.* (1991), utilizes solar radiation to raise the temperature of the soil above 31–32°C, a temperature at which many organisms are killed. Soil solarization is a hydrothermal process and hence, depends for its success on moisture for maximum heat transfer to soil organisms. In appropriate circumstances it has been used to successfully control pest fungi, bacteria, weeds and nematodes (see DeVay *et al.*, 1991) and shows great potential, especially for countries and regions with high intensity of sunlight, but also appropriate levels of soil moisture, air temperatures, soil colour and structure (DeVay, 1991).

3.5.4 Barriers and trap crops

Barriers and trap crops refer to the use of susceptible plant species to divert pests from their primary host. Obviously the trap crops must be attractive to the target pest and tolerant of heavy attack. The plants can either be sprayed with pesticides or destroyed, necessarily before the pests reproduce or the main crop becomes available to them (Coaker, 1987). They have been extensively used in the control of plant parasitic nematodes and of insect pests. Castor is grown as a trap crop in cotton, tobacco and chillies to reduce damage by *Spodoptera littura* (Jayaraj and Santharam, 1985) and okra is grown in the vicinity of cotton to reduce attack by flea beetles (Ezueh, 1991).

Trap crops have also been successfully used to reduce the incidence of virus spread via insect vectors. Large monocultures of susceptible crops normally provide a favourable environment for the transmission of virus from one plant to another and from one planting to another. The planting of barriers of alternative susceptible crops can significantly reduce the ratio of virus spread in a field (Thottappilly *et al.*, 1990) (e.g. Lima and Gonclaves (1985) reported barrier crops useful in controlling aphid-transmitted viruses).

Other types of barriers include reflective mulches which act as a repellent to aphid vectors in certain crops (Simons, 1982; Kring and Schuster, 1992) and polyethylene nets used for the protection of field vegetables against insect attack. The nets have been shown to offer protection to cauliflowers against *Delia radicum*, *Brevicoryne brassicae* and Lepidoptera, and to Japanese radishes against *Delia radicum* (Ester *et al.*, 1994). Although costs of control may be high, this may be offset by higher profit from improved quality of produce.

3.5.5 Planting characteristics

In general, a well-prepared seed-bed will allow plants to develop good root systems and become healthy and vigorous. Aspects of planting or sowing which can affect pest incidence and development include the quality of the seed-bed, depth of sowing, seed density/spacing and planting time. Of most relevance are plant density and planting time. It is important to manipulate plant densities and row spacings to achieve rapid canopy shading by crops. Plant characteristics that are associated with competitive ability against weeds include plant height, leaf shape and size and leaf area index (Shenk, 1994). However, plant characteristics and spacings that favour control of one pest may act to the detriment of the control of another. Dense stands of a crop often result in high humidities which encourages many diseases. In addition if the distance between individual plants is reduced this increases the rate of spread of some aerial spores (Parry, 1990). In contrast to this, it has been demonstrated in Nigeria, Malawi and Uganda that to escape from infestations of groundnut rosette virus-carrying aphids, groundnut should be sown early with a close spacing (Wightman *et al.*, 1990). Given the degree of variation in response between different pests and their hosts every situation and pest-cropping system has to be evaluated on an individual basis and selection of appropriate planting densities based on the density that provides the best yields while providing economical control of the key pests.

Optimum sowing times are dependent on a range of factors including time of harvest of the previous crop, soil type, the rotation and crop variety. In many cases there are some yield benefits to early sowing and if this can be combined with a situation where there is desynchronization between the hostplant development and pest population build-up (Ezueh, 1991) then reduction in pest numbers may provide an additional bonus. Date of sowing has proved to be effective in the control of some soybean defoliators (Jackai *et al.*, 1990) while cotton sown in August avoids attacks by flea beetles and pink bollworm (El Amin and Ahmed, 1991). As with planting density, experiments need to be carried out to optimize crop yield in relation to levels of pest control and each pest-cropping system evaluated on an individual basis. This of course makes area-wide recommendations more difficult.

3.5.6 Phytosanitation and certification

Phytosanitation is concerned with control of sources of infestation and inoculum, through the maintenance of tidy fields and farms, the disposal of unwanted crops and crop residues and the elimination of weeds or volunteers that can harbour pests. Since many pests can survive dormant periods in alternative hosts or crop residues, etc. it is important that these be disposed of to reduce colonization in the next season. Weeds for instance, are important sources for non-persistent, semi-persistent and circulative viruses and their elimination can contribute to virus control (Zitter and Simons, 1980). The pink bollworm *Pectinophora gossypiella* can be controlled by uprooting and destroying the cotton plant after harvest. This measure kills the diapausing larvae and reduces subsequent population levels (Hill and Waller, 1982). The practice of roguing a crop, which involves removing and destroying plants as soon as they become diseased, can reduce the rate of pest population development. However, such practices can become labour intensive and hence, costly. The use of fire to burn stubble from a harvested field is one of the oldest weed-control practices known. Fire kills many weeds, diseases and insects and in addition returns bound nitrogen and phosphates to the soil and increases soil pH (Shenk, 1994). However, it can also lead to the loss of soil organic matter and soluble nutrients and increase soil erosion on sloping land.

Certification involves use of seed which is known to be pest free. Seed material is usually produced either in an area where climatic conditions make pest attack unlikely or under stringent pest management conditions. Certification is especially important in situations where it is known that seed or tubers are a major source of inoculum. The seed potato certification scheme in Europe represents one of the most extensive and long-lived certification programmes known.

3.5.7 Water management

In situations where crops are not solely rainfed, the management of water may be used as a means of pest control. Drip irrigation will favour a crop rather than the weeds and will provide the crop with a competitive advantage. Flooding is also an effective practice for the control of a number of weeds and of course is usefully employed in this way with vast areas of rice paddy (Shenk, 1994). The flooding of land shortly after preparation for 2–3 days has also been found to be an effective means of killing soil dwelling pupae of *Heliothis* and *Spodoptera* species in cotton (El Amin and Ahmed, 1991). However, in general water supplies are managed primarily to promote healthy and vigorous plant growth with pest control usually only considered of secondary importance.

3.6 INTERFERENCE METHODS

The use of semiochemicals, the sterile insect technique and use of insect growth regulators are loosely grouped together under the common theme of interference methods. All of the techniques interfere with normal physiological function or behaviour of insect pests. Semiochemicals are used to affect insect mating or aggregative behaviour, insect growth regulators affect the normal growth and metamorphosis of insects, while the sterile insect technique interferes with reproduction.

3.6.1 Semiochemicals

Semiochemicals are substances or mixtures of substances emitted by one species that modify the behaviour of receptor organisms of like or different species (Tinsworth, 1990). The best-known examples of semiochemicals used in pest control are those associated with the use of sex pheromones. These are however, just one class of behaviour-modifying chemicals, the whole range of which are having an increasingly diverse and significant role to play in the development of measures for the control of insect pests (Inscoe *et al.*, 1990). Table 3.6 provides a list of definitions of the different types of semiochemical, among which only the synomones and apneumones have yet to find a role in pest management.

Nordlund (1981) defined a pheromone as a substance which is secreted by an organism to the outside that causes a specific reaction in a receiving

Table 3.6 The different types of semiochemical involved in insect communication. (After Nordlund, 1981)

Pheromone
A substance that is secreted by an organism to the outside and causes a specific reaction in a receiving organism of the same species

- Sex pheromone – a substance generally produced by the female to attract males for the purpose of mating
- Aggregation pheromone – a substance produced by one or both sexes, and bringing both sexes together for feeding and reproduction
- Alarm pheromone – a substance produced by an insect to repel and disperse other insects in the area. It is usually released by an individual when it is attacked

Allelochemical
A substance that is significant to organisms of a species different from its source, for reasons other than food

- Allomone – a substance produced or acquired by an organism that, when it contacts an individual of another species evokes in the receiver a behavioural or physiological reaction that is adaptively favourable to the emitter but not the receiver
- Kairomone – a substance produced or acquired by an organism that, when it contacts an individual of another species evokes in the receiver a behavioural or physiological reaction that is adaptively favourable to the receiver but not the emitter
- Synomone – a substance produced or acquired by an organism that, when it contacts an individual of another species evokes in the receiver a behavioural or physiological reaction that is adaptively favourable to both emitter and receiver
- Apneumone – a substance emitted by a non-living material which evokes a behavioural or physiological reaction that is adaptively favourable to a receiving organism but detrimental to an organism of another species that may be found in or on the non-living material

organism of the same species. Pheromones are usually classified as sex, aggregation and alarm pheromones (Table 3.6). The number of pheromones identified and their application have greatly increased since the first chemical structure of a pheromone was announced in 1959 (Inscoe *et al.*, 1990) (Figure 3.1), the growth in numbers of sex and aggregation pheromones or parapheromones accounting for the majority of these. The sex pheromones of Lepidoptera have been widely used in traps for monitoring population levels; one of the best examples of which is probably the pea moth (*Cydia nigricana*) trap monitoring system (Macauley *et al.*, 1985). However, there are numerous examples (see Ridgway *et al.*, 1990; McVeigh *et al.*, 1993) and the principles of pheromone trap monitoring are now well established (Wall, 1990). In an increasing number of cases it has been possible to use pheromones to directly control pests through mating disruption (e.g. Critchley *et al.*, 1985; Cassagrande, 1993; de Vlieger and Klijnstra, 1993),

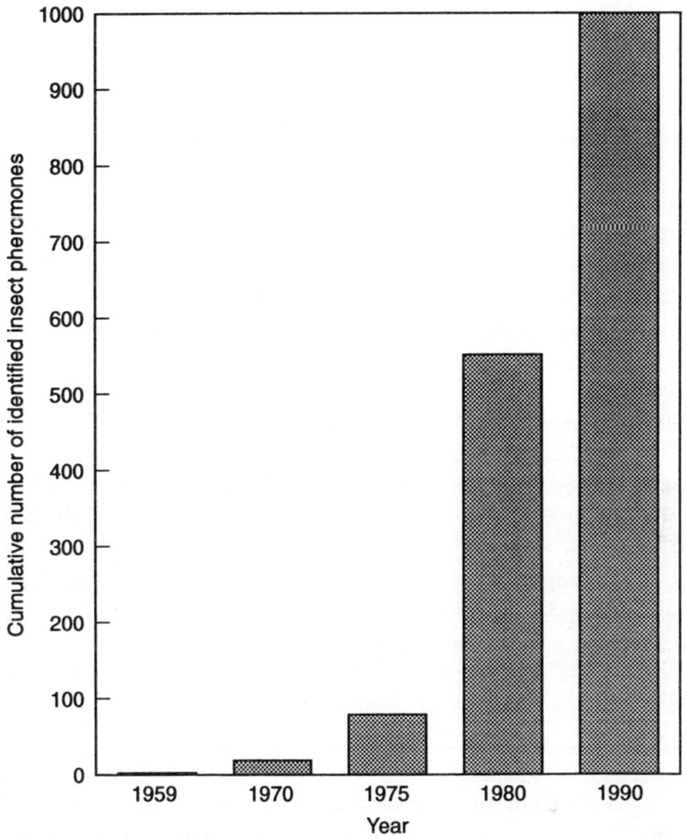

Figure 3.1 The cumulative growth in numbers of insect pheromones which have been isolated and identified since 1959.

lure and kill techniques (e.g. Otieno *et al.*, 1988; Montiel, 1992) and mass trapping (systems available for 12 Lepidoptera, four Coleoptera, two Diptera and one Hymenoptera; Inscoe *et al.*, 1990).

Control and monitoring systems for coleopteran pests are more often based on aggregation pheromones which are often combined with secondary plant substances, and cause both sexes to aggregate on the host plant. They are currently used to control forest pests (e.g. Bedard and Wood, 1980; Baake and Strand, 1981; Borden, 1990) but aggregation pheromones can also be used for monitoring beneficial insect populations (Pickett, 1988b).

Insect alarm pheromones can also be used to good effect in pest management. In the case of the honey bee *Apis mellifera* the alarm pheromone has been used to repel the bees from oilseed rape before insecticide application (Free *et al.*, 1985; Pickett, 1988a). Many species of aphid produce an alarm pheromone, (E)-β-farnesene, which is normally released when aphids are attacked, increasing their mobility and hence, chances of escape from natural enemies. This reaction to the pheromone has been utilized ingeniously to increase aphid contact with synthetic pyrethroids (Griffiths and Pickett, 1987) and with the fungal spores of *Verticillium lecanii* (Hockland *et al.*, 1986) and thereby increasing the effectiveness of these control measures.

Pheromones of invertebrate pests can be closely related or even identical to secondary plant substances. For instance, (E)-β-farnesene, the aphid alarm pheromone, occurs commonly in plants but does not interfere with aphid colonization because of an accompanying biosynthesis in the plant of a potent inhibitor (Pickett, 1985). However, many other secondary plant substances do produce allomonic effects, usually in the form of anti-feedants.

The best-known anti-feedant is azadirachtin produced by the neem tree (*Azadirachta indica*) but there are many others. The brassicas produce mustard oils, mostly comprising of organic isothiocyanates; cultivars of oilseed rape *Brassica rapae* low in the isothiocyanate sinigrin are more readily eaten by a range of herbivores (Pickett, 1985). The water pepper *Polygonium hydropiper* produces a simple compound polygodial which when applied in the field reduced barley yellow dwarf virus incidence through deterring aphid colonization (Pickett *et al.*, 1987), one of the first examples of a plant-derived anti-feedant being used successfully against an arable pest. Plant allomones also include compounds which attract pollinating insects and other beneficials. The attraction of parasitoids to crop volatiles (Powell and Zhang, 1983) could be of use in pest management to monitor populations of natural enemies or potentially, to retain them in the crop to attack colonising pests (Pickett, 1988b). Under laboratory conditions a plant-derived substance β-caryophythene has been shown to lower the dispersal of aphids when attacked by the predator *Chrysoperla carnea*. This compound could potentially be applied directly to crops to improve predation of aphids (Pickett, 1988b).

Kairomones are substances which benefit the receiver rather than the emitter. They are crop or animal volatiles that are used by the pest to detect its host. A classical example of this has been the studies conducted to identify the kairomone in cattle that attracts the tsetse fly (*Glossina* spp.). The work was pioneered by Vale (1974a,b) and has resulted in the development of lures based on acetone and carbon dioxide and a control technique utilizing insecticide-impregnated screens baited with the lure (see Hall, 1990). Identification of kairomones in crop plants can be potentially equally advantageous. It is generally acknowledged that the presence of plant extracts can improve the attraction of pheromones and catches in traps, e.g. grandlure and monoterpines from cotton are used in traps to catch bollworms (Dickens, 1989). If crop kairomones can be identified then the opportunity exists to breed crop plants with lower levels of the chemical and thereby decrease crop attractiveness for the pest (Pickett, 1985).

The biosynthesis of semiochemcials is relatively simple and often involves processes already taking place in the crop plants or related species. Indeed, semiochemicals are often the same as or related to secondary plant metabolites; hence, breeding or genetic manipulation could be used to supplement secondary metabolites in plants as a means of effective control (Pickett *et al.*, 1989). Genetic manipulation could also be used to transfer genes for semiochemical production from insects into plants where it could be employed to protect plants as a form of anti-xenotic resistance (Pickett, 1988a). One thing is very clear however, that semiochemicals will have a diverse and important role in future pest management strategies as greater understanding is obtained of their properties and functions.

3.6.2 Sterile insect techniques

The sterile insect technique (SIT) involves the laboratory mass rearing, irradiation and sterilization of large numbers of individuals of an insect species which are then released to mate with the natural wild populations. Since females that mate with sterile male insects produce no offspring, the target pest population is gradually eliminated over several generations. The technique has been most successfully used against the cattle screw worm *Cochliomyia hominivorax* in the USA (Davidson, 1974; Drummond *et al.*, 1988). The spectacular success of this programme created widespread interest in the approach but despite considerable investment in research there have been few programmes that have been as successful as the screw worm programme where SIT has been able to eliminate pests without recourse to other control measures – in contrast to, for example, the Mediterranean fruit fly sterile insect programme (Saul, 1989).

The technique involves the mass rearing of the insect, the production of sterile insects through cytoplasmic incompatibility, hybrid sterility or through dominant lethal mutations and then the release of these sterile insects (Boller, 1987; Lindquist and Busch-Petersen, 1987). However, a large amount of information is required before suitable candidates for SIT can be identified (Boller, 1987) including: (i) knowledge of economic or public health impact of the pest; (ii) in-depth understanding of the pest's ecology and behaviour; (iii) an ability to successfully sterilize the insect without reducing its natural competitiveness; (iv) availability of a technique for economic mass rearing, handling and release of the insects; and (v) procedures for monitoring the quality of the reared insect.

With these requirements it is not unreasonable to expect few species to be suitable for control by SIT. Despite the fact that the SIT has many of the attributes required of components for insect pest management systems, i.e. it is an ecologically sound technique, species-specific, it has no impact on other components of an ecosystem, it can provide long-term suppression of a pest, it is however, only ever likely to maintain a small niche in the gamut of control measures used in IPM.

3.6.3 Insect growth regulators

The term insect growth regulators (IGRs) was designed to describe a new class of biorational compounds in the late 1960s (Staal, 1975). IGRs may be defined as materials which are natural biochemicals or exogenously applied chemicals that cause morphological and physiological changes during the growth or development of insects (Chamberlain, 1975). The group includes insect hormones, growth regulators, developmental inhibitors and synthetic hormone mimics. The majority of the research to date has concentrated on juvenile hormones and their synthetic analogues, juvenoids, although some effort has been devoted to identification and utilization of anti-juvenile hormone agents.

The chemical structure of the first juvenile hormone JH1 was elucidated in 1967 (Staal, 1975). Since then a number of juvenile hormone analogues have been developed that are successfully used for control of a range of pests, particularly public health and stored product pests such as mosquitoes (Gratz, 1985; Otieno *et al.*, 1988), cockroaches (Staal *et al.*, 1985; Ross and Cochran, 1990) and flies, fleas and pharaoh ants (Staal, 1987). These juvenoids have a number of favourable characteristics which make them attractive for use in IPM programmes: (i) their spectacular activity on specific target species (within a sensitive window); (ii) little immediate cross-resistance can be expected to exist; (iii) the toxicity for non-target pests is extremely low; and (iv) they are generally not repellent (Staal, 1987). However, to balance against these favourable attributes are three fundamental drawbacks with their use. These are: (i) a general lack of field stability; (ii) the absence of direct mortality (except for kinoprene); and (iii) narrow windows of sensitivity in the last larval instar and in the eggs. Hence, the restricted use of IGRs to public health and stored product pests. However, most of the drawbacks of juvenoids do not apply to anti-juvenile hormone agents, particularly the sensitivity window which is much greater, affecting the next moult, making them more attractive for use in field crops. Anti-juvenile hormone agents have great potential but breakthroughs regarding activity and selectivity will only be forthcoming after further advances are made in the knowledge of regulation, production, transport and metabolism and reception of endogenous juvenile hormones (Staal, 1986).

Anti-juvenile hormone agents are found as secondary chemicals in a wide range of plants, as are phytoecdysones and phytojuvenoids (Bowers, 1985). Their relative abundance in different plant families (111 plant families have been shown to contain at least 69 phytoecdysone steroids) with moulting hormone activity (Bergamasco and Horn, 1983) suggests a role as plant defensive compounds. The identification of these compounds lends possibilities for exploitation through conventional plant breeding, genetic manipulation and through their use as models for development of synthetic analogues. The whole subject of juvenile hormones and their analogues provides a number of exciting opportunities for future exploitation and utilization in integrated pest management programmes.

REFERENCES

Agricola, U., Agounke, D., Fischer, H.U. and Moore, D. (1989) The control of *Rastrococcus invadens* Williams (Hemiptera: Pseudococcidae) in Togo by the introduction of *Gyranusoidea tebgyi* Noyes (Hymenoptera: Encyrtidae). *Bulletin of Entomological Research*, **79**, 671–8.

Ahlstrom-Olsson, M. and Jonasson, T. (1992) Mustard meal mulch – a possible cultural method for attracting natural enemies of brassica root flies into brassica crops, in *Proceedings Working Group Meeting – 'Integrated Control in Field Vegetable Crops'*, XV/4, (eds S. Finch and J. Freuler), IOBC/WPRS, pp. 171–5.

Akobundu, I.O. and Poku, J.A. (1987) Weed control in soybeans in the tropics, in *Soybeans for the Tropics*, (eds S.R. Singh, K.O. Rachie and K.E. Dashiell), John Wiley & Sons, New York, pp. 69–77.

Ayres, P.G. and Paul, N.D. (1990) The effects of disease on interspecific plant competition, in *The Exploitation of Micro-organisms in Applied Biology*, The Association of Applied Biologists, Warwick, pp. 155–62.

Baake, A. and Strand, L. (1981) Pheromones and traps as part of an integrated control of the spruce bark beetle. Some results from a control program in Norway in 1979 and 1980. *Rapport fra norsk institutt for skogforskning*, **5/81**, 1–39.

Bailey, J.A. (1990) Improvement of mycoherbicides by genetic manipulation, in *The Exploitation of Micro-organisms in Applied Biology*, The Association of Applied Biologists, Warwick, pp. 33–8.

Baker, R.T. (1985) Biological control of plant pathogens: definitions, in *Biological Control in Agricultural IPM Systems*, (eds M.A. Hoy and D.C. Herzog), Academic Press, Orlando, pp. 25–39.

Barlow, F. (1985) Chemistry and formulation, in *Pesticide Application: Principles and Practice*, (ed. P.T. Haskell), Oxford Science Publications, Oxford, pp. 1–34.

Beck, S.D. (1965) Resistance of plants to insects. *Annual Review of Entomology*, **10**, 207–32.

Bedard, W.D. and Wood, L.D. (1980) Suppression of *Dendroctonius brevicornis* by using a mass-trapping tactic, in *Management of Insect Pests with Semiochemicals – Concepts and Practice*, (ed. E.R. Mitchell), Plenum Press, New York, London, pp. 103–14.

Bedford, G.O. (1981) Control of the rhinocerus beetle by baculovirus, in *Microbial Control of Pests and Plant Diseases, 1970–1980*, (ed, H.D. Burges), Academic Press, Orlando, pp. 409–26.

Bence, J.R. (1988) Indirect effects and biological control of mosquitoes by mosquito fish. *Journal of Applied Ecology*, **25**, 505–21.

Bergamasco, R. and Horn, D.H.S. (1983), in *Endocrinology of Insects*, (eds G.H. Downer and H. Laufer), Alan R Liss, New York, pp. 627–54.

Boller, E.F. (1987) Genetic Control, in *Integrated Pest Management*, (eds A.J. Burn, T.H. Coaker and P.C. Jepson), Academic Press, London, pp. 161–87.

Borden, J.H. (1990) Use of semiochemicals to manage coniferous tree pests in Western Canada, in *Behavior-Modifying Chemicals for Insect Management*, (eds R.L. Ridgway, R.M. Silverstein and M.N. Inscoe), Marcel Dekker Inc., New York, Basel, pp. 281–315.

Bowers, W.S. (1985) Phytochemical disruption of insect development and behavior, in *Bioregulators for Pest Control*, (ed. P.A. Hedin), American Chemical Society, Washington DC, pp. 225–36.

Buckner, C.H. (1966) The role of vertebrate predators in the biological control of forest insects. *Annual Review of Entomology*, **11**, 449–70.

Caseley, J.C. (1994) Herbicides, in *Weed Management of Developing Countries*,

FAO Plant Production and Protection Paper 120, (eds R. Labrada, J.C. Caseley and C. Parker), FAO, pp. 183–223.

Cassagrande, E. (1993) The commercial implementation of mating disruption for the control of rice stemborer, *Chilo suppressalis*, in rice in Spain, in *Working Group 'Use of Pheromones and Other Semiochemicals in Integrated Control'*, IOBC wprs Bulletin, (eds L.J. McVeigh, D.R. Hall and P.S. Beevor), IOBC, pp. 82–9.

Cate, J.R. (1990) Biological control of pests and diseases: integrating a diverse heritage, in *New Directions in Biological Control*, (eds R.R. Baker and P.E. Dunn), Alan R Liss, New York, pp. 23–43.

Chamberlain, W.F. (1975) *Journal of Medical Entomology*, **12**(4), 395.

Chiverton, P.A. and Sotherton, N.W. (1991) The effects on beneficial arthropods of the exclusion of herbicides from cereal crop edges. *Journal of Applied Ecology*, **28**, 1027–39.

Coaker, T.H. (1987) Cultural methods: the crop, in *Integrated Pest Management*, (eds A.J. Burn, T.H. Coaker and P.C. Jepson), Academic Press, London, pp. 69–88.

Cock, M.J.W. (1985) The use of parasitoids for augmentative biological control of pests in the People's Republic of China. *Biocontrol News and Information*, **6**, 213–33.

Cock, M.J.W. (1994) Biological weed control, in *Weed Management for Developing Countries*, FAO Plant Production and Protection Paper 120, (eds R. Labrada, J.C. Caseley and C. Parker), FAO, pp. 173–80.

Cowgill, S.E., Wratten, S.D. and Sotherton, N.W. (1993) The selective use of floral resources by the hoverfly *Episyrphus baleatus* (Diptera: Syrphidae) on farmland. *Annals of Applied Biology*, **122**(2), 223–31.

Cremlyn, R. (1978) Pesticides in the Environment, in *Pesticides Preparation & Mode of Action*, John Wiley, Chichester, New York, pp. 210–21.

Critchley, B.R., Campion, D.G., McVeigh, L.J., McVeigh, E.M., Cavanagh, G.G., Hosny, M.M., Nasr, El-Sayed, A., Khidr, A.A. and Naguib, M. (1985) Control of pink bollworm, *Pectinophora gossypiella* (Saunders) (Lepidoptera: Gelechiidae), in Egypt by mating disruption using hollow-fibre, laminate-flake and microencapsulated formulations of synthetic pheromone. *Bulletin of Entomological Research*, **75**, 329–45.

Cuijpers, T.A.M.M., Steeghs, N.W.F. and Smits, P.H. (1994), in *International Symposium on Crop Protection*, 3 May 1994, p. 40.

Davidson, G. (1974) *Genetic Control in Insect Pests*, Academic Press, London.

de Vlieger, J.J. and Klijnstra, J.W. (1993) Mating disruption of codling moth and fruit tree leafrollers in apple orchards with TNO dispensers, in *Working Group 'Use of Pheromones and Other Semiochemicals in Integrated Control'*, IOBC wprs Bulletin, (eds L.J. McVeigh, D.R. Hall and P.S. Beevor), IOBC, pp. 99–103.

DeBach, P. (1964) *Biological Control of Insect Pests and Weeds*, Chapman & Hall, London.

DeBach, P. (1974) *Biological Control by Natural Enemies*, Cambridge University Press, Cambridge.

Dent, D.R. (1991) *Insect Pest Management*, CAB International, Wallingford.

Dent, D.R. (1993) The use of *Bacillus thuringiensis* as an insecticide, in *Exploitation of Microorganisms*, (ed. D. Garth Jones), Chapman & Hall, London, pp. 19–32.

DeVay, J.E. (1991) Historical review and principles of soil solarization, in *Soil Solarization*, (eds J.E. DeVay, J.J. Stapleton and C.L. Elmore), FAO, pp. 1–15.

DeVay, J.E., Stapleton, J.J. and Elmore, C.L. (1991) *Soil solarization. Proceedings of the First International Conference on Soil Solarization, Amman, Jordan, 19–25 Feb 1990*, FAO, Rome.

Dickens, J.C. (1989) Green leaf volatiles enhance aggregation pheromone of boll weevil, *Anthonomus grandis*. *Entomologia Experimentalis et Applicata*, **52**, 191–203.

Dodd, A.P. (1959) The biological control of prickly-pear in Australia. *Monographie Biologiae*, **8**, 565–77.

Drummond, R.O., George, J.E. and Kunz, S.E. (1988) *Control of Arthropod Pests of Livestock: A Review of Technology*, CRC Press Inc., Boca Raton, Florida.

El Amin, E.T.M. and Ahmed, M.A. (1991) Strategies for integrated cotton pest control in the Sudan. 1 – Cultural and legislative measures. *Insect Sci. Application*, **12**(5/6), 547–52.

Emge, R.C., Melching, J.S. and Kingsolver, C.H. (1981) Epidemiology of *Puccinia chondrillina*, a rust pathogen for the biological control of rush skeleton weed in the United States. *Phytopathology*, **71**, 839–43.

Ester, A., Embrechts, A., Vlaswinkel, M.E.R. and de Moel, C.P. (1994) Protection of field vegetables against insect attacks by covering the plot with polyethylene nets, in *International Symposium on Crop Protection*, Gent University 3 May 1994, University of Gent, p. 64.

Evans, H.C. and Ellison, C.A. (1990) Classical biological control of weeds with micro-organisms: past, present, prospects, in *The Exploitation of Micro-organisms in Applied Biology*, The Association of Applied Biologists, Warwick, pp. 39–49.

Evans, H.F. (1990) The use of bacterial and viral control agents in British forestry, in *The Exploitation of Micro-organisms in Applied Biology*, The Association of Applied Biologists, Warwick, pp. 195–203.

Ezueh, M.I. (1991) Prospects for cultural and biological control of cowpea pests. *Insect Sci. Application*, **12**(5/6), 585–92.

Ferron, P. (1981) Pest control by the fungi *Beauveria* and *Metarhizium*, in *Microbial Control of Pests and Plant Diseases, 1970–1980*, (ed. H.D. Burges), Academic Press, Orlando, pp. 465–82.

Free, J.B., Pickett, J.A., Ferguson, A.W., Simpkins, J.R. and Smith, M.C. (1985) Repelling foraging honeybees with alarm pheromones. *Journal of Agricultural Science, Cambridge*, **105**, 255–60.

Gednalske, J.V. and Walgenbach, D.D. (1984) Effect of tillage practices on the emergence of *Smicronyx fulvus* (Coleoptera: Curculionidae). *Journal of Economic Entomology*, **77**, 522–4.

Georgis, R. (1990) Commercialization of steinernematid and heterorhabditid entomopathogenic nematodes, in *Brighton Crop Protection Conference – Pests and Diseases*, BCPC Farnham, pp. 275–80.

Gratz, N.G. (1985) Control of dipteran vectors, in *Pesticide Application: Principles and Practice*, (ed. P.T. Haskell), Oxford Science Publications, Oxford, pp. 273–300.

Greathead, D.J. (1984) Biological control constraints to agricultural production, in *Advancing Agricultural Production in Africa*, (ed. D.L. Hawksworth), Commenwealth Agricultural Bureaux, pp. 200–6.

Greathead, D.J. and Waage, J.K. (1983) *Opportunities for Biological Control of Agricultural Pests in Developing Countries*, The World Bank, Washington.

Greaves, J.H. (1989) *Rodent Pests and their Control in the Near East*, FAO, Rome.

Greaves, J.H. and Jones, P.J. (1985) Vertebrate pest control, in *Pesticide Application: Principles and Practice*, (ed. P.T. Haskell), Oxford Science Publications, Oxford, pp. 334–77.

Greaves, M.P. and MacQueen, M.D. (1990) The use of mycoherbicides in the field, in *The Exploitation of Micro-organsisms in Applied Biology*, The Association of Applied Biologists, Warwick, pp. 163–8.

Gregoire, J.-C., Baisier, M., Merlin, J. and Naccache, Y. (1990) Interactions between *Rhizophagus grandis* (Coleoptera: Rhizophagidae) and *Dendroctonus micans* (Coleoptera: Scolytidae) in the field and the laboratory. Their application for the biological control of *D. micans* in France, in *The Potential for Biological Control of Dendroctonus and Ips Bark Beetles,* (eds D. Kulhavy and M.E. Miller), The Stephen Austin University Press, Nagocdoches, pp. 95–108.

Griffiths, D.C. and Pickett, J.A. (1987) Novel chemicals and their formulation for aphid control, in *Proceedings of the 14th International Symposium on Controlled Release of Bioactive Materials, Toronto, 1987,* (eds P.I. Lee and B.A. Leonhardt), The Controlled Release Society Inc., Toronto, pp. 243–4.

Habeck, M.H., Lovejoy, S.B. and Lee, J.G. (1993) When does investing in classical biological control research make economic sense? *Florida Entomologist,* **76**(1), 96–101.

Hall, D.R. (1990) Use of host odor attractants for monitoring and control of tsetse flies, in *Behavior-Modifying Chemicals for Insect Management,* (eds R.L. Ridgway, R.M. Silverstein and M.N. Inscoe), Marcel Dekker Inc., New York, pp. 517–30.

Haskell, P.T. (1985) *Pesticide Application: Principles and Practice,* Oxford University Press, Oxford.

Hill, D.S. and Waller, J.M. (1982) *Pests and Diseases of Tropical Crops,* Longman, London.

Hockland, S.H., Dawson, G.W., Griffiths, D.C., Marples, B., Pickett, J.A. and Woodcock, H.N. (1986) The use of aphid alarm pheromone (E-β-farnesene) to increase effectiveness of the entomophilic fungus *Verticillium lecanii* in controlling aphids on chrysanthemums under glass, in *Fundamental and Applied Aspects of Invertebrate Pathology,* (eds R.A. Samson, J.M. Valk and D. Peters), Foundation of the Fourth International Colloquium of Invertebrate Pathology, Wageningen, p. 252.

Hominick, W.M. (1990) Entomopathogenic rhabditid nematodes and pest control. *Parasitology Today,* **6**, 148–52.

Hoy, J.B., Kauffman, E.E. and O'Berg, A.G. (1972) A large-scale field test of *Gambusia affinis* and chlorpyrifos for mosquito control. *Mosquito News,* **32**, 161–71.

Huffaker, C.B. (1977) Augmentation of natural enemies in the People's Republic of China, in *Biological Control by Augmentation of Natural Enemies,* (eds R.L. Ridgway and S.B. Vinson), Plenum Press, New York, pp. 329–39.

Inscoe, M.N., Leonhardt, B.A. and Ridgway, R.L. (1990) Commercial availability of insect pheromones and other attractants, in *Behavior-Modifying Chemicals for Insect Management,* (eds R.L. Ridgway, R.M. Silverstein and M.N. Inscoe), Marcel Dekker Inc., New York, pp. 631–715.

Jackai, L.E.N., Panizzi, A.R., Kundu, G.G. and Srivastava, K.P. (1990) Insect pests of soybean in the tropics, in *Insect Pests of Tropical Food Legumes,* (ed. S.R. Singh), John Wiley & Sons, New York, Chichester, pp. 91–156.

Jayaraj, S. and Santharam, G. (1985) Ecology-based integrated control of *Spodoptera litura* (F.) on cotton, in *Microbial Control and Pest Management,* (ed. S. Jayaraj), TNAU, Coimbatore, pp. 256–64.

Jervis, M. and Kidd, N. (1995) *Insect Natural Enemies. Practical Approaches to their Study and Evaluation,* Chapman & Hall, London.

Johnson, D.A. and Gilmore, E.C. (1980) Breeding for resistance to pathogens in wheat, in *Biology and Breeding for Resistance to Arthropods and Pathogens in Agricultural Plants,* Proceedings of an International Short Course in Host Plant Resistance, Texas A & M University, 22 July–2 August 1979, (ed. M.K. Harris), pp. 263–75.

Jones, T.A. (1988) A probability method for comparing varieties against checks. *Crop Science,* **28**(6), 907–12.

Kaaya, G.P. (1993) Biological control: an environmentally safe alternative to the use of chemical pesticides, in *Community-Based and Environmentally Safe Pest Management*, (eds R.K. Saini and P.T. Haskell), ICIPE Science Press, Nairobi, pp. 15–29.

Katan, J. (1981) Solar heating (solarisation) of soil for control of soilborne pests. *Annual Review of Phytopathology*, **19**, 211–36.

Katan, J.A., Greenberger, H.A. and Grinstein, A. (1976) Solar heating by polyethylene mulching for the control of diseases caused by soil-borne pathogens. *Phytopathology*, **66**, 683–8.

King, E.G., Hopper, K.R. and Powell, J.E. (1985) Analysis of systems for biological control of crop arthropod pests in the U.S. by augmentation of predators and parasites, in *Biological Control in Agricultural IPM Systems*, (eds M.A. Hoy and D.C. Herzog), Academic Press, Orlando, pp. 201–28.

Kirkwood, R.C. (1987) Uptake and movement of herbicides from plant surfaces and the effects of formulation and environment upon them, in *Pesticides on Plant Surfaces*, (ed. H.J. Cottrell), John Wiley & Sons, New York, Chichester, pp. 1–25.

Koike, S.T., Smith, R.F and Schulbach, K.F. (1992) Resistant cultivars, fungicides combat downy mildew of spinach. *California Agriculture*, **46**(2), 29–31.

Kring, J.B. and Schuster, D.J. (1992) Management of insects on pepper and tomato with UV-reflective mulches. *Florida Entomologist*, **75**(1), 119–29.

Lacey, A.J. (1985) Weed control, in *Pesticide Application: Principles and Practice*, (ed. P.T. Haskell), Oxford Science Publications, Oxford, pp. 456–85.

Lima, J.A.A. and Gonclaves, M.F.B. (1985) O uso de barreira viva visando o controle do 'Cowpea aphid-borne mosaic virus' em cultura de feijao-de-corda (*Vigna unguiculata*). *Fitopatologia Brasileira*, **10**, 321.

Lindquist, D.A. and Busch-Petersen, E. (1987) Applied insect genetics and IPM, in *Integrated Pest Management, Protection Integrée, Quo Vadis? An International Perspective, Parasitis 86*, (ed. V. Delucchi), Geneva, pp. 237–55.

Macauley, E.D.M., Etheridge, P., Garthwaite, D.G., Greenway, A.R., Wall, C. and Goodchild, R.E. (1985) Prediction of optimum spraying dates against pea moth (*Cydia nigricana*) using pheromone traps and temperature measurements. *Crop Protection*, **4**, 85–98.

Manners, J.G. (1982) *Principles of Plant Pathology*, Cambridge University Press, Cambridge.

Martin, J.N., Leonard, W.H. and Stamp, D.L. (1976) *Principles of Field Crop Production*, Collier Macmillan, New York, Chichester.

Matthews, G.A. (1992) *Pesticide Application Methods*, 2nd edn, Longman Scientific & Technical, London.

McVeigh, L.J., Hall, D.R. and Beevor, P.S. (1993) *Working Group 'Use of Pheromones and Other Semiochemicals in Integrated Control'*. Proceedings of IOBC/WPRS Working Group Meeting. Chatham, UK, 11–14 May 1993. *IOBC/WPRS Bulletin*, **16**(10), 372.

Mitchell, T.D. (1985) Goats in land and pasture, in *Goat Production and Research in the Tropics*, (ed. J.W. Copland), ACIAR, Canberra, pp. 115–16.

Mohyuddin, A.I. and Shah, S. (1977) Biological control of *Mythimna separata* (Lep.: Noctuidae) in New Zealand and its bearing on biological control strategy. *Entomophaga*, **22**(4), 331–3.

Montiel, M.B. (1992) The influence of the Project ECLAIR 209 on the development of new oliveculture, in *Research Collaboration in European IPM Systems* BCPC Monograph No. 52, (ed. P.T. Haskell), BCPC, Brighton, pp. 77–80.

Moore, D. (1993) Biological control and the rural community, in *Community-Based and Environmentally Safe Pest Management*, (eds R.K. Saini and P.T. Haskell), ICIPE Science Press, Nairobi, pp. 111–22.

Munaan, A. and Wikardi, E.W. (1986) Towards the biological control of coconut insect pests in Indonesia, in *Biological Control in the Tropics*, (eds M.Y. Hussein and A.G. Ibrahim), Penerbit University, Pertainan, Malaysia, pp. 149–57.

Neuenschwander, P. and Herren, H.R. (1988) Biological control of the cassava mealybug, *Phenacoccus manihoti*, by the exotic parasitoid *Epidinocarsis lopezi* in Africa. *Phil. Transactions of the Royal Society of London*, **318**, 319–33.

Nordlund, D.A. (1981) Semiochemicals: a review of terminology, in *Semiochemicals, Their Role in Pest Control*, (eds D.A. Nordlund, R.L. Jones and W.J. Lewis), John Wiley & Sons, New York, pp. 13–23.

Norris, D.M. and Kogan, M. (1980) Biochemical and morphological bases of resistance, in *Breeding Plants Resistant to Insects*, (eds F.G. Maxwell and P.R. Jennings), John Wiley & Sons, New York, pp. 23–62.

Oatman, E.R., Gilstrap, F.E. and Voth, V. (1976) Effect of different release rates of *Phytoseiulus persimilis* (Acarina: Phytoseiidae) on the twospotted spider mite on strawberry in Southern California. *Entomophaga*, **21**, 269–73.

Otieno, W.A., Onyango, T.O., Pile, M.M., Laurence, B.R., Dawson, G.W., Wadhams, L.J. and Pickett, J.A. (1988) A field trial of the synthetic oviposition pheromone with *Culex quinquefasciatus* Say (Diptera: Culicidae) in Kenya. *Bulletin of Entomological Research*, **78**, 463–78.

Parry, D.W. (1990) How do we control disease?, in *Plant Pathology in Agriculture*, (ed. D.W. Parry), Cambridge University Press, Cambridge, pp. 86–158.

Pickett, J.A. (1985) Production of behaviour-controlling chemicals by crop plants. *Phil. Transactions of the Royal Society of London*, **310**, 235–9.

Pickett, J.A. (1988a) The future of semiochemicals in pest control. *Aspects of Applied Biology*, **17**, 397–406.

Pickett, J.A. (1988b) Integrating use of beneficial organisms with chemical crop protection. *Phil. Transactions of the Royal Society of London*, **318**, 203–11.

Pickett, J.A., Dawson, G.W., Griffiths, D.C. *et al.* (1987) Development of plant-derived antifeedants for crop protection, in *Pesticide Science and Technology*, (eds R. Greenhalgh and T.R. Roberts), Blackwell Science, Oxford, pp. 125–8.

Pickett, J.A., Wadhams, L.J. and Woodcock, C.M. (1989) Chemical ecology and pest management: some recent insights. *Insect Science Application*, **10**(6), 741–50.

Popiel, I. and Hominick, W.M. (1992) Nematodes as biological control agents: Part II, in *Advances in Parasitology*, Academic Press Limited, Orlando, pp. 381–433.

Powell, W. and Zhang, Z. (1983) The reactions of two cereal aphid parasitoids, *Aphidius uzbekistanicus* and *A. ervi* to host aphids and their food plants. *Physiological Entomology*, **8**, 439–43.

Püntner, W. (1981) *Manual for Field Trials in Plant Protection*, Ciba-Geigy, Basel.

Richards, M.G. and Rodgers, P.B. (1990) Commercial development of insect biocontrol agents, in *The Exploitation of Micro-organisms in Applied Biology*, The Association of Applied Biologists, Warwick, pp. 245–53.

Richardson, P.N. (1990) Uses for parasitic nematodes in insect control strategies in protected crops, in *The Exploitation of Micro-organisms in Applied Biology*, The Association of Applied Biologists, Warwick, pp. 205–10.

Ridgway, R.L., Silverstein, R.M. and Inscoe, M.N. (1990) *Behavior-Modifying Chemicals for Insect Management*, Marcel Dekker Inc., New York.

Robinson, R.A. (1987) *Host Management in Crop Pathosystems*, MacMillan Publishing Co., New York.

Rodriguez-Kabana, R. and Curl, E.A. (1981) *Annual Review of Phytopathology*, **18**, 311.

Rogers, C.E. (1985) Cultural management of "*Dectus texanus*" (Coleoptera: Cerambycidae) in sunflower. *Journal of Economic Entomology*, **78**, 1145–8.

Ross, M.H. and Cochran, D.G. (1990) Response of late-instar *Blattella germanica*

(Dictyoptera: Blattellidae) to dietary insect growth regulators. *Journal of Economic Entomology*, **83**(6), 2295–305.

Russell, G.E. (1978) *Plant Breeding for Pest and Disease Resistance*. Studies in the Agricultural and Food Sciences, Butterworth, London, Boston.

Ryder, M.H. and Jones, D.H. (1990) Biological control of crown gall, in *Biological Control of Soil-Borne Plant Pathogens*, (ed. D. Hornby), CAB International, Wallingford, pp. 45–63.

Saul, S.H. (1989) Genetics of the Mediterranean fruit fly (*Ceratitis capitata*) (Wiedemann), in *Genetical and Biochemical Aspects of Invertebrate Crop Pests*, (ed. G.E. Russell), Intercept, Hampshire, pp. 1–36.

Schroth, M.N. and Hancock, J.G. (1985) Soil antagonists in IPM systems, in *Biological Control in Agricultural IPM Systems*, (eds M.A. Hoy and D.C. Herzog), Academic Press, Orlando, pp. 415–31.

Shenk, M.D. (1994) Cultural practices for weed management, in *Weed Management for Developing Countries*, FAO Plant Production and Protection Paper 120, (eds R. Labrada, J.C. Caseley and C. Parker), FAO, pp. 163–70.

Simmonds, F.J. (1968) Economics of biological control. *PANS*, **14**, 207–15.

Simons, J.N. (1982) Use of oil sprays and reflective surfaces for control of insect-transmitted plant viruses, in *Pathogens, Vectors, and Plant Diseases. Approaches to Control*, (eds K.F. Harris and K. Maramorosch), Academic Press, Orlando, pp. 71–3.

Singh, D.P. (1986) Breeding for resistance to diseases and pests, in *Breeding for Resistance to Diseases and Insect Pests*, Springer-Verlag, Berlin.

Sotherton, N.W., Boatman, N.D. and Rands, M.R.W. (1989) The 'conservation headland' experiment in cereal ecosystems. *The Entomologist*, **108**, 135–43.

Staal, G.B. (1975) Insect growth regulators with juvenile hormone activity. *Annual Reviews of Entomology*, **20**, 417–60.

Staal, G.B. (1986) Anti juvenile hormone agents. *Annual Review of Entomology*, **31**, 391–429.

Staal, G.B. (1987) Juvenoids and anti juvenile hormone agents as insect growth regulators, in *Integrated Pest Management, Protection Integrée: Quo Vadis? An International Perspective, Parasitis 86*, (ed. V. Delucchi), Geneva, pp. 277–92.

Staal, G.B., Henrick, C.A., Grant, D.L. *et al.* (1985) Cockroach control with juvenoids, in *Bioregulators for Pest Control*, (ed. P.A. Hedin), American Chemical Society, Washington DC, pp. 201–18.

Stinner, B.R. and House, G.J. (1990) Arthropods and other invertebrates in conservation-tillage agriculture. *Annual Review of Entomology*, **35**, 299–318.

Strange, R.N. (1993) *Plant Disease Control. Towards Environmentally Acceptable Methods*, Chapman & Hall, London.

Summerell, B.A. and Burgess, L.W. (1989) Factors influencing survival of *Pyrenophora tritici-repentis*: stubble management. *Mycological Research*, **93**, 38–40.

Sumner, D.R., Doupnik, B. and Boosalis, M.G. (1981) Effects of reduced tillage and multiple cropping on plant diseases. *Annual Review of Phytopathology*, **19**, 167–87.

Sütterlin, S. and van Lenteren, J.C. (1994) Biological control of whitefly in *Gerbera*: a success story, in *International Symposium on Crop Protection*, 3 May 1994, University of Gent, p. 32.

Takahashi, F. (1964) Reproduction curve with two equilibrium points: a consideration in fluctuation of insect populations. *Research in Population Ecology*, **6**, 28–38.

TeBeest, D.O. (1993) Biological control of weeds with fungal plant pathogens, in

Exploitation of Microorganisms, (ed. D.G. Jones), Chapman & Hall, London, pp. 1–17.

Thomas, M.B., Wratten, S.D. and Sotherton, N.W. (1991) Creation of 'Island' habitats in farmland to manipulate populations of beneficial arthropods: predator densities and emigration. *Journal of Applied Ecology*, **28**, 906–17.

Thottappilly, G., Rossel, H.W., Reddy, D.V.R., Morales, F.J., Green, S.K. and Makkouk, K.M. (1990) Vectors of virus and mycoplasma diseases: an overview, in *Insect Pests of Tropical Food Legumes*, (ed. S.R. Singh), John Wiley & Sons, New York, pp. 323–42.

Tinsworth, E.F. (1990) Regulation of pheromones and other semiochemicals in the United States, in *Behavior-Modifying Chemicals for Insect Management*, (eds R.L. Ridgway, R.M. Silverstein and M.N. Inscoe), Marcel Dekker Inc., New York, Basel, pp. 569–603.

Trujillo, E.E. (1976) Biological control of hamakua pamakani with plant pathogens. *Proceedings of the American Phytopathological Society*, **3**, 298.

Trujillo, E.E. (1985) Biological control of hamakua pamakani with *Cercosporella* sp. in Hawaii, in *Proceedings of the VI International Symposium on Biological Control of Weeds*, (ed. E.S. Delfosse), Agriculture Canada, Vancouver, pp. 661–71.

Tuleen, D.M., Frekeriksen, R.A. and Vudhivanich, P. (1980) Cultural practices and the incidence of sorghum downy mildew in grain sorghum. *Phytopathology*, **70**, 905–8.

Vale, G.A. (1974a) New field methods for studying the responses of tsetse flies (Diptera: Glossinidae) to hosts. *Bulletin of Entomological Research*, **64**, 199–208.

Vale, G.A. (1974b) The responses of tsetse flies (Diptera: Glossinidae) to mobile and stationary baits. *Bulletin of Entomological Research*, **64**, 545–88.

van den Bosch, R. and Messenger, P.S. (1973) *Biological Control*, International Textbook Company Ltd, Aylesbury.

van Lenteren, J.C. (1986) Parasitoids in the greenhouse: successes with seasonal inoculative release systems, in *Insect Parasitoids*, (eds J. Waage and D. Greathead), Academic Press, London, pp. 341–74.

Vandermeer, J. (1989) *The Ecology of Intercropping*, University Press, Cambridge.

Wall, C. (1990) Principles of monitoring, in *Behavior-Modifying Chemicals for Insect Management*, (eds R.L. Ridgway, R.M. Silverstein and M.N. Inscoe), Marcel Dekker Inc., New York, Basel, pp. 9–23.

Waller, J.M. (1985) Plant disease control, in *Pesticide Application: Principles and Practice*, (ed. P.T. Haskell), Oxford Science Publications, Oxford, pp. 425–55.

Way, M.J. and Cammell, M.E. (1985) Biological considerations, in *Pesticide Application: Principles and Practice*, (ed. P.T. Haskell), Oxford Science Publications, Oxford, pp. 68–94.

Whitehead, A.G. and Bridge, J. (1985) Control of plant nematodes, in *Pesticide Application: Principles and Practice*, (ed. P.T. Haskell), Oxford Science Publications, Oxford, pp. 398–426.

Wightman, J.A., Dick, K.M., Rango Rao, G.V., Shanower, T.G. and Gold, C.G. (1990) Pests of groundnut in the semi-arid tropics, in *Insect Pests of Tropical Food Legumes*, (ed. S.R. Singh), John Wiley & Sons, New York, Chichester, pp. 243–322.

Wilkins, R.M. (1990) *Controlled Delivery of Crop-Protection Agents*, Taylor & Francis Ltd, London.

Winstanley, D. and Crook, N.E. (1990) The potential for improving a granulosis virus for the control of codling moth, in *The Exploitation of Micro-organisms in Applied Biology*, The Association of Applied Biologists, Warwick, pp. 11–16.

Winstanley, D. and Rovesti, L. (1993) Insect viruses as biocontrol agents, in *Exploitation of Microorganisms*, (ed. D.G. Jones), Chapman & Hall, London, pp. 104–36.

Zandstra, B.H. and Motooka, P.S. (1978) Beneficial effects of weeds in pest management – a review. *PANS*, **24**(3), 333–8.

Zhang, Z.Q. (1992) The use of beneficial birds for biological pest control in China. *Biocontrol News and Information*, **13**, 11N–16N.

Zitter, T.A. and Simons, J.N. (1980) Management of viruses by alteration of vector efficiency and by cultural practices. *Annual Review of Phytopathology*, **18**, 289–310.

CHAPTER 4

Defining the problem

D.R. Dent

4.1 INTRODUCTION

Many agricultural systems are now in a continuous state of change. As the agricultural systems change, new problems emerge which have to be addressed. The solution to the problem, if adopted, then contributes to a further modification of the system which itself may produce problems of a different kind. Not all modifications will create new problems, and as a system develops and evolves it should gradually move towards a more balanced state which meets the economic needs of the human population it sustains. However, if the economic needs of the population are not met then the agricultural system will continue to be modified, usually through means of increased intensification.

Changes in an agricultural system that give rise to pest problems may warrant the development of a pest management programme leading to a new pest management system. If the IPM system is adopted by farmers then it may have 'knock-on' effects for another component of the agricultural system. If this effect is a negative one then a new problem will have been created. Examples of this kind of negative knock-on effects of one component of a system on another are common throughout the history of pest control, e.g. widespread use of pesticides against one pest causing an outbreak of a secondary pest. Zadoks (1993) provides a more complex illustration of the problem. Improvements in soil tillage, water management, plant nutrition (especially high nitrogen levels) and plant breeding have combined to allow an increase in plant density. This however, has led to higher humidity within the crop canopy and to concomitant changes of pest species. Zadoks notes that this development is not irreversible since new idiotypes of crop plants, new crop structures (as affected for example by the distance between rows) and new machinery may be developed that will address the problem. Such continuous modifications reflect the need for a

Integrated Pest Management Edited by David Dent. Published in 1995 by Chapman & Hall, London. ISBN 0 412 57370 9

more holistic approach to problem solving which looks beyond the immediate issue (perhaps isolated in a specific discipline) to the root or cause of the problem (which may result from changes evident only from a perspective across disciplines).

Time is also an important component to take into consideration. The relevance of changes to a system may in some circumstances only be appreciated over a long period of time. Although research and development (R&D) is a continuous process the introduction and adoption of the results will often be manifest in more discrete stages. To understand how a number of modifications have combined to create a problem it may be necessary to evaluate changes over an extended period of time.

It is essential that the whole dimension of a problem be identified, whether from a transdisciplinary or historical perspective, but preferably a combination of both. Failure to do this will inevitably lead to the development of inappropriate solutions. Time spent on the systematic, directed and focussed definition of a pest problem, identifying the key components of the process affecting pest status, damage and control (Norton and Mumford, 1993), is a vital step in the process of devising appropriate solutions to a pest problem. Combined with a precise statement of current research capability, such an approach provides a solid basis for the targeting of research, development and advisory effort, concentrating on the key questions associated with feasible and acceptable pest management options.

This chapter considers the initial stages, the trigger events (including funding) that lead to the start of a programme; the analysis of the problem in historical, ecological, yield loss and socioeconomic terms and the means by which the present status of research can be evaluated. All these components combine to provide the information required for setting the goals of an IPM programme/system.

4.2 TRIGGER EVENTS AND FUNDING

The initiation of an IPM programme will be dependent on a combination of two factors, first an occurrence of events that triggers a need for development of an IPM system and second, the availability of funding to make it possible to undertake the necessary R&D. Trigger events may be market, technology or politically led, while the investment in the development of programmes or particular aspects of a pest control strategy may be provided by either the public (including international aid) or private sector, or a combination of both.

Market-led events which may trigger the initiation of an IPM programme are those that depend on farmer and/or consumer demand. Farmers will look to adopt new strategies for pest control, such as IPM, if: (i) their existing control measures fail, e.g. failure of pesticides due to the development of pest resistance; (ii) the market value of a crop product improves to the extent that the potential loss to a pest increases; or (iii) if a new production technology increases crop susceptibility and hence, the losses incurred to pests (Norton, 1982b). Farmers will also adopt alternative strategies when new markets for their crop products become available. For instance, the increased consumer demand and a premium price for organically grown, pesticide-free food products may encourage farmers to seek alternative control measures to pesticides. In situations where the market-led events cause dramatic losses to farmers or where staple food crops are put at risk then the funding for the initiation of a pest management programme is likely to be swift and substantial. In less serious circumstances funding would normally be provided piecemeal, dealing, not at the IPM programme level, but with specific projects aimed at solving a particular aspect of the problem, for instance, more money for development of a specific and new technological innovation.

Basic research provides unpredictable sources of technological innovation. Technology-led events refer to the development and availability of these innovations, and the subsequent impact they may have on farmer adoption of a particular pest control strategy. The most obvious and dramatic example of the introduction of such technology and its impact on pest control, was the availability in the 1940s of organochlorine insecticides, and the 2,4-D and MCPA herbicides. These two classes of pesticide completely transformed the way farmers viewed crop protection and revolutionized farming agronomic practice, particularly in cereals in Europe. In the case of herbicides, their introduction reduced the need for conventional tillage practices and promoted the use of direct drilling of seed. Since then, there have been various other innovations which have provided farmers with a range of control measures (e.g. sex pheromones, mating disruption; *Bacillus thuringiensis*, a microbial insecticide; selective systemic fungicides; sterile insect technique) for use in pest management. The most recent technology which will undoubtedly have a significant impact on future pest management strategies, is that of gene manipulation. This will have an important influence on development of a range of control measures, particularly microbial pesticides and pest-resistant crop plants. How such organisms and crops will be integrated into overall pest management strategies has yet to be adequately addressed.

Political-led events refer to the situations such as regional or national pest problems (e.g. locust swarms), or to pressures from the general public that oblige governments to implement appropriate policy changes. Growing concern among the general public for environmental issues, particularly pesticide misuse, has prompted some governments to formally and explicitly advocate the use of IPM as an environmentally friendly form of crop protection (Table 4.1). Under such conditions the case for funding IPM programmes will at least be heard more sympathetically and the agencies empowered by government to allocate resources will give proper attention

Table 4.1 Integrated pest management as official policy; (+) explicit statement; (−) implicit statement. (From Zadoks, 1993)

1985	India	Ministerial declaration	+
1985	Malaysia	Ministerial declaration	+
1986	Germany	Parliamentary decision, Plant Protection Act	+
1986	Indonesia	Presidential decree	+
1986	Philippines	Presidential declaration	−
1987	Denmark	Parliamentary decision	−
1987	Sweden	Parliamentary decision	−
1991	The Netherlands	Cabinet decision: Multi-Year Plan for Crop Protection	−
1992	United Nations World's Heads of State, Conference on Environment Development	Agenda 21, Rio de Janeiro	+

to the needs of IPM. Government agencies responsible for funding R&D programmes will still have limited budgets with which to meet their own priorities. These tend to require a maximum return for a minimal investment (Dent, 1993) and to favour projects or programmes that show 'timeliness and promise' (SERC, 1991) or that underpin near market R&D (e.g. MAFF, 1991). These priorities will in their turn discriminate against some IPM projects in favour of others, for instance, the development of pest control 'products' as opposed to 'techniques' (Dent, 1993: section 6), that may ultimately act to the detriment of IPM as a whole.

Another tendency in research and in research funding, that runs counter to the need for the development of 'complete' integrated, pest management programmes, has been adherence to a policy of research specialization. Most scientists are specialists (Edwards, 1990) and specialists tend to be held in higher regard than generalists (Tait, 1987). In the field of pest management this means concentration of research effort into individual, specific control measures usually along the lines of rigid discipline boundaries. There are too few scientists trained at the level of pest management having knowledge and expertise that crosses traditional disciplines, or the abilities to integrate and manage multidisciplinary teams of scientists (Dent, 1991, 1992). For this reason, most research remains *ad hoc* efforts by individual pest control specialists, each developing so-called integrated pest management programmes independently of one another (Pimentel, 1985). There is however, hope that the situation will change. Greater emphasis is now being placed on the development of collaborative research programmes in order to make more effective use of limited resources (Dent, 1992). It also seems to have been realized that there are some subjects that can only be effectively studied at the programme rather than the project level (Phillips, 1989). Programmes have certain advantages over projects. They tend to: (i) facilitate and stimulate the development of coherent research; (ii) offer greater surety of keeping a research team together allowing the benefits of their collective and shared expertise to bear fruit; (iii) provide a good environment for training and developing young scientific talent (especially in the skills of interdisciplinary research); and (iv) shift more of the management decisions to the 'local' level, where they ought to be (Phillips, 1989). The funding for research programmes is generally significantly greater than for projects and tends to support a larger team of scientists for a longer period. It is important that funding arrangements match the changing needs of the science; in pest management there is now a great need for funding at the programme level to allow the development of interdisciplinary, collaborative IPM programmes.

The degree of funding for a programme will always ultimately determine what can be achieved and the methods that are used to achieve it. Under most present arrangements one of the skills required in the application for grants is to properly match the required funding to the levels of work necessary to achieve stated goals. This can create difficulties, particularly with complex research programmes when grants are awarded before a complete and detailed analysis of the problem. As the programme gets under way scrutiny of the problem reveals constraints and dimensions which need to be addressed, but that were not costed in the original proposal. In an ideal situation, a feasibility assessment should be carried out as a preliminary to all IPM R&D programmes. The assessment should involve detailed study of the factors which are likely to influence the success of the programme (Table 4.2) (Mumford and Norton, 1987). IPM research should only proceed if those favourable conditions exist, or are likely to, by the time the research and development are completed. However, feasibility studies would require *a priori* funding, and as yet no funding agency has had the foresight to support such assessments.

Table 4.2 Some factors in favour of conducting research and development on pest control programmes. (From Mumford and Norton, 1987)

Severe pest damage
High crop prices
Narrow pest spectrum
Consistent pest problem
High current control costs
Poor effectiveness of current controls
Low environmental acceptability of current controls
Risk aversion in farmers
Low expected cost of R&D
High chance of technical success in R&D
Politically powerful farmers', consumers', or environmental lobby
Extension or marketing system to disseminate new IPM
Farmers with capital and skill to implement new IPM or capacity for government-run regional programme

4.3 HISTORICAL ANALYSIS

In the absence of a feasibility study the first essential step at the start of a programme is to define the current situation and identify the factors that contributed to its development. This may be more easily stated than achieved however, simply because the factors contributing to a particular problem, e.g. a pest outbreak, may be complex and the result of a particular combination of circumstances, manifested over a long period of time. Initial attention tends to focus on recent events, on the pest and the measures used to control it. To determine the cause of an outbreak however, it may be necessary to consider the problem from a broader historical perspective. In situations where a commonly used control measure has failed, such as the failure of pesticide use through the development of pest resistance, the initial reaction is to find an alternative but similar type of control measure, e.g. another pesticide. However, the actual cause of the pest problem may be more complex and quite unrelated to the present failure of the pesticide. Simply replacing it will not necessarily provide a long-term solution. For instance, when blackgrass (*Alopecurus myosuroides*) developed resistance to the herbicides chlorotoluran and isoproturon, the resistant grass populations were found mostly in fields where intensive winter cereal cropping and non-ploughing cultivation techniques had been practised (Moss and Cussans, 1991). Blackgrass is an annual weed propagated solely by seeds which mostly germinate in the autumn from September to November; consequently, *A. myosuroides* is mainly associated with autumn-sown crops. The trend in the 1970s to grow more autumn- than spring-sown cereals with minimum cultivation, necessitated an increased frequency of herbicide application. The combination of these factors, autumn-sown crops and minimum tillage and the subsequent increased herbicide use contributed to the development of herbicide resistance in the 1980s. Given this information the most effective strategy to contain existing resistant blackgrass populations was considered to be a combination of ploughing and the inclusion in the arable crops of a rotation on which effective herbicides can be used (Moss and Cussans, 1991).

Many present day pest problems which develop in agriculture are caused by modification of the agroecosystem through changed agronomic practices (Pimentel, 1977, 1991). One of the major insect pests of maize (corn) in the USA is the corn root worm complex, with yield losses four times higher in 1977 than they were in 1955, largely because of the replacement of maize rotations with continuous maize crops (Pimentel *et al.*, 1977). The rotations were used to break the breeding cycle of the pest, reducing the carry over from one year to the next. In other situations, outbreaks can be attributed to past changes in cropping practice through the introduction of monocultures (Cromartie, 1991), new crops (Pimentel, 1991) or crop cultivars, e.g. the introduction of high-yielding, susceptible cereal cultivars in the 'Green Revolution' (Lipton and Longhurst, 1989), and the subsequent reliance on pesticide use in attempts to redress the problem. In order to identify constraints and opportunities for resolving problems, it is necessary to map out the sequence of interacting events which have contributed to them. A technique that is particularly useful for achieving this is the historical profile (Norton, 1987, 1990, 1993; Norton and Mumford, 1993).

An historical profile is constructed by first identifying the major factors that directly or indirectly influence the pest problem. Then on the basis of statistical information and expert opinion, the changes in these components

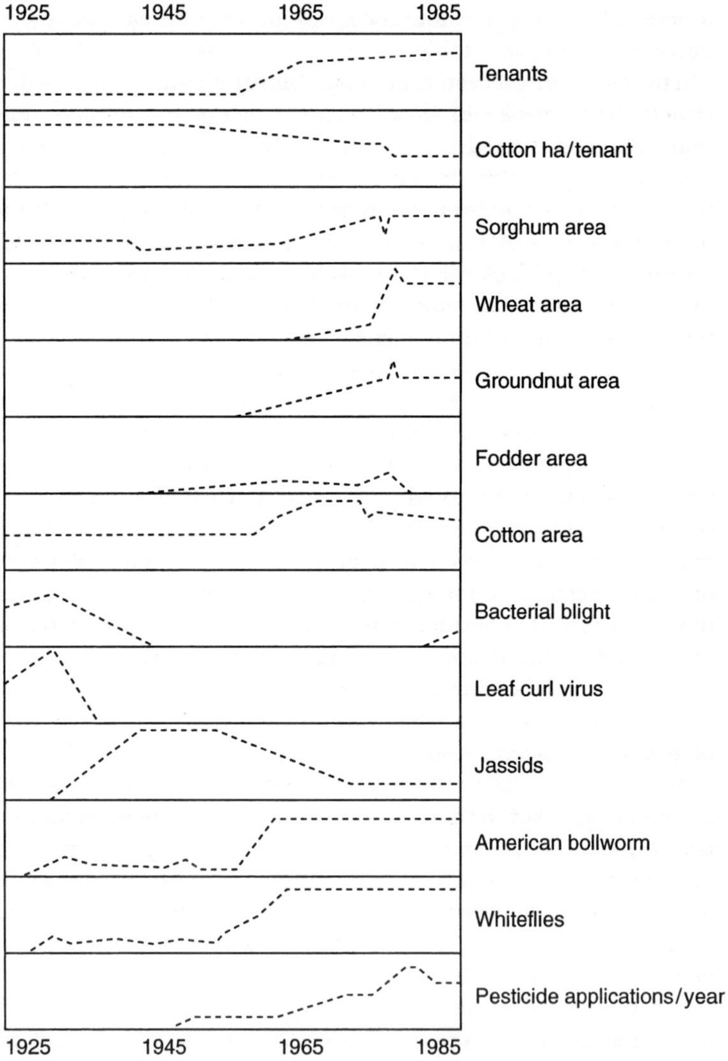

Figure 4.1 Historical profile for pest problems in the Sudan Gezira. (After Griffiths, 1984).

are graphed over a relevant time period to the present (Norton, 1990) (Figure 4.1). Historical profiles are usefully constructed in a workshop environment. These bring together relevant people (for example, scientists, extension workers, farmers, agro-industry, sociologists, economists) who can provide a range of different viewpoints and perspectives (Norton, 1993). Although, the composition of the group can be relatively important, the objectives of the workshop can be set to achieve what is feasible for a particular group (Chadwick and Marsh, 1993; Norton, 1993). The historical profile acts to structure and focus the discussion to obtain fast results; it defeats the object if the analysis becomes too deep, since the analysis then becomes an end in itself.

The technique serves four very important functions: (i) it provides a structured means of bringing together diverse information about factors that have contributed to the development of a pest problem; (ii) used in a workshop environment it acts as a focus to bring expertise and knowledge to bear on the problem; (iii) it raises questions and hypotheses; and (iv) it provides an explicit basis for thinking about future developments (Norton, 1987). This last function is particularly important because it may provide indications of plausible solutions to the problem, or at least identify the options available, and hence form a basis for the goals of a programme.

4.4 SOCIOECONOMIC ANALYSIS

Over a decade ago, a farmer would have been viewed merely as a recipient of the benefits derived from pest management research. The control measures advocated by the scientists would be based on their perception of the farmer's problem and would have provided technological solutions that the scientists considered appropriate and robust. Thankfully the failings of such narrow approaches have been realized and attempts are now made to involve the farmer not only in the initial planning stages of a programme but also in the research, development and implementation processes. This is a significant advance that means the farmer's circumstances, in terms of their perceptions, needs, objectives, and resource constraints, are taken into account; the development of the pest management system is also placed in the context of the overall farming systems and the social and political forces acting on them. Such an approach, almost by its nature, necessitates a multidisciplinary perspective and hence, involvement of scientists, socioeconomists, extension workers, representatives of agribusiness and the farmers, thus promoting the integration of effort, often so lacking in the development of pest management systems.

During the initial stages of programme and/or systems development baseline information is required on current pest control practices, costs and returns, and other internal aspects of the farmer's circumstances. This information is of vital importance, primarily because research effort must be effectively targeted, but also to allow comparison and assessment of the success and benefits of proposed strategies, to identify the gaps in research that farmers perceive as important, and to define farmer information requirements (Reichelderfer *et al.*, 1984), the latter indicating the need, or otherwise, for work in the areas of extension, education and training as opposed to primary research. The means by which such information has been traditionally collected is through the use of an exploratory survey, which could involve either postal questionnaires (e.g. Mumford, 1982a) or direct interviews and questioning of farmers (Byerlee *et al.*, 1980; Collinson, 1980; Shaner *et al.*, 1982). The essential objective of these exploratory surveys is to gather as much information as possible in a very short time through a series of interviews with a large number of farmers (Byerlee *et al.*, 1980). The emphasis is on the researchers to identify the questions that are relevant and the farmers to provide the answers. More recently exploratory surveys have been superseded by rapid rural appraisals (RRA) where the interview tends to be less structured. The intention with RRA is to reverse the one-sided relationships between the specialists and the farmers so that the specialists also learn from the interaction. The RRA should provide an opportunity for mutual learning, exchange of ideas, skills and knowledge (Chambers *et al.*, 1989). Mathema and Galt (1989) identified three components to RRA: the key informant survey, farm household interviews and farmer group interviews. The key informant survey involves interviews with knowledgeable people relevant to the problem, e.g. a village elder or knowledgeable farmer. The researchers do not discuss the informant's particular farm but consider the farming system in general. The individual household interview is also an informal interview, held in the farmer's home and an open-ended discussion takes place around key questions and prompts. The farmer group interviews on the other hand, are mainly used as a means of checking whether information obtained about the local farming system stands up to scrutiny (Mathema and Galt, 1989). Communication with farmers, in these ways, brings the researchers into direct contact and enables them to observe first hand the farmers' crop husbandry and pest control practices. It will clearly highlight the constraints under which the farmers work, their own perceptions of the problem and the solutions the farmers consider most appropriate.

The farmers' ability to control pests will be constrained by a whole variety of factors. These may include access to relevant crop protection information, the availability of off-farm inputs, access to credit facilities, or limitations caused by poor infrastructure. All these factors and others will need to be considered in order that a true assessment can be made and steps taken to either remove the constraint, or to develop pest management systems which take it into account.

The use by a farmer of a pest control measure which has to be purchased as an off-farm input will be dependent on its availability (Norton, 1985). An input such as an insecticide (a pest control product; section 7.6.1) must be present in the retailers at the appropriate time and at a suitable cost (Kenmore *et al.*, 1985; Ruthenberg and Jahnke, 1985; Waibel, 1993). The farmer may have insufficient capital to purchase these inputs and hence, is dependent on the availability of credit or the maintenance of subsidies by government. Credit may be available from a variety of sources, such as commercial or agricultural banks, merchants' cooperatives or money lenders (Abbott and Makeham, 1979; Makeham and Malcolm, 1986; Ellis, 1992) but the form in which it is available will affect the flexibility a farmer has in adopting different pest control options. Farmers are sometimes provided with a technology package at the time they draw their loan which removes the farmers' freedom of choice, generalizes pest management decision making and precludes farmer experimentation (Goodell, 1984). There are also constraints imposed when quick mid-season loans are unavailable for need-based use of control measures, such as pesticides (Kenmore *et al.*, 1985). If the bank or the retailer are distant from the farm, with poor roads and means of communication, a farmer may be unable to make use of loans or control measures even when they are available. Restrictions placed on the availability of control measures by poor infrastructure can effectively prevent adoption of most IPM systems. The accessibility of extension information and/or other means of communicating relevant pest management information also can have a significant impact on adoption of different control measures.

Norton (1982a) classified information relevant to pest management decision making under four headings: (i) fundamental; (ii) historical; (iii) real-time; and (iv) forecast information. Farmers will need to have basic information about the basic biology and ecology of the pests in their crops and of the measures available for their control. Historical information will be built up through experience and will provide the farmer with a measure of perspective against which to gauge the likelihood and severity of pest attack. This information will be compared with real-time data on pest status and damage in the current crop and decisions about control will be largely based on this comparison. The final category, forecast information, is based on the other categories of information and will provide estimates of future levels of attack and damage. The farmer may provide for their own information needs, or rely on an extension advisor, but the source must be trusted and the information relevant and reliable (Lawson, 1982).

A combination of available information and experience contribute to an individual's perception of a problem (Mumford, 1982a). However, an individual's perception is not always a good representation of the real problem either. How farmers perceive pest problems and the value they place on various control measures is clearly of great importance to those involved in developing pest management systems. There are several possible ways in which perceptions of pest hazards can be formed including: (i) direct or indirect 'experience' of a pest on a crop; (ii) 'experience' of the same or similar pests on another crop; (iii) 'experience' of different pests on the same crop; and (iv) awareness of solutions to pest problems on crops in general (Mumford, 1982b). A number of these 'perceptions' have been identified in actual farmer practice. A survey by Tait (1977) found that the use of fungicides and insecticides varied much more between farmers than between crops on the same farm, despite differences in actual pest problems on these crops. The use of insecticides and fungicides on potatoes was also generally associated even though the conditions necessary for prevalence of the aphid insect pest were directly opposite for those of the fungal pathogen. Results from a survey by Mumford (1982a) appeared to explain the phenomenon. Different crops on the same farm tended to be treated similarly (pesticide application tended to be uniformly high or low among different crops) and individual crops are often treated similarly for weeds, insects and diseases. Hence, if a crop is treated with one type of pesticide such as a herbicide, it is more likely to be treated with another, either an insecticide or a fungicide (Mumford, 1982b). It would seem that if a crop is perceived to be under threat from one pest then an individual is more likely to be conscious of the need to protect the crop from other pests. Equally likely is the scenario that the farmer's perception of a threat to one crop affects his perception of threat from the same pest in another crop.

The important role of socioeconomic analyses during the preliminary stages of a pest management programme should not be underestimated. Of all the factors likely to affect the success of the programme and the adoption of the IPM system, the socioeconomic component holds one of the most crucial keys. The programme managers must be sure that they are familiar with and understand the perceptions, needs, objectives (see below) and constraints faced by the farmers, the channels of communication they use and the types of information they require.

Given sufficient information from an RRA then it may be possible to place farmers into a 'recommendation domain'; groups of farmers whose climate, soils, pest problems and socioeconomic constraints are sufficiently similar that a single recommendation is applicable to the entire group (Byerlee *et al.*, 1980; Reichelderfer *et al.*, 1984). In this way an IPM strategy can be devised for a specific recommendation domain, providing opportunities for better targeting of research.

4.5 RESEARCH STATUS ANALYSIS

Research in pest management is most frequently aimed at one of two objectives: improving our understanding of the underlying biological and ecological processes associated with the pest problems or developing specific control measures (Norton, 1987). A third, too often overlooked, area of study is that of yield loss assessment, which serves to provide information useful in the allocation of resources for research, control and extension (Walker, 1987). Description of these three areas of research: (i) biological and ecological; (ii) technical; and (iii) yield loss assessments provide the bases for choosing between control options, a framework for research and for incorporation of new information (Heong, 1985). The techniques which can be used to assist in this description for each of the different aspects are considered in the following sections. A knowledge of the current status of research in each of these areas is necessary for a comprehensive analysis of the problem and for identifying potential options for developing solutions.

4.5.1 Biological and ecological analysis

The definition used in Chapter 1 considers IPM as a pest management system '. . . in the context of the associated environment and population dynamics of the pest species . . .'. From this and definitions cited by others (Rabb and Guthrie, 1970; Flint and van den Bosch, 1981) it is obvious that IPM is essentially an ecological approach to pest management based on a sound knowledge and understanding of the biological and ecological factors influencing pest population dynamics. The emphasis on pest ecology should provide a basis for a more rational and sustainable approach to pest management, through utilization of more environmentally friendly control measures. However, any population ecologist will readily point out that the factors influencing pest population dynamics are often diverse and complex. There are many factors both biotic and abiotic that can influence the population dynamics of a pest species. These may include environmental factors, temperature, light, humidity, rainfall, soil nutrients, etc.; hostplant condition and cultivar; growth stage, to name but a few. However, not all the information available may be relevant to the particular problem. It is necessary from the point of view of programme or system development to identify the key components of the pathosystem in a concise, but easily interpreted/understood description – remembering that any description must be relevant and comprehensible to those collaborating from a number of different disciplines.

In the normal course of events a scientist would seek to collate and synthesize relevant information by writing a review of the literature. Such reviews tend to make great use of and assume knowledge of generally accepted concepts, ideas, terminology associated with a particular discipline that are readily understood within a discipline but not necessarily by others from other disciplines (section 5.6.2). Thus, while reviews might provide a good basis for collating and synthesizing information within a discipline they are a less effective way of communicating information between disciplines, especially where the purpose is to put over only key components and interactions. A more rational basis for communication of this type of information in multidisciplinary situations is to make use of the descriptive analysis techniques proposed and utilized by Norton and Mumford (1993) and colleagues. These include, in particular, the use of a life-history diagram, interaction matrices and time profiles.

A life-history diagram or life-cycle model (Figure 4.2) produced for each key pest in the system will depict each of the development stages and the potential points in the cycle for different types of natural mortality factors or put weak points for introduction of managed control measures.

These diagrams provide an explicit but simple means of representing the pest's life-cycle that can be readily understood by specialist and non-specialist alike. An interaction matrix takes this description one stage further (Holt *et al.*, 1987; Norton, 1990; Norton and Mumford, 1993).

An interaction matrix can be constructed for each pest under study. The key components in the pathosystem are identified; these may include climatic factors, management factors, crop factors, natural enemies and life-history characteristics. A matrix is then constructed which allows the primary effect of any column component on any row component to be shown (Norton, 1990) (Figure 4.3). A relationship between components can be indicated with a dot and the cell is left blank where no direct relationship is thought to exist. Taking any particular column the dots in the column indicate the primary effects that component has on the system. Similarly for any row, the dots indicate the various factors that can have an effect on that component. Interactions may occur between these primary effects and they can lead on to secondary and subsequent effects as one proceeds through the matrix, generally from the top left-hand side to the bottom right-hand side (Norton, 1990).

As well as producing a useful tool for determining key components at an initial stage of a programme, interaction matrices have also been used

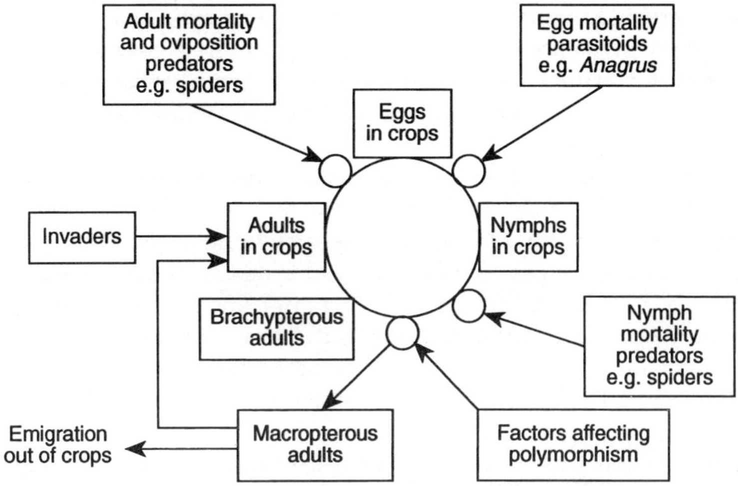

Figure 4.2 Life-history diagram of the brown plant-hopper *Nilaparvata lugens*. (After Heong, 1985).

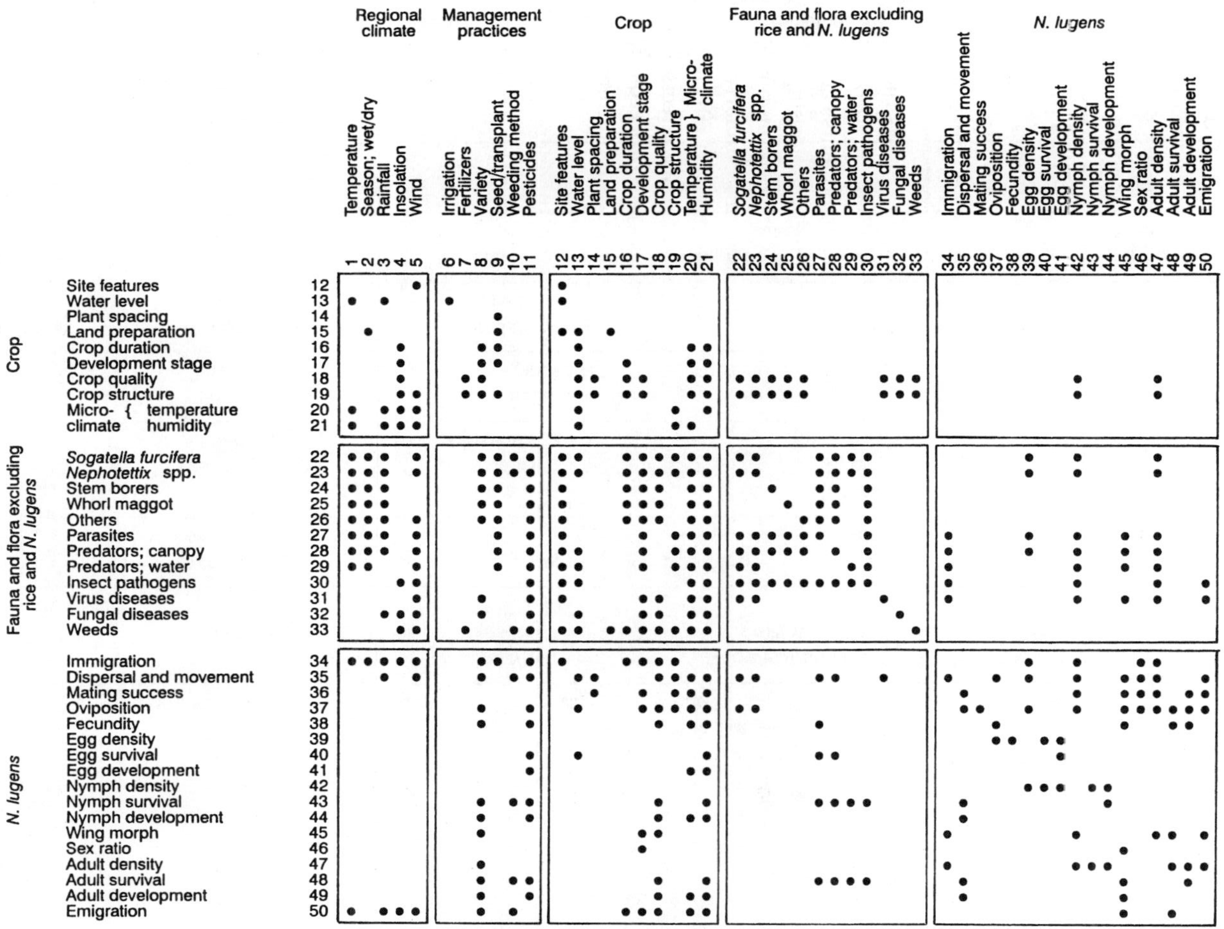

Figure 4.3 An interaction matrix for *Nilaparvata lugens*. (After Holt *et al.*, 1987)

System component	Winter March	April	May	June	July	August	Sept/Oct
Tree	Wood/buds	Wood/leaf/buds blossom BB GC PB PF		Wood/leaf/buds fruitlet		Wood/leaf/buds fruit	Wood/buds
Codling/tortrix	larva bark	pupa		adult/egg	larva fruit	adult/egg	larva fruit
Winter moth	adult/egg bark	larva buds, leaves blossoms		pupa soil			adult
Sawfly	larva soil	pupa soil	adult/egg blossom	larva fruitlet soil			
Rosy apple aphid	egg bark	nymph/adult(2 gens) leaves		on plantain			adult/egg bark
Woolly aphid	nymph bark	adult bark		nymph/adult (3 gens) leaves			nymph bark
Red spider mite	egg bark	juvenile stage/adult/egg (5 gens) leaves					egg bark
Mildew	mycelium buds	mycelium/conidium (many gens) blossoms, leaves, buds, leaves					mycelium buds
Scab	perithecium leaves in soil	ascospore aerial	mycelium/conidium (many gens) blossoms/leaves, fruitlet/leaves fruit/leaves				perithecium leaves in soil

Phytophthora	mycelium/ canker soil/wood	mycelium/conidium/canker soil/wood soil, fruit/wood		canker wood
Nectria	canker/ascospore wood	conidia/canker wood	conidia/canker fruit/wood	canker wood
Gloeosporium	canker wood/leaves	canker wood	mycelium/canker leaves/wood fruit/wood	canker wood/leaves
Fireblight	canker wood	bacterial growth/canker blossoms, leaves/wood, leaves/wood		canker wood
Phytoseiid mites	adult bark	egg/juvenile stages/adult (4 gens) tree		adult bark
Anthocorids	adults bark	egg/nymph/adult (2 gens) tree		adult bark
Mirids	egg bark	nymph tree	adult tree	egg bark

Figure 4.4 A time profile for components of an orchard pest system. The stages of blossom development are BB, bud-burst; GC, green cluster; PB, pink bud; PF, petal fall. (After Norton, 1982a).

in the design of simulation models (Holt *et al.*, 1987; Day and Collins, 1992).

The life-system diagrams and interaction matrices tend to be constructed for each pest in a system. Time profiles and damage matrices place more emphasis on the host crop, and its complete complex of pests. A time profile can be used to list the components of the pathosystem showing the life-stages and location of the major pests on the host throughout the cropping season (Norton, 1979, 1982b) (Figure 4.4) or to describe the crop management events and their timing (Norton and Mumford, 1993). A damage matrix (Figure 4.5) is used to indicate which components of yield (e.g. vegetative buds, leaves, wood, fruit buds, blossom, etc.) are affected by the major pests (Norton, 1982b), i.e. it classifies pest problems according to the type of damage they cause. The time profiles and the damage matrix provide a simple but effective means of describing the structure of the pathosystem and identifying where information is lacking or unavailable.

4.5.2 Control measures

There are essentially three steps involved in the description of control measures. First, identifying control measures currently used by the farmers; second, identifying the current state of research on all available measures; and third, a consideration of the feasibility of all available control measures in terms of those in current use, those available for implementation and those potentially available from current R&D.

The control measures used by farmers may be embodied in the cropping system, through diversification, tillage practice or crop rotation, or may involve the use of specific control agents (Norton, 1976). These different measures and actions are undertaken at different periods during the cropping season. Each of the actions, which will occur in sequence throughout the season represents a decision point for the farmer, i.e. where to cultivate or use minimum tillage, which rotation to use, which pesticide to apply and so on. Obviously there will be a series of such decisions throughout the season, the combination of options selected influencing subsequent choices and so on. A pictorial representation of this process in the form of a decision tree (Figure 4.6) provides a useful means of identifying options available to farmers and the times at which pest management decisions are made. Obviously as the season progresses the number of options available to the farmer diminishes (Norton, 1976; Norton and Mumford, 1993). Decision trees are now being used more often and examples are appearing in the scientific literature (Mumford, 1978; Moss, 1980; Heong, 1985; Norton, 1985). A similar technique is that of the decision profile, a specific form of seasonal profile which places the different activities on a time axis and allows comparison with other events and the development of the crop (Norton and Mumford, 1993). Both these techniques, the decision tree and the decision profile are valuable tools for descriptive analysis, not only identifying the options used but also providing insights into those that are

Tree components	Codling/ tortrix	Winter moth	Sawfly	Apple aphid	Woolly aphid	Mites	Mildew	Scab	*Nectria*	*Phytophthora*	*Gloeosporium*	Fire blight
Vegetative buds				*			*					
Leaves				*	*	*	*	*				
Wood									*	*	*	*
Fruit buds		*		*			*					
Blossom trusses	*	*		*	*	*	*					
Fruitlet	*		*					*				
Fruit	*							*	*	*	*	*
Fruit in storage								*	*	*	*	*

Figure 4.5 A damage matrix showing which components of yield are affected by the major insect pests and diseases in an apple orchard. * denotes a damaging effect. (After Norton, 1982a).

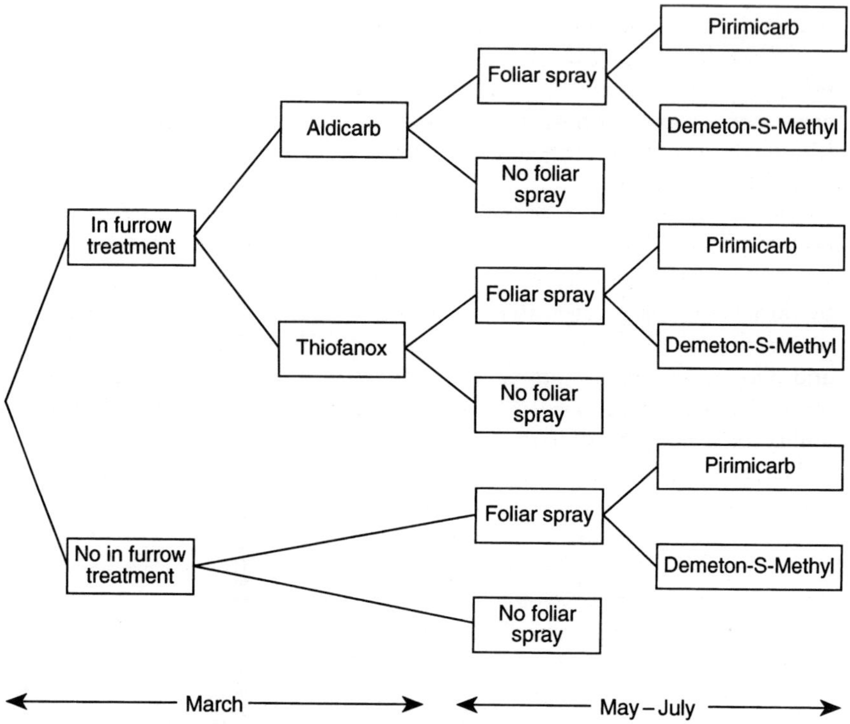

Figure 4.6 A decision tree for sugar-beet aphid–virus control in England, 1982. (After Mumford, 1981).

a workshop environment to force a team to consider all the possible options, providing a framework for deciding which options and strategies require further investigation (Barlow *et al.*, 1979; Norton and Mumford, 1993).

Describing the availability of control measures must include an assessment of the status of research programmes involved in the development of such measures. Only by such an assessment can an overview be obtained of those measures that are nearing completion, those that are available requiring implementation, or those still showing promise but unlikely to be available during the duration of the funding for the current programme. Being able to predict the likely availability and relevance of new technologies is an important component of the descriptive processes that has important implications for the design and planning aspects of programme development.

Research on control measures, especially well-established forms of control, e.g. pesticide development, hostplant resistance, pheromone technologies, tends to follow a roughly similar sequence or pathway (e.g. Figures 4.7 and 4.8). Such pathways arise through accumulated experience of the work of many scientists and often provide a theoretical optimum pathway that is devised only with hindsight. However, such pathway diagrams, if constructed for each of the different control measures currently under investigation, can be used as a benchmark against which to judge their progress. For instance, if hostplant resistance is considered technically feasible for inclusion into a pest management system then the availability of a resistant cultivar for use within the current funding period will be dependent on the stage reached along the research pathway. If trials are still being undertaken for field evaluation of resistance, then the whole sequence of stages to bring the resistance to the level of a registered cultivar and use by farmers is many stages away (Figure 4.8). Use of such research pathway diagrams provide a simple, visual means of describing the research process and allow a systematic appraisal of progress. In this way they provide a useful indicator of the status of the research and the likelihood of availability of a new technology.

The availability of a control measure provides only one of a number of

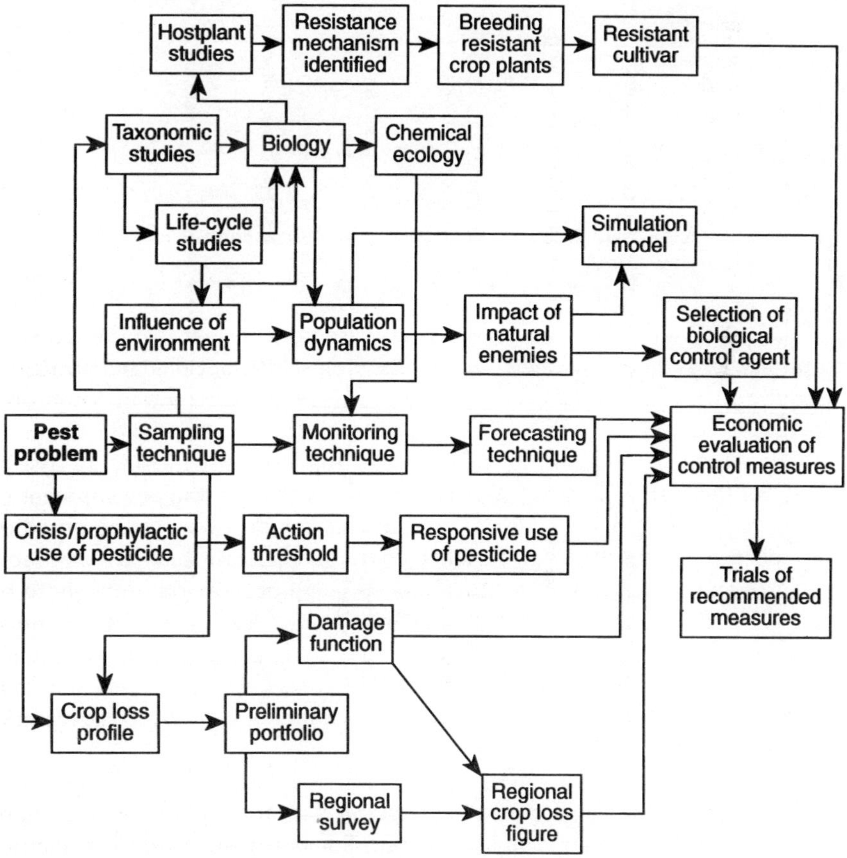

Figure 4.7 A hypothetical example of a research pathway for development of an IPM programme.

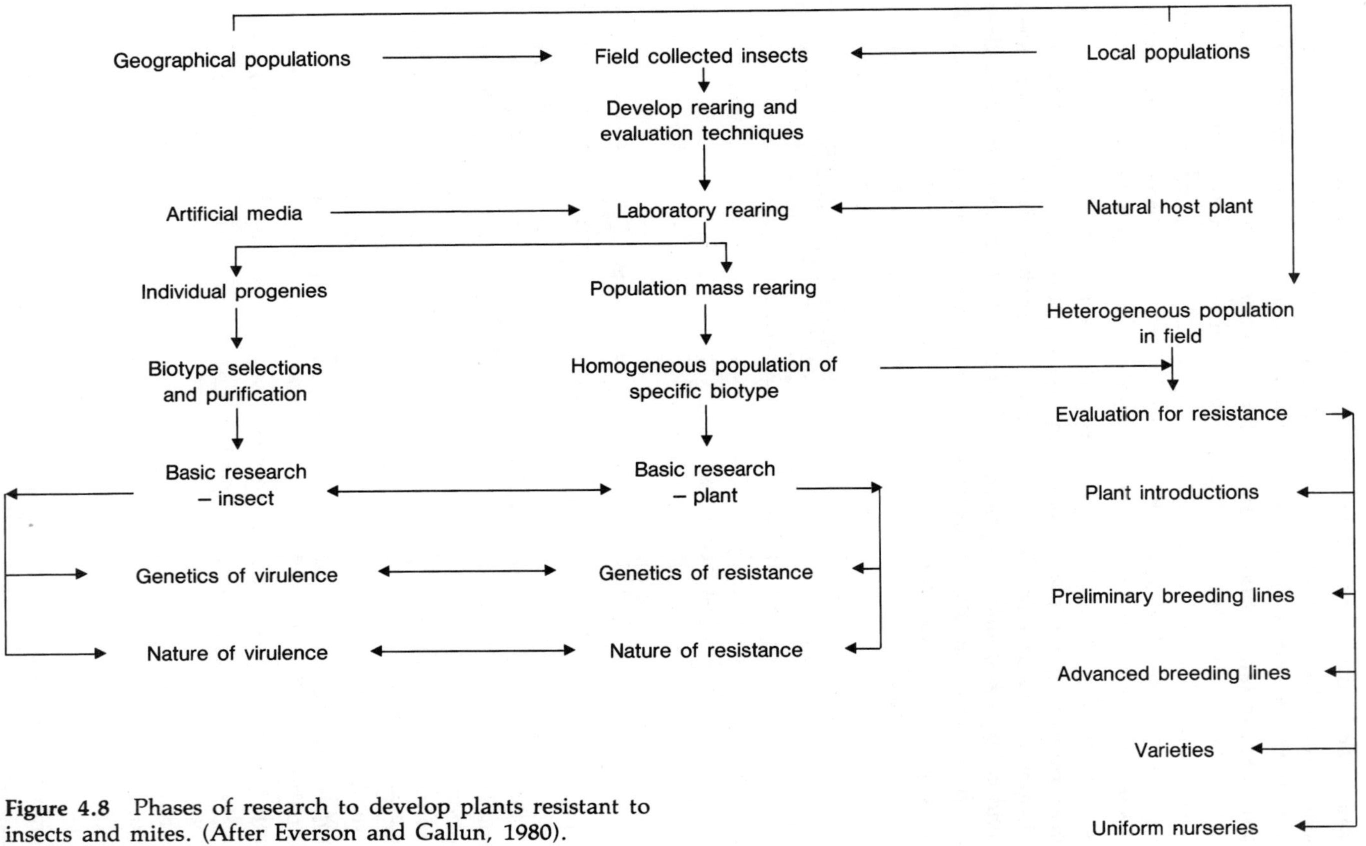

Figure 4.8 Phases of research to develop plants resistant to insects and mites. (After Everson and Gallun, 1980).

which the acceptability, desirability and feasibility of measures is assessed criteria on inclusion in an IPM programme and/or system. For a control measure to be applicable it should not only be available but technically possible, practically feasible, economically desirable, environmentally acceptable, politically advantageous and socially acceptable (Norton, 1987; Bomford, 1988; Norton and Mumford, 1993). A technique that can be used to help assess whether or not different control measures meet these criteria is the feasibility table (e.g. Table 4.3). A list of potential control measures should be produced (Norton and Mumford (1993) recommend use of a brainstorming session for this purpose) and these listed on the vertical axis of a table. The various criteria that might be used to assess feasibility (see above; but there may be others) are listed across the top of the table. Each of the control measures is then considered in relation to the number of criteria it fulfils. Obviously those control measures that fulfil only some of the criteria are rejected whereas those fulfilling all or most of the criteria will be included in a programme or investigated further.

The appraisal of control measures may be taken one stage further by constructing a control attribute table (Table 4.4). These can provide more specific information concerning four main attributes of control measures: (i) cost; (ii) control characteristics; (iii) hazard; and (iv) infrastructure requirements. The information may identify where research effort may be concentrated to improve a particular control measure by, for instance, improving ease of use or reliability.

4.5.3 Yield loss assessment

A knowledge of the extent of yield losses caused by various pests in a cropping system provides baseline information which can be used to determine priorities for R&D. It can also be used to assess the likely cost of control (where sufficient information is available). The relative importance of different pests in a system can be determined through the construction of a crop loss profile (Chiarappa, 1981). In most situations however, priorities for R&D are defined on an *ad hoc* basis, most frequently in response to a particular pest outbreak and without full recognition of the individual factors and their interacting contribution to crop loss. Quantification of crop losses are largely ignored because of the intrinsic difficulties in obtaining meaningful yield loss data. Yield loss assessment methods (for reviews see Chiarappa, 1971, 1981; Bardner and Fletcher, 1974; Walker, 1983, 1984; Dent, 1991) that are available are largely complex, expensive and require extensive research input. Unless experimental assessments of yield loss have already been carried out in advance then during the initial stages of a programme, yield loss information will have to be based on data from 'indirect' assessments.

Indirect measures of yield loss may be obtained from a variety of sources. Van der Graff (1981) cites the results of pesticide or varietal field trials, statements from experts or general enquiries among those most affected, as useful sources of such indirect yield loss information. The last two methods

Table 4.3 A feasibility table for wild duck control in Australia. (After Bomford, 1988)

Control options	*Feasibility/acceptability criteria*					
	Technically possible	*Practical with farmer resources*	*Economically desirable*	*Environmental acceptability*	*Political acceptability*	*Social acceptability*
1. Grow another crop	Yes	No				
2. Grow decoy crop	Yes	Yes	?	Yes	Yes	Yes
3. Predators and disease	No					
4. Sowing date	Yes	Yes	?	Yes	Yes	Yes
5. Sowing technique	Yes	Yes	?	Yes	Yes	Yes
6. Field modifications	Yes	Yes	?	Yes	Yes	Yes
7. Drain or clear daytime refuges	Yes	No				
8. Shoot	Yes	Yes	?	Yes	?	Yes
9. Prevent access, netting	Yes	Yes	No			
10. Decoy birds or free feeding	Yes	Yes	?	Yes	Yes	Yes
11. Repellants	Yes	No				
12. Deterrents	Yes	Yes	?	Yes	Yes	Yes
13. Poisons	Yes	Yes	?	No		
14. Resowing or transplanting seedlings	Yes	Yes	?	Yes	Yes	Yes

Table 4.4 Control attributes table – a hypothetical example

Control measure	*Chemical insecticide*	*Microbial insecticide*
Cost: H, high; M, medium; L, low		
Product purchase	M	M
Application	L	L
Equipment/machinery	H	H
Operating/maintenance	M	M
Time	L	L
Labour	L	L
Control characteristics: 1, good; 5, poor		
Reliability	2	4
Specificity	3	2
Ease of use	2	2
Efficacy		
Speed of action	2	4
Toxicity to target organ	1	2
Compatibility (natural mortality factors)	5	2
Duration of control	3	4
Hazard: H, high; M, medium, L, low		
Operator	H	L
Family	H	L
Consumer	L	L
Environment	H	L
Infrastructure support requirements		
Extension services	H	M
Diagnostic laboratories	M	M
Credit facilities	M	M

however, provide only qualitative data. Observers questioned on their appraisal of a situation often differ considerably in their assessments. Their perception of the problem also tends to be very time and location specific. Information may be biased because the observers report a view on the basis of the reason for which they think they are being questioned. If based on only a few individuals such approaches to assessing yield losses should be used only as a last resort. Where an opportunity exists for introducing a more structured approach, then the opinions of a range of 'experts' can be utilized through use of the Delphi method (Reichelderfer *et al.*, 1984).

The Delphi method was invented by Olaf Helmer (1969) as a technique for eliciting expert judgements about phenomena which are not conducive to normal means of objective measurement (Saaty and Kearns, 1985). A multidisciplinary pool of 'experts' is asked to answer anonymously a questionnaire in which they have to provide estimates of crop losses that they believed could be attributed to a pest or a range of pests. Each then receives an average of the estimates provided by the others, and has an opportunity to modify their own estimate if they desire. The process is repeated a number of times with anonymous feedback so that the experts can change their views in the light of what the others think. Eventually most of the group would reach a consensus, at which point the group meets to discuss its assessment and its underlying assumptions (Dale, 1978). The method is considered to be better than a face-to-face debate simply because it allows the experts to consider the views of others, anonymously and without prejudice, and to change their own estimates without embarrassment. There are however, a number of problems with use of the Delphi technique. Saaty and Kearns (1985) cite the following: (i) the superficial nature of its interdisciplinary perspective (because it degenerates to an arbitrary consensus that fails to capture the diversity of knowledge held by all participants); (ii) its failure to investigate any underlying assumptions on which each expert makes an assessment; and (iii) the conclusions reached are shaped and pre-determined by the questions that have been asked. The last point should not cause too much concern in a yield loss assessement because the yield loss estimates are very definite questions for which answers are required. The other two concerns have also been dealt with in a revised method, referred to as the Policy Delphi Method. This method makes use of experts and individuals affected by the outcome of the judgements; in our case dealing with pest management this would mean the farmers. The problem of the artificial consensus is solved by modifying the feedback mechanism so that the group response accentuates conflict or polarization of views instead of emphasizing the central tendency. Also, greater opportunities are provided for exchange of ideas and debate.

Despite its shortcomings the Delphi method, or modifications of it, has great potential for use in estimating yield losses in situations where experimental and survey data are not available. It is a technique that has not yet been used in pest management problems (Heong, 1985) but it could be easily built into the initial stages of any pest management programme and provide important guidelines for establishing priorities for research.

Indirect estimates of yield may also be obtained from pesticide and varietal field trials data (Van der Graff, 1981). Although carried out for reasons other than yield loss evaluation, where trials provide general data on yield reduction by comparison of a treated or untreated plot, the results should provide an indication of pest-associated losses. However, some caution will need to be applied in their interpretation especially where the varieties, cultivation practices and locations are not those associated with damage and yield loss in farmers' fields.

The lack of good data on yield losses will hinder the ability of an IPM programme to develop appropriate control strategies and to assess their economic value. Every effort should be made to obtain even the most rudimentary experimental yield loss data or information, but where this is not possible or where relevant information is unavailable, then use of the Delphi method is recommended.

4.6 GOALS AND STRATEGIES

Once the existing situation has been analysed from an historical perspective, in terms of research status and from a socioeconomic standpoint, it is necessary to identify the goals and objectives of the programme that will be required to develop the solutions to the pest management problem. The setting of goals is an important element of the design and planning of a programme. If the goals are inappropriate then the programme will inevitably fail in its attempts to solve the real problems which have been identified. The word 'goal' is used here as a generic term and as a designation for the concrete results of visions and levels of ambition, and for the benchmarks against which the eventual success of the programme will be judged (Karlöf, 1987). The 'vision' aspect of a goal refers to the mental forecast that creates the future from a montage of current facts, hopes, aspirations and opportunities. The level of ambition reflects the degree of desire and the motivation to perform. The two factors 'vision' and 'level of ambition' combine to provide a 'goal'.

Goals are intimately linked to the actions necessary to acheive them (i.e. the implementation of a strategy necessary to achieve the stated goals), in a hierarchial way, so that the goal of one level becomes the strategy of the

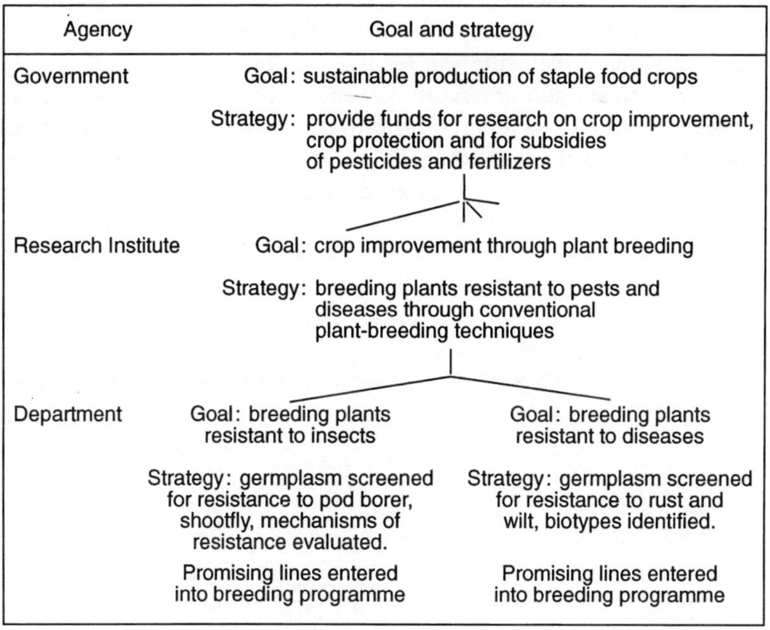

Figure 4.9 A hypothetical example of the hierarchical nature of goals and strategies.

level above it. Conversely, the strategy at one level becomes the goal of the level below it (Karlöf, 1987) (Figure 4.9). In practice this would mean, for instance, that a government's goal of self sufficiency in production of staple food crops, achieved through a strategy of providing funds for crop improvement, crop protection and subsidies for pesticides and fertilizers, would provide public research institutes with the goals of improving crop production and protection methods (particularly through use of fertilizers and pesticides) (Figures 4.9 and 4.10). These goals would then be translated by various departments into more specialist strategies such as breeding plants resistant to pests, or genetic manipulation to incorporate resistance genes. In turn, these strategies would provide the goals for an individual research team, and so on down to the level of a goal, and strategy for each individual scientist. Hence, the goals and strategies adopted by the government will ultimately have an impact on the goals and strategies adopted in the development of a particular IPM programme. This will also be true of goals and strategies of the extension service.

Part of an extension service's work will involve planning on the basis of local needs but they will also be expected to implement plans and objectives at the national level. National programmes will provide a framework within which the extension agent plans his or her local programmes and establishes priorities which must be followed (Oakley and Garforth, 1985). Provided the goals set for the extension services are compatible with those for R&D, then the implementation of an IPM system may be encouraged. Of course, if they are incompatible then they may act as a constraint, impeding systems implementation (Lim *et al.*, 1981).

Agencies, such as agrochemical companies, or other crop protection-based companies, may have an input into the development of an IPM programme, either directly through collaborative R&D or indirectly by providing products or services for use in an IPM system. The primary goals

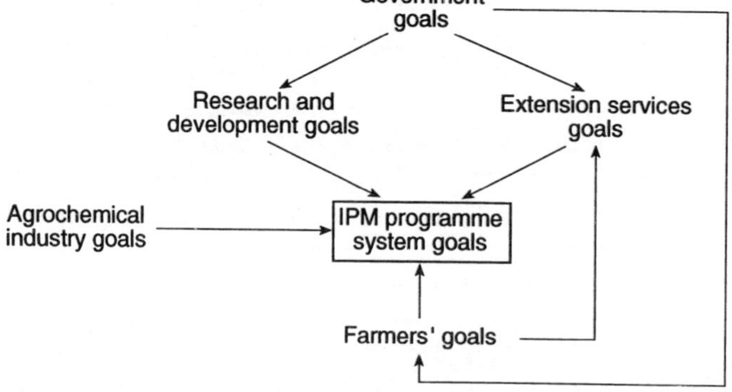

Figure 4.10 The goals of various agencies which impinge on the development of an IPM programme.

of such companies will be to gain satisfactory profits and sustain or increase their share of the market (Norton, 1976). There may be conflict of interest here, especially where promotional campaigns and/or package deals run counter to the goals of an IPM system. More often now, however, commercial companies collaborate in IPM programmes, especially where the potential exists for exploiting new technology (e.g. Haskell, 1992), even despite apparent differences in the respective goals of the different partners (Jones and Esterbaranz, 1992), with the industrialist looking to obtain a saleable product from the work and the researcher carrying out work in order to obtain recognition through publications, higher degrees, etc.

The most important goals for consideration during the development of an IPM programme are those of the farmer. Ultimately it is the farmers who will utilize the resulting IPM system and if such a system does not meet their objectives then inevitably it will be rejected.

Farmers are a very diverse group. They differ widely in the area and type of land farmed, cropping systems, wealth and farming objectives. The objectives of subsistence farmers are generally held to be different from those of commercial farmers. The former are concerned with guaranteeing a food supply and the latter, profit. However, most farmers will have in common an overall objective to undertake activities that will make themselves and their families as content as possible (Reichelderfer *et al.*, 1984). In order to deal with the complexity of the problem of satisfying farmers' goals it is necessary to break these goals into their constituent components, ambition and vision. It is widely accepted that farmers are intent on making their income as large as possible, with income considered not just in terms of cash, but also the value placed on food for home consumption and the capacity for future production (level of ambition). If their goal is to maximize profit then this will be reflected in their choice of control strategy. You would expect the farmer to select and use a strategy with the highest expected value. However, there are other goals which may seem more appropriate to the farmer, such as adopting strategies that minimize net costs (Valentine *et al.*, 1976), minimize the costs of the worst outcome or the probability of a particular disaster level (Mumford and Norton, 1984). The vision component of a farmer's goals is based on three factors: (i) the farmer's evaluation of outcomes in terms of food or net revenue; (ii) his attitude to risk and uncertainty; and (iii) his time horizon and preferences (Norton, 1977). Studies have concentrated on the idea of farmer's attitude to risk, and the effect this has on his crop protection decisions (Norton, 1976, 1977; Lane, 1981; Mumford, 1981). Decisions are said to be subject to risk when the resulting outcome is not determinate but one of a range of possible outcomes (Norton, 1976). Studies have tended to confirm the idea that farmers are generally risk averse; that is, they make decisions based on the need to insure against a poor or unsatisfactory outcome (Mumford, 1981, 1982b).

Norton, (1976, 1982b) introduced the idea that farmers may in fact adopt what is referred to as a 'satisficing' approach. This is an approach where the farmers learn through trial and error the strategies that prove satisfactory (Mumford and Norton, 1984). Since a farmer is likely to be faced with the same sort of problem from season to season and year to year, each decision made will be influenced by previous experience and previous decisions, incorporating (i) and (iii) aspects of 'vision'. This will inevitably lead to the farmer adopting 'standard operating procedures' which reduce the effort required in making frequent pest control decisions. These standard operating procedures will reflect the farmer's goals, which will also remain fairly constant over time.

The goals of an IPM programme and the resulting system will be dependent on the goals of not only the farmers but also agencies such as the government, the research community, the extension service and the agrochemical industry. The IPM programme will need to marry the essential elements of each of these agencies' goals and reflect their individual aspirations in setting its own programme and system goals. If a programme does not achieve this and the goals of IPM are at odds with those of other agencies, then it can only impede the progress of the programme and systems' development. The setting of appropriate goals is of the utmost importance since the direction and performance level of the IPM team depend on it. They need to be defined with care, understanding and tact.

Goal setting is hard work, but since it is so vital to the success of the programme it is worth spending the time necessary to get it right. The process requires that you mentally start at the finish and then work backwards. The clearer one can be about the end result (even though it may change) the more effective will be the plans to achieve it. Funding may have been obtained for an R&D programme to develop technologies appropriate for use by farmers, or the intention could be to develop an operational IPM system involving a programme of R&D as well as design and implementation of an appropriate delivery system. The best way to perhaps try to capture the goal is by asking questions such as, 'How will we know we have finished?', 'What will the end result/product actually look like?'. Discuss proposed goals with the end user, the farmers; they will be able to tell you whether or not the goals are appropriate.

An effective goal has five characteristics. It is specific, measurable, agreed upon, realistic, and time-framed (Randolph and Posner, 1988). A goal must be well defined, with no ambiguity, specifically addressing the issue in question. It should be clear enough so that anyone with a basic knowledge of the subject can read it, understand it and know what it is you are trying to accomplish. A goal should be measurable in terms of the activities carried out to achieve it. For instance, a component of a goal may be 'to develop an IPM technology transfer package'. At the end of the programme and during its development it should be possible to assess whether such a package has resulted or will result from the work. However, in this instance it would also be necessary to qualify what was meant by a 'technology transfer package' – since this could mean anything! There must be agreement about what is actually meant and whether or not the end result will solve the problem or respond to the need that led to the inititation of the programme. Agreement about the goal(s) is also required from all participants because disagreement will only lead to disenchantment and lack of commitment to the work. This will be particularly evident where participants believe goals are not realistic. A goal must be attainable and everyone within the programme will need to be convinced of this to ensure necessary levels of commitment are maintained. The time given to complete a programme may be dependent on availability of funds. The goals must be achievable within the given time frame. If a programme covers unfamiliar subjects or problems it will be necessary to build in time for learning and gaining familiarity. All of these five characteristics must be encompassed by the goal, if it is to be effective.

REFERENCES

Abbott, J.C. and Makeham, J.P. (1979) *Agricultural Economics and Marketing in the Tropics*, Longman Group Ltd, Harlow, Essex.

Bardner, R. and Fletcher, K.E. (1974) Insect infestations and their effects on the growth and yield of field crops: a review. *Bulletin of Entomological Research*, **64**, 141–60.

Barlow, N.D., Norton, G.A. and Conway, G.R. (1979) A systems analysis approach to orchard pest management. Mimeographed report, Environmental Management Unit, Dept of Zoology and Applied Entomology, Imperial College, London.

Bomford, M. (1988) Effect of wild ducks on rice production, in *Vertebrate Pest Management in Australia*, (eds G.A. Norton and R.P. Pech). CSIRO – Division of Wildlife and Ecology, Canberra, pp. 53–7.

Byerlee, D., Collinson, M.P., Perrin, R.K., Winkelmann, D.L., Biggs, S., Moscardi, E.R., Martinez, J.C., Harrington, L. and Benjamin, A. (1980) *Planning Technologies Appropriate to Farmers: Concepts and Procedures*, CIMMYT, El Batan, Mexico.

Chadwick, D.J. and Marsh (1993) *Crop Protection and Sustainable Agriculture*, John Wiley & Sons, New York.

Chambers, R., Pacey, A. and Thrupp, L.A. (1989) *Farmer First: Farmer Innovation and Agricultural Research*, Intermediate Technology Publications, London.

Chiarappa, L. (1971) *Crop Loss Assessment Methods, FAO Manual on the Evaluation and Prevention of Losses by Pests, Diseases and Weeds*, CAB, Wallingford.

Chiarappa, L. (1981) Establishing the crop loss profile, in *Crop Loss Assessment Methods – Supplement 3*, (ed. L. Chiarappa), CAB, Wallingford, pp. 21–4.

Collinson, M.P. (1980) Farming system research in the context of an agricultural research organisation, in *Farming Systems in the Tropics*, 3rd edn, (ed. H. Ruthenberg), Clarendon Press, Oxford, pp. 381–9.

Cromartie, W.J. (1991) Environment control of insects using crop diversity, in *Handbook of Pest Management in Agriculture*, (ed. D. Pimentel). CRC Press, Boca Raton, Florida, pp. 183–216.

Dale, E. (1978) *Management: Theory and Practice*, 4th edn, McGraw-Hill, Singapore.

Day, R.K. and Collins, M.D. (1992) Simulation modeling to assess the potential value of formulation development of lamdba-cyhalothrin. *Pesticide Science*, **15**, 45–61.

Dent, D.R. (1991) *Insect Pest Management*, CAB International, Wallingford.

Dent, D.R. (1992) Scientific programme management in collaborative research, in *Research Collaboration in European IPM Systems*, BCPC Monograph No. 52, (ed. P.T. Haskell), BCPC, Brighton, pp. 69–76.

Dent, D.R. (1993) Products versus techniques in community-based pest management, in *Community-Based and Environmentally Safe Pest Management*, (eds R.K. Saini and P.T. Haskell), ICIPE Science Press, Nairobi, pp. 169–80.

Edwards, S. (1990) Scientific research moves towards the 21st century. *Science and Public Affairs*, **4**(2), 131–8.

Ellis, F. (1992) *Agricultural Policies in Developing Countries*, University of Cambridge, Cambridge.

Everson, E.H. and Gallun, R.L. (1980) Breeding approaches, in *Breeding Plants Resistant to Insects*, (eds F.G. Maxwell and P.R. Jennings). John Wiley & Sons, New York, pp. 513–33.

Flint, M.L. and van den Bosch, R. (1981) *Introduction to Integrated Pest Management*, Plenum Press, New York.

Goodell, G. (1984) Challenges to international pest management research and extension in the Third World: do we really want IPM to work? *Bulletin of the Entomological Society of America*, (Fall), 18–26.

Griffiths, W.T. (1984) A review of the development of cotton pest problems in the Sudan Gezira. PhD Dissertation, University of London.

Haskell, P.T. (1992) *Research Collaboration in European IPM Systems, BCPC Monograph no. 52*, BCPC, Farnham, Surrey.

Helmer, O. (1969) *Analysis of the Future: The Delphi Method*, Rand Corporation, Santa Monica.

Heong, K.L. (1985) Systems analysis in solving pest management problems, in *Integrated Pest Management in Malaysia*, (eds B.S. Lee, W.H. Loke and K.L. Heong). Malaysian Plant Protection Society, Kuala Lumpur, pp. 133–49.

Holt, J., Cook, A.G., Perfect, T.J. and Norton, G.A. (1987) Simulation analysis of brown planthopper (*Nilaparvata lugens*) population dynamics on rice in the Philippines. *Journal of Applied Ecology*, **24**, 87–102.

Jones, O.T. and Esterbaranz, J.P. (1992) The role of industry in IPM systems development, in *Research Collaboration in European IPM Systems. BCPC Monograph No. 52*, (ed. P.T. Haskell), BCPC, Brighton, pp. 65–8.

Karlöf, B. (1987) *Business Strategy in Practice*, John Wiley & Sons, New York.

Kenmore, P.E., Heong, K.L. and Putter, C.A. (1985) Political, social and perceptual aspects of integrated pest management programmes, in *Integrated Pest Management in Malaysia*, (eds B.S. Lee, W.H. Loke and K.L. Heong), MAPPS, Malaysia, pp. 47–67.

Lane, A.B. (1981) Pest control decision making in oilseed rape. PhD Dissertation, Imperial College, University of London.

Lawson, T.J. (1982) Information flow and crop protection decision making, in *Decision Making in the Practice of Crop Protection*, (ed. R.B. Austin). BCPC Publications, Farnham, pp. 21–32.

Lim, G.S., Heong, K.L. and Ooi, P.A. (1981) Constraints to integrated pest control in Malaysia, in *IOBC Special Issue: Conference on Future Trends of Integrated Pest Management*, May/June 1980, IOBC, Paris, pp. 61–6.

Lipton, M. and Longhurst, R. (1989) *New Seeds and Poor People*, Unwin Hayman Ltd, London.

MAFF (1991) Ministry of Africulture, Fisheries and Food. Studentship Booklet. HMSO, London.

Makeham, J.P. and Malcolm, L.R. (1986) *The Economics of Tropical Farm Management*, University of Cambridge, Cambridge.

Mathema, S.B. and Galt, D.L. (1989) Appraisal by group trek, in *Farmer First*, (eds R. Chambers, A. Pacey and L.A. Thrupp), Intermediate Technology Publications, London, pp. 68–73.

Moss, S.R. (1980) Some effects of burning cereal straw on seed viability, seedling establishment and control of *Alopecurus myosuroides* Huds. *Weed Research*, **20**, 271–86.

Moss, S.R. and Cussans, G.W. (1991) The development of herbicide-resistant populations of *Alopecurus myosuroides* (Black grass) in England, in *Herbicide Resistance in Weeds and Crops*, (eds J.C. Caseley, G.W. Cussans and R.K. Atkin), Butterworth Heinemann, Oxford, pp. 45–55.

Mumford, J.D. (1978). Decision making in the control of sugar beet pests, particularly viruliferous aphids. PhD Dissertation, University of London.

Mumford, J.D. (1981) A study of sugar beet growers' pest control decisions. *Annals of Applied Biology*, **97**, 243–52.

Mumford, J.D. (1982a) Farmers' perceptions and crop protection decision making, in *Decision Making in the Practice of Crop Protection*, (ed. R.B. Austin), BCPC Publications, Farnham, pp. 13–19.

Mumford, J.D. (1982b) Perceptions of losses from pests of arable crops by some farmers in England and New Zealand. *Crop Protection*, **1**(3), 283–8.

Mumford, J.D. and Norton, G.A. (1984) Economics of decision making in pest management. *Annual Review of Entomology*, **29**, 157–74.

Mumford, J.D. and Norton, G.A. (1987) Economic aspects of integrated pest management, in *Integrated Pest Management, Protection Integreé: Quo Vadis? An International Perspective*, (ed. V. Delucchi), Parasitis 86, Geneva, pp. 397–408.

Norton, G.A. (1976) Analysis of decision making in crop protection. *Agro-Ecosystems*, **3**, 27–44.

Norton, G.A. (1977) Background to agricultural pest management modelling, in *Proceedings of Conference on Pest Management*, (eds G.A. Norton and C.S. Holling), IIASA, Laxenburg, Austria, pp. 161–76.

Norton, G.A. (1979) Background to agricultural pest management modelling, in *Pest Management: Proceedings of an International Conference*, (eds G.A. Norton and C.S. Holling), Pergamon, Oxford, pp. 161–76.

Norton, G.A. (1982a) Crop protection decision making – an overview, in *Decision Making in the Practice of Crop Protection*, (ed. R.B. Austin). BCPC Publications, pp. 3–11.

Norton, G.A. (1982b) A decision-analysis approach to integrated pest control. *Crop Protection*, **1**(2), 147–64.

Norton, G.A. (1985) Economics of pest control, in *Pesticide Application: Principles and Practice*, (ed. P.T. Haskell), Clarendon Press, Oxford, pp. 175–89.

Norton, G.A. (1987) Pest management and world agriculture – policy, research and extension, in *Papers in Science, Technology and Public Policy*, No. 13: Pest Management and World Agriculture – Policy; Research and Extension, Imperial College of Science & Technology, University of London, The Science Policy Research Unit, University of Sussex, The Technical Change Centre, London, pp. 1–28.

Norton, G.A. (1990) Decision tools for pest management: their role in IPM design and delivery. (Paper presented to the FAO/UNEP/USSR Workshop on Integrated Pest Management).

Norton, G.A. (1993) Agricultural development paths and pest management: a pragmatic view of sustainability, in *Crop Protection and Sustainable Agriculture*, (eds D.J. Chadwick and Marsh). John Wiley & Sons, pp. 100–15.

Norton, G.A. and Mumford, J.D. (1993) *Decision Tools for Pest Management*, CAB International, Oxford.

Oakley, P. and Garforth, C. (1985) *Guide to Extension Training*, FAO, Rome.

Phillips, D. (1989) Funding the UK science base: modes of support. *Science and Public Affairs*, **4**(1), 59–63.

Pimentel, D. (1977) Ecological basis of insect pest, pathogen and weed problems, in *The Origins of Pest, Parasite, Disease and Weed Problems*, (eds J.M. Cherrett and G.R. Sagar), Blackwell, Oxford, pp. 3–31.

Pimentel, D. (1985) Using genetic engineering for biological control: reducing ecological risks, in *Engineered Organisms in the Environment: Scientific Issues*, (eds H.O. Halvorson, D. Pramer and M. Rogul), American Society for Microbiology, Washington, DC, pp. 129–40.

Pimentel, D. (1991) Diversification of biological control strategies in agriculture. *Crop Protection*, **10**(1), 243–53.

Pimentel, D., Shoemaker, C., La Due, E.L., Rovinsky, R. and Russell, N. (1977) *Alternatives for Reducing Insecticides on Cotton and Corn Economic and Environmental Impact*, Environmental Research Laboratory, EPA, Athens.

Rabb, R.L. and Guthrie, F.E. (1970) *Concepts of Pest Management*, North Carolina State University Press, Raleigh.

Randolph, W.A. and Posner, B.Z. (1988) *Effective Project Planning and Management*, Prentice Hall, London.

Reichelderfer, K.H., Carlson, G.A. and Norton, G.A. (1984) Economic guidelines for crop pest control. FAO Plant Production and Protection Paper 58.

Ruthenberg, H. and Jahnke, H.E. (1985) *Innovation Policy for Small Farmers in the Tropics*, Oxford University Press, Oxford.

Saaty, T.L. and Kearns, K.P. (1985) *Analytical Planning*, Pergamon Press, Oxford.

SERC (1991) *Science and Engineering Council, Research Grants*.

Shaner, W.W., Phillips, P.F. and Schmehl, W.R. (1982) *Farming Systems Research and Development Guidelines for Developing Countries*, Westview Press, Colorado.

Tait, E.J. (1977) The use of forecasting as a method of rationalising pesticide application, in *Proceedings 1977 British Crop Protection Conference*, BCPC, Brighton, pp. 235–9.

Tait, E.J. (1987) Planning an integrated pest management system, in *Integrated Pest Management*, (eds A.J. Burn, T.H. Coaker and P.C. Jepson). Academic Press, London, pp. 189–207.

Valentine, W.J., Newton, C.M. and Talevio, R.L. (1976) Compatible systems and decision models for pest management. *Environmental Entomology*, **5**, 891–900.

Van der Graff, N.A. (1981) Increasing reliability of crop loss information: the use of 'indirect' data, in *Crop Loss Assessment Methods – Supplement 3*, (ed. L. Chiarappa), CAB, Oxford, pp. 65–9.

Waibel, H. (1993) Government intervention in crop protection in developing countries, in *Crop Protection and Sustainable Agriculture*, (eds D.J. Chadwick and J. Marsh), John Wiley & Sons, Chichester, pp. 76–93.

Walker, P.T. (1983) Assessment of crop losses, in *Pest and Vector Management in the Tropics with Particular Reference to Insects, Ticks, Mites and Snails*, (eds A. Youdeowei and M.W. Service), Longman, Essex, pp. 75–83.

Walker, P.T. (1984) The quantification and economic assessment of crop losses due to pests, diseases and weeds, in *Advancing Agricultural Production in Africa*, Commonwealth Agricultural Bureaux, pp. 175–81.
Walker, P.T. (1987) Losses in yield due to pests in tropical crops and their value in policy decision-making. *Insect Science Application*, 8(4/5/6), 665–71.
Zadoks, J.C. (1993) Crop protection: why and how, in *Crop Protection and Sustainable Agriculture*, (eds D.J. Chadwick and J. Marsh), John Wiley & Sons, Chichester, pp. 48–60.

CHAPTER 5

Programme planning and management

D.R. Dent

5.1 INTRODUCTION

The initial stages of the programme should have provided the IPM team with all relevant available information, collated and synthesized in an appropriate form and on the basis of which the programme/system goals will have been set. The next stage in the procedure is that of planning and management, i.e. identifying the means by which the goals will be achieved and the process of implementing the necessary strategy to achieve them.

The first step in this process is to devise the type and form of IPM system that will meet the required goals. Questions need to be asked such as, who is the system for, on what scale will it be applicable, what control measures are to be utilized and in what way will they be applied? What are the perceived benefits to be, and over what time scale will these occur? These are all very important questions which need to be answered in the process of devising an IPM system (section 5.2). Once a decision on these matters have been made it is then a question of considering how the programme is to be organized and managed to ensure that the IPM system is developed in the required way.

Since the resources available will ultimately affect what can and cannot be done, an evaluation of resources at the disposal of the team is also required (section 5.3). On the basis of this an evaluation will need to be made concerning the establishment of appropriate organizational structures for the allocation of responsibilities and tasks, and the management of activities (section 5.4). The techniques of programme planning will be required to give the programme direction and purpose, while monitoring techniques will need to be put in place to enable evaluation of progress

Integrated Pest Management Edited by David Dent. Published in 1995 by Chapman & Hall, London. ISBN 0 412 57370 9

towards the stated goals (section 5.5). Proper management will contribute to good communications, team dynamics and integration of effort (section 5.6). The contribution of these separate but interacting components should ensure the development of a targeted IPM system.

5.2 DEVISING AN IPM SYSTEM

The ultimate goal of an IPM programme is to develop an IPM system. Hence, before the needs of an IPM programme can be considered the form and components of the IPM system have to be defined. The means by which an IPM system is implemented is also relevant to system design but this is considered separately in Chapter 7. This section is concerned with the means by which an IPM system is devised as part of the process of defining programme needs.

The first question which has to be addressed is 'Who is the IPM system for?' This question will have been largely dealt with in the initial stages of programme planning where farmers should have been classified according to 'decision domains', i.e. those farmers that, in decision-making terms for pest management, can be treated as a single homogeneous group (section 4.4). What needs to be decided at this stage is the number of farmers, the district size or regional area over which the IPM system eventually will be implemented. In pest management terms this will be primarily dependent on the area over which the level of pest infestation can be considered relatively uniform (Carlson and Rodriguez, 1984; Reichelderfer *et al.*, 1984). The more mobile the pest and uniform the level of damage caused, the larger the area over which the system can be implemented. If the area is large then an area-wide or regional IPM system becomes possible.

In theory there is much to commend the use of regional pest management systems (Southwood and Norton, 1973). These include: (i) greater accountability taken of the impact of the IPM system so that it meets long term goals; (ii) a wider choice of control measures used (since a number of measures can only be successfully employed over a wide area, e.g. classical biological control, sterile insect technique); and (iii) an economy of scale may be obtained from widespread use of a control measure, such as an aerial application of insecticides. These advantages are even more evident in situations where information on pest mobility is readily available, there are high costs for individual farm pest-suppression methods, and where improved monitoring and pest control technology is available specifically for area-wide pest management systems (Carlson and Rodriguez, 1984). The technologies that are now available for use on a regional scale include mating disruption, for example *Cydia molesta* in Australia (Vickers *et al.*, 1985); *Lymantria dispar* (Kolodny-Hirsch and Schwalbe, 1990) as well as more familiarly, the use of sterile insect technique (SIT) (*Dacus dorsalis*) (Steiner *et al.*, 1965; Ushio *et al.*, 1984; Koyama *et al.*, 1982); *Anthonomus grandis* (Ridgway *et al.*, 1990); Mediterranean fruit fly, (Saul, 1989); and biological control through classical use of introductions (e.g. *Phenacoccus manihoti* in Africa; Neuenschwander and Herren, 1988).

The problems arise in area-wide systems when there is a variability in demand for pest control measures across farms. This may occur either because of variable levels of pest infestation or because the farmers switch to growing alternative, substitute crops which are not susceptible to the pest in question. Problems can also develop because use of an area-wide pest control system changes overall crop production levels or crop prices, or because the system itself develops high information or administrative costs (Reichelderfer *et al.*, 1984). If farmers choose not to participate in a regional scheme then the costs of participation for the remaining farmers can become prohibitively high. In addition, with measures such as biological control, a minimum level of participation may be required in order to preserve (through appropriate practices) populations of the released natural enemies. These latter problems may be avoided through the public funding of systems since this will tend to involve little direct contribution towards costs by individual farmers. However, where area-wide control systems cannot be publicly funded and/or in situations where pest infestation or cropping systems are more variable, then farm-based IPM systems will need to be devised.

5.2.1 Models of farmer participation

These two different scales of approach, regional and farm-based pest management systems can be categorized as a function of the degree to which farmers participate in their implementation. Andrews *et al.* (1992) consider two models, the 'Farmer bypass model' and the 'Farmer as Protagonist model'. The former model applies to a regional pest management system where there is minimal farmer and extension involvement. They have straightforward research programmes uncomplicated by outreach and extension. This contrasts with the Farmer as Protagonist model which recognizes the ecological and socioeconomic heterogeneities of farming systems that require both site-specific decision-making and control measures. Each of the two models can be further subdivided into participatory and non-participatory approaches to research and implementation.

Non-participatory research–non-participatory implementation refers to situations such as classical biological control and the use of quarantine procedures. The research and implementation can be carried out totally independently requiring no input from the farmers who are seen simply as beneficiaries of the outcome. The *participatory research–non-participatory implementation* is an uncommon model because it involves farmer participation in establishing research strategy and in the development of the technology but then 'sitting back' to enjoy the benefits as the pest control measure is applied on their behalf. Examples of this type of approach would be a situation in which farmers were involved in establishing the need for control measures such as classical biological control or SIT and before release assisted in creating a suitable environment for establishment of released insects. In the case of biological control introductions this may mean farmers engaging in habitat modification or desisting in pesticide application, while for SIT it could involve farmers applying pesticides to reduce natural pest population levels to aid the process of population suppression.

The most common, more conventional model is the *non-participatory research–participatory implementation* approach. This, essentially top-down approach, assumes that scientists involved in fundamental research pass ideas and information to applied scientists who then develop techniques and products to pass on to extension workers who finally introduce these to the farmer. This model has undoubtedly been very successful in the development and adoption of pest control products (section 7.6.1) such as chemical pesticides and microbial pesticides, packaged in simple ways to promote dissemination and adoption. The industrial/private sector have found it a particularly amenable approach for the promotion and sale of their agrochemical products. However, despite its intensive use the *non-participatory research–participatory implementation* approach has a number of limitations. For example, it is less effective when information rather than inputs are to be disseminated and it has not proven useful for techniques such as conservation of natural enemies and habitat modification (Andrews *et al.*, 1992) (section 7.6.1). The *non-participatory research–participatory implementation* approach tends also to be less effective in situations where farming and cropping practices are heterogeneous and will fail in situations where the scientists do not take consideration of all the key factors as they carry out their research. Despite these drawbacks, this model has been widely applied. More recently however, an increasing emphasis has been given to the need to involve farmers in development in general (Chambers, 1983; Chambers *et al.*, 1989) and particularly in the design of IPM systems (Norton, 1990; Norton and Mumford, 1993). This has led to the realization that the alternative approach based on *participatory research–participatory implementation* will probably yield research pertinent to the needs of the farmer and hence, will mean that there is more likelihood that the products of such research will be adopted by the farmers.

Integrated pest management systems designed for individual farmers would ultimately seem to fit the *participatory research–participatory implementation* model more readily than the *non-participatory–participatory implementation* model. The latter model has been shown to be effective where a simple message needs to be delivered; IPM however, is a much more complex concept which is less amenable to simple dissemination approaches. For instance, a pesticide is highly toxic to the pest, visibly effective, apparently easy to apply and is reliable. It is also based on a simple concept, i.e. a problem with an associated action that directly redresses the problem. By contrast IPM systems are more complex, are often counter-intuitive, involving approaches which have an indirect rather than an obviously direct impact on the pest problem, and they may not provide visibly dramatic results (Goodell, 1984). IPM systems will tend not to be dependent on use of a single control measure but on a number of interacting measures, the sum effect of which provides the required level of control. In similar situations in rural development, where the introduction of a new production system is dependent on the interaction of several practices and inputs for its success, then the system is provided as a technical package (Adams, 1982; Ruthenberg and Jahnke, 1985; Sumberg and Okali, 1989). These packages, which in the case of crop production might include a new variety, higher plant density or altered planting arrangement, fertilizer and pesticide use are provided as an all-or-nothing proposition (Sumberg and Okali, 1989). In developing countries such packages of inputs are tied to agricultural credit programmes. The package is purchased at the time the borrowers draw their loan (Goodell, 1984). By availing themselves of this credit at lower interest rates the farmers are effectively investing in inputs such as pesticides before they even know whether they will have a pest problem. Such compulsory packaged systems lend themselves to corruption, preclude farmer experimentation and discourage alternative forms of control (Goodell, 1984, 1989). Clearly similar types of packages for IPM are undesirable, but since IPM is based on the utilization of a number of compatible, often interacting control measures, which applied in combination provide sufficient levels of pest control but in isolation are rarely worthwhile, then some form of package for IPM would seem inevitably necessary. The latest ideas from rural development specialists argue for a more open, interactive approach in which a 'basket' of alternative technologies are presented and the farmers then selects those which most readily fit their requirements (Chambers *et al.*, 1989). Such an approach, although in principle providing a logical way to proceed, would, for IPM, place an unrealistic demand on research. It would require scientists to develop a combination of technologies that work in different permutations. Such aims are at present unrealistic for IPM.

The *participating research–participatory implementation* approach should, however, suit the more complex needs of IPM, where on-farm research is used to identify the constraints to performance of packages of new improved technologies (Sumberg and Okali, 1989). On-farm trials are viewed as a means of validating a given package and the role of the farmer in this research is to identify any remaining constraints in the package before more general dissemination. As an IPM system becomes more complex in design and implementation then the role of the farmer in the development of the system becomes increasingly critical, and hence the greater emphasis that needs to be placed on use of the *participatory research–participatory implementation* model.

On the assumption that the model utilized will include both *participatory/*

non-participatory/research – implementation combinations of the farmer as *protagonist model*, then the next step in devising an IPM system is to consider the constituent components of any IPM system and the choices that may be made concerning their inclusion and use.

5.2.2 Constituent components of an IPM system

There are three different components that can potentially contribute to an IPM system: (i) products and devices and accompanying procedures; (ii) control techniques or practices; and (iii) information. A product may be a pesticide, a natural enemy packaged for augmentative release, a pheromone formulation used for mating disruption, while a device may be a trap for monitoring or a lure and kill target surface. Each would normally be supplied with instructions providing details of the conditions for most appropriate use and application. The control techniques and practices include cultural/agronomic practices that influence pest infestation levels such as planting times and density, cropping pattern, cultivation, and habitat modification, e.g. sowing flowering plants as food source for parasitoids. Control techniques differ from products and devices in that the latter are purchased as off-farm inputs while the former tend to involve on-farm resources, (although not always). Extra labour may be required to implement innovative techniques (section 7.6.1). In situations where no off-farm inputs are required then the new approaches, techniques, etc., are presented to the farmers as 'information'.

Norton and Mumford (1982) distinguish between four categories of information relevant to pest management: (i) fundamental; (ii) historical, (iii) real-time; and (iv) forecast information (section 4.4). Fundamental information relates to the basic technical, biological and ecological processes which affect the damage caused by pests and the effectiveness of control measures (Norton, 1982). It would include the knowledge of the fundamental principles of IPM and basic concepts and rules such as the economic threshold. The type of information that would be required for dissemination would depend on the extent of knowledge and understanding of the farmers in the target group. In terms of disseminating new information it may need to include a means of pest identification and description of life-cycles as well as the information on which control measures to use and how to achieve optimum performance. Recommendations arising from research would fall into this latter category. For instance, research may show that a number of pesticides applied at reduced rates to a particular plant growth stage controls pests adequately and allows development of natural enemy populations; this information could be presented to the farmer as a series of recommendations, i.e. a list would be provided of more environmentally friendly pesticides, their recommended application rates and details of timing of application.

Historical information, in the form of a record of past pest outbreaks and damage can be used to indicate trends in pest development and allow some assessment of the probability of forthcoming attacks to be made. This category of information has less relevance to this stage of devising an IPM system than the other categories. It is however, particularly important in terms of the farmer's own perception of the pest problem. Each farmer will have past experiences which will influence their perception of the risk and the need for control, the success of previous control measures and the need for alternatives (section 4.5). An understanding of these perceptions will need to be built into any framework of information that is presented to the farmer.

Real-time information is primarily concerned with the need to provide individual farmers with information on current pest status in terms of levels of pest infestation and damage. Relevant data may be collected by on-farm monitoring schemes or by regional surveillance. Information on other variables such as weather, crop development and prices also fall into this category.

The fourth category of information is referred to as forecast information, which is dependent on the other three types of information (Norton, 1982; Norton and Mumford, 1982). Forecast information involves providing estimates of the future state of important variables such as levels of pest attack and damage. Some form of modelling technique, a regression model, or a more complex simulation model is usually employed (Norton and Mumford, 1982) (Chapter 7). These four categories of information, the products and the techniques are the components which potentially are available for construction of an IPM system. The choice of control measures and the approach taken will ultimately affect the relative mix of products, techniques and the information that will constitute the IPM system.

5.2.3 Attributes of an IPM system

The aim should be to devise an IPM system which is sufficiently robust to maintain control over a prolonged period (Matthews, 1984). To achieve this an IPM system will require a number of attributes (Table 5.1). First and foremost the system must be effective. For the farmer this means that the IPM system should be at least as good as the conventional control measure. A farmer is only likely to accept reduced effectiveness, that is an increased level of damage, if there is some incentive or indirect advantage such as a premium price for 'organically' grown produce. The system must be economic: any increased costs of inputs whether products or labour costs with the new system must be offset by increased production or some other means of improvement. No farmer will adopt and sustain use of uneconomic pest management practices. They will also be less inclined to adopt an IPM system that is too complex or inflexible. IPM systems must be designed to be as simple as possible, utilizing the minimum number of control measures compatible with maintaining pest populations at appro-

Table 5.1 Suitable attributes for an IPM system

An IPM system should
• provide effective control of pest complex
• be economically viable
• be simple and flexible
• utilize compatible control measures
• be sustainable
• have minimum harmful impact on the environment, the producer and consumer

priate levels. The number of decision points should be kept to a minimum (and made as early in the season as possible, since the number of options available for control are reduced as the season progresses) and where feasible the inclusion of complicated procedures or application methods should be avoided.

A package of control measures should be flexible and not totally preclude use of only some of the individual measures, allowing for some farmer selection and experimentation. The individual control measures should of course be compatible and optimize natural mortality factors. It is important during the design of an IPM system to consider the level of control which is required and the best mix of control measures that will achieve this with minimal antagonism. Also, recourse should constantly be made to basic principles, identifying how control measures are likely to affect initial levels of infestation and/or reduce the intrinsic rate of increase (section 2.6.1). Too little attention is often given to these basic points in the design stage which can lead to problems later on during implementation. Where possible it is advisable to evaluate the effects of different control measures through use of simulation models, since these can provide useful pointers, and prevent investment of resources into potentially unprofitable areas. Finally, the IPM system should be sustainable, have minimum impact on the environment and present no hazard to the farmers, their families or the consumers of the crop products.

There are few that would dispute the general value of the attributes of an IPM system listed in Table 5.1 and outlined above. The important point is to draw attention to the more systematic way in which an IPM system can be devised. At the end of this design phase the programme should have available (at least) an explicitly defined conceptual model of the system indicating how it is thought each of the components will interact (this should form the basis of a simulation model of the system), a decision tree indicating the sequence and number of management decisions that will be made and a preliminary assessment of expected economic costs and benefits of the proposed system. Organized in this way the programme managers can then consider their resource requirements.

5.3 PROGRAMME AND SYSTEMS RESOURCE REQUIREMENTS

IPM programmes and systems development can only take place within the context of available resources, whether these are time, human, institutional or financial. The type and extent of resources will always influence the size and scope of what can be undertaken and ultimately what is achieved. Hence, there is always a problem of properly matching the funding to the work required to achieve stated goals. Difficulties tend to most often arise where grants are awarded before a full, detailed analysis of the problem, which subsequently reveals additional dimensions that should have been included in the initial proposal. The programme may then have to adjust budgets and priorities to ensure they address these new aspects, or otherwise carry on with the original proposal in the knowledge that their work is excluding some important dimension. The ability to accurately define resource needs in relation to stated goals is an essential component of good programme/systems management. While it is not always possible to accurately assess resource needs at the planning stage (because it is impossible to accurately predict the future) the use of the following equation can provide a logical basis for decisions. It is particularly useful for predicting expected levels of expenditure and time required to complete projects.

$$\text{Expected value} = \frac{\text{Optimistic estimate} + 4\ (\text{most likely estimate}) + \text{pessimistic estimate}}{6}$$

For instance, if you needed to estimate expected consumable costs on a project then an optimistic estimate might be £3000, a pessimistic estimate £4500, then you would budget for £3900. This kind of calculation is not required for every single activity but it is useful for assessing the 'cost' of critical activities of which you have little experience (Randolph and Posner, 1988). As a general principle it is always wise to build extra resources into estimates. This is known as slack, and allows a project to be manoeuvred back to track in the event of unforeseen problems or costs. The following sections briefly describe the different types of resources (Figure 5.1) and their role in the development of IPM programmes and systems.

Quite often the largest recurrent cost to a programme or system will be the salaries of the specialists involved (Pearson, 1988). Given this it is very important that the correct decisions are made concerning the relative balance of specialists, their level of expertise, competence and ability to complete their work. In the first instance, the manager may have to determine the most appropriate balance of scientists and extension workers involved given the relative emphasis on programme or systems development. Obviously this balance may change as an R&D programme progresses and greater emphasis is placed on systems development. Conversely if the development of an IPM system provides the major thrust of the work

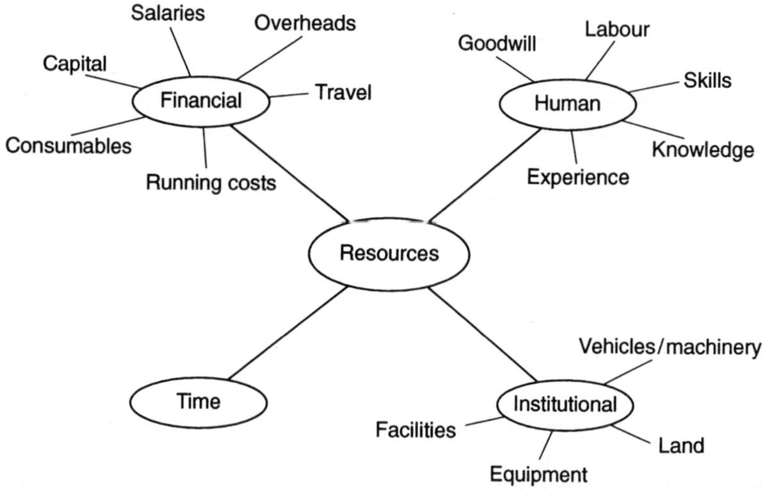

Figure 5.1 Resources available in the development of an IPM programme.

and research is only required to fill in minor 'research gaps' then extension workers may provide the bulk of the manpower involved.

The numbers of individuals involved from each discipline must match the priority attached to the task and the difficulty of carrying it out. It is also particularly important for the manager to be aware of the existence of the different research paradigms to ensure that those scientists selected adhere to a disciplinary paradigm most relevant to the needs of the programme (Chapter 7). Hence, programme managers need a broad knowledge of IPM and of the individual scientists involved in the relevant disciplines. This places a great burden on a programme manager because few individuals can have the breadth of knowledge and the necessary contacts across the different disciplines (especially in different organizations) to ensure that well-informed choices are made. It is more likely that a more pragmatic approach is taken based on established links and contacts rather than the recruitment of the most appropriate specialist for the work (Dent, 1992). Such pragmatic approaches are understandable but should be undertaken with caution, since ultimately the poor choice of personnel may serve only to weaken the work of the programme. Likewise, there is a tendency, especially in universities, to employ younger scientists, often as postgraduate students studying for a PhD degree, rather than more experienced scientists or post-doctoral scientists. While this may be seen as a convenient way of saving money in the short term, again, it may cause difficulties, particularly because the goals of obtaining PhDs and the type of research required for achieving these are not always coincident with the goals of an IPM programme. Also, the relative inexperience of the postgraduate scientist may tend to reduce the likelihood of pertinent and successful research. This is not to say however, that by involving more experienced scientists there will not be problems or deficiencies in the work that is carried out. With more experienced scientists there is always a concern that they will have less time to commit to a programme (given their greater diversity of interests and administrative responsibities).

From a management perspective it is a good idea to determine in advance the proportion of total man-hours that those involved can commit to the programme. The younger more inexperienced scientists will undoubtedly be able to commit a greater proportion of their time to programme research than their elder more experienced counterparts. The dynamism, enthusiasm and commitment of younger scientists may also be deemed to outweigh the value of experience. They also tend to be more open to new ideas and approaches, and have the flexibility essential to developing interdisciplinary research teams (section 5.3).

Interdisciplinary research may also place an extra burden on resources compared with research in each constituent discipline. For instance, a pathologist may devise an experimental field trial which involves only one or two system parameters and will design the experiment to meet the demands of statistical significance. If an economist were also then involved in the trial there would be a need to estimate a larger number of parameters to carry out, for example, a cost/benefit analysis. The requirement for a much larger experiment and hence, more resources, automatically brings one face-to-face with the economics of resource allocation in the programme (Headley, 1985). In the polarization of tasks in interdisciplinary research it is inevitable that some sacrifices will be necessary to achieve some success without violating the budget.

Experiments may be carried out in farmers' fields, experimental research station fields, or in the laboratory. Those carried out in farmers' fields will require a close association, involvement and goodwill between the programme team and the farmers, and also that the funds are available to either rent the farmers' land or to compensate for any losses caused by the failure of experiments. It is now generally acknowledged that more value can be attached to the signficance of pest management programmes developed on-farm than in experimental research station fields, largely because of the anomalies of station fields and crop-growing conditions (Reed *et al.*, 1985). However, not all research can be carried out in farmers' fields. Adaptive research may be well suited to this, but more fundamental research, or research where the scientist is simply attempting to demonstrate biological feasibility, will need to be conducted in a laboratory or experimental fields of a research station, where the necessary facilities and equipment are available.

Collaborating organizations may already have all the necessary equipment and facilities for carrying out a research project. Indeed this may be used as a criterion for their participation. Often however, funding for a new project is seen as an opportunity to upgrade facilities or to purchase new pieces of equipment. A manager needs to be aware of the costs and the potential for lost time and research output due to delays in purchase, installation and training associated with use of new equipment. There is also a learning curve during which time the value of research obtained through use of the new equipment may be below required standards. A manager must identify potential situations in which significant project time may be lost.

Travel costs, for example, those associated with visits by scientists to and from experiments in farmers' fields, have a time element associated with them. If a scientist has to drive long distances to field sites then the time taken driving is time in which no research is being conducted. In this context time has an associated cost that has implications for the programme. Managers may wish to quantify this time element as a means of gauging the value of the research. It may be more appropriate for a scientist to spend less time driving to a nearer, but less suitable field site than to waste time driving to a more distant but more appropriate field site. Time is a valuable resource and should not be wasted. It may be necessary, on occasion, for the programme or systems manager to balance scientific expedience with economic prudence on the basis of an economic assessment of time. This can be done by calculating on an hourly or daily basis the cost of the scientists' time to the programme on the basis of their gross salary, an appropriate proportion of running costs and overheads as well as consumables. The final hourly figure (Table 5.2) represents the cost to the programme of each hour of work undertaken by the scientist. Such figures tend to place the value of work into a context that everyone can appreciate and hence is a useful measure/criterion by which to assess the cost of various research tasks.

The costs of a programme are ultimately assessed in financial terms. Most programmes tend to break their costs/expenditure into the categories in Figure 5.1; salaries, capital, consumables, travel, overheads and running

Table 5.2 Evaluating an hourly rate for work

1. Calculate the annual sum of money under the following categories:

	£
Annual salary	
Taxes (+ National Insurance)	
Employer's National Insurance	
Professional overheads (heating, electricity, rent)	
Miscellaneous costs (visits, conferences, vehicles)	

2. Estimate value of weekly rate by dividing by number of weeks worked per year
3. Estimate value of hourly rate by dividing (2) by number of hours worked per week

costs. Salaries usually represent the biggest cost to a programme and are self-explanatory. Capital budgets are those items of expenditure which are allocated to the purchase of equipment and furnishings, while consumables include recurring items such as experimental materials, e.g. Petri dishes, cage materials, fertilizer, pesticides. The overheads of a programme are the costs that the participating organizations charge for the administration of the programme. It is usually estimated as fixed proportion of the total budget. The figure is meant to cover, the costs associated with the administration of salaries, personnel and purchasing, in addition to the cost of a contribution towards general expenditure associated with the provision of electricity, water, etc. The programme is unlikely to receive any of the income under this heading, although the situation does vary enormously between different organizations. Running costs are the funds required to actually run the programme; mainly items that cannot be placed under consumables categories, e.g. vehicle and equipment maintenance.

Proper use of resources, whether time, institutional, functional or human resources is a vital aspect of the development of any IPM programme or system. In the first instance, it is a matter of correctly matching the funds to the tasks but later during the course of the programme it is essential that the resources be properly managed. Depending on the size of the programme the manager may need to delegate this responsibility to an administrator.

5.4 ORGANIZATIONAL STRUCTURES

Appropriate organizational structures are required by all organizations, whatever their size, to ensure the most suitable division of labour, and use of resources, to provide channels of communication and contact between participants, and ultimately the management of activities to ensure organizational goals are achieved (Heap, 1989). Particular attention needs to be given to the establishment and use of appropriate structures when it is necessary to integrate the inputs from a multidisciplinary team or to promote interdisciplinary collaboration. The latter requires a more considered approach than is necessarily available from the more common organizational structures used in business and commerce.

5.4.1 Open and closed systems

Organizational structures, in their most basic form, are either open or closed systems. A closed system is used for example, where a product passes through a number of processes, moving through various phases of development until its final form, e.g. the processes involved in developing chemical or microbial insecticides (Graham-Bryce, 1981; Baldwin, 1986), or production of biological control agents for mass release (Scopes and Biggerstaff,

1971). Closed systems thus obviously have some relevance to the development of specific pest control products but they are less applicable to the general development of pest management programmes. Of greater relevance are the 'open' or diversified structures which involve the division of an organization into separate functional groups each dealing with an aspect of pest management or a specific control measure. The functional groups may be socioeconomists, plant pathologists, weed scientists or even more specifically, fungicide or herbicide specialists, scientists involved in identifying hostplant resistance mechanisms or pheromone chemists. Each group works to solve a particular problem through activities which contribute to the overall development of the programme. The functional groups may belong to more than one organization and be separated geographically, but they can nevertheless still constitute a single programme with a common goal.

A programme will basically consist of the different functional groups (the operating system carrying out the activities contributing to the overall goals), and supported by a management and administrative system which will direct, motivate and administer resources as required. The structure may be organized mechanistically or organically (Heap, 1989). A mechanistic approach involves the use of formal relationships and communication channels based on a hierarchical form. It is supported by written rules and procedures and a formal reporting process. The organic approach to open structures is much more informal and allows lateral interaction and communication. Such structures are considered more flexible and can be more responsive to unexpected changes and new ideas. A combination of the more positive aspects of both the organic and mechanistic approaches is found in a third alternative, the matrix organizational structure. This involves the use of functional groups, but running across these is a project-based structure. The project, which in this case would be a contribution to an IPM programme, involves the secondment of individuals on a temporary basis from the normal functional group to serve on the project. The matrix organization forms a two-way flow of authority and responsibility (to the programme and to the functional group or department) and hence, will mean that the individuals concerned have two managers to which they report. Problems arise with this approach where there is a conflict of interest between the two managers. For the matrix organization to be successful each person or functional group must agree to cooperate and coordinate efforts so that employees are not asked to do two different things at the same time. The programme manager, who may have a team drawn from any number of functional groups and/or organizations, must have an input into the evaluation of each individual's performance. Although this may be difficult to organize across institutes, it is essential if a manager is to be able to motivate team members and ensure their commitment to the programme (Randolph and Posner, 1988).

5.4.2 Sociocognitive frameworks

The matrix organization emphasizing, as it does, the involvement of various functional groups in a programme or project is essentially a multidisciplinary approach, well suited to pest management problems. Where, however, there is interest in moving beyond that of a basic multidisciplinary approach to the integration of effort and interdisciplinary research then organizational structures need to be considered at the sociocognitive level (Rossini *et al.*, 1978). The term 'sociocognitive' is used because interdisciplinary research is team research that must entail social interaction among team members in order to produce convergence of disciplinary perspectives. In one of the few studies carried out to evaluate methods used by scientists to integrate their research, Rossini *et al.* (1978) identified four sociocognitive frameworks: (i) common group learning; (ii) modelling; (iii) negotiation among experts; and (iv) integration by leader.

Common group learning is an approach where the programme output is taken from the portion of each team member's knowledge which is common to all, i.e. the intersection of the individual's knowledge. The limits of the problem are bounded and divided into areas for consideration based on availability of expertise within the group. Each individual will carry out a preliminary analysis of their particular aspect of the programme which is then discussed and criticized as a group. The analyses are then rewritten by a different individual who is usually not an expert. The output is considered to be the common intellectual property of the group (Figure 5.2).

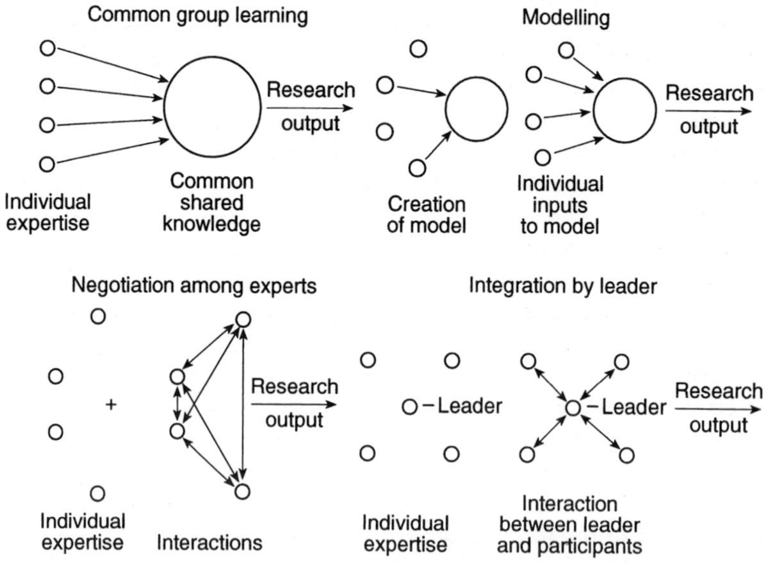

Figure 5.2 Organizational frameworks for interdisciplinary research. (After Swanson, 1979).

The modelling framework provides a common focus for the team. Not all the members of the team need participate in the construction of the model but they should all agree on its form. The model may be in a simple conceptual form or a full mathematical model depending on the problem, and the abilities and perspectives of the team members. Models tend to narrow the focus of interest both by excluding non-essential relationships (which may be desirable) and by excluding relevant aspects of the world that do not fit within their framework (which is undesirable). For this reason a model based sociocognitive framework works best with very specific, narrowly focused problems.

In the studies conducted by Rossini and colleagues, common group learning and modelling were the most commonly used frameworks; negotiation among experts was the least used. Here the problem is bounded as a group and the work divided among the team on the basis of their disciplinary expertise. Each member analyses their aspect of the problem, incorporating any complex and esoteric theories and approaches they consider appropriate. Integration is then achieved by negotiation, mainly at the boundary region and where analyses overlap. Where this occurs analyses will be redone to reflect the findings of the other experts.

The integration by leader approach places a great burden on the leader, who acts as the sole integrator and interacts with each member (members do not interact among themselves). The leader needs to understand and assimilate each member's contribution to the programme (Figure 5.2). The research output will be heavily biased by the leader's experience, knowledge and abilities.

Each of the frameworks discussed above has advantages and disadvantages and will be more suited to some situations and not to others. There also may be situations which call for a combination of frameworks (Dent, 1991) (Figure 5.3). The common group-learning approach is limited to the techniques, concepts, theories and models that are familiar to each member of the group. This tends to decrease the depth of analysis. The approach has the advantage however, of placing the burden of confronting an expert upon the whole group; a depersonalization which is shared by the modelling framework as it forces individuals to meet the information needs of the model. In contrast, negotiation among experts depends on confrontation, particularly at the boundary between regions of expertise, where more equal conditions exist. It is an approach more contrary to standard research training than any of the other frameworks, since it involves internal tampering with disciplinary analysis, a recipe for confrontation. The framework based on the integration by leader is weakened by its demands on the leader. It tends to play down depth of analysis simply because a single individual cannot be expected to grasp the details of highly specialized analyses outside their own area of expertise (Rossini *et al.*, 1978). There will be a tendency for the research output to be multidisciplinary rather than interdisciplinary (Swanson, 1979).

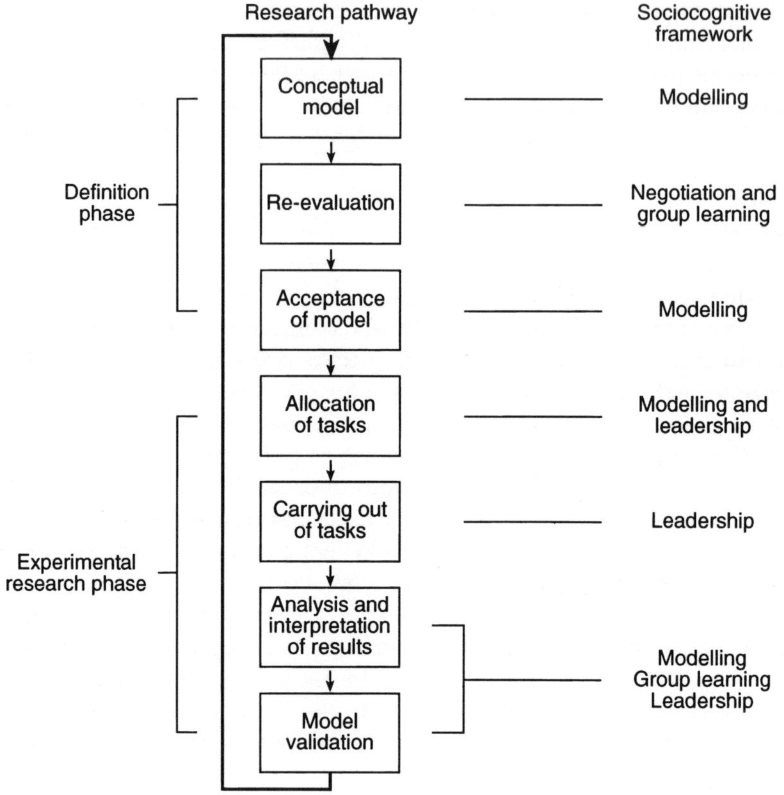

Figure 5.3 A scheme for the type of sociocognitive framework appropriate for each stage of a research programme. (After Dent, 1991).

5.4.3 Institutional effects

The institutes involved in a programme will also have a significant influence on the integration and success of the research undertaken. Different organizations have different structures, some of which are more conducive than others to the needs of interdisciplinary work. University departments with their emphasis on disciplinary expertise and reward systems that favour specialization rather than generalization and interdisciplinary activities, fare poorly as organizations that function to facilitate interdisciplinary participation (Rossini *et al.*, 1978; Tait, 1987; Dent, 1991, 1992). Small contract research organizations having flexibility, more appropriate internal structures (e.g. flexible accounting mechanisms), incentive and reward schemes, are by far the most conducive to interdisciplinary research programmes (Rossini *et al.*, 1978). The larger the organization and the more inflexible its internal structures, the less amenable is it likely to be to interdisciplinary team work.

Where a number of organizations are involved in developing a single IPM programme (e.g. Haskell, 1992) then the physical separation of functional groups may impede integration. Such difficulties faced by physical separation may be compounded by organizational distances, i.e. their interests, goals, procedures and professional orientation. However, all of these 'distances' can be overcome or at least mitigated by systematically ensuring good communication (section 5.5).

Setting up and maintaining an appropriate organizational structure is important if the disparate expertise of socioeconomists, scientists, extension workers, farmers and industrialists are to be brought together in order to develop integrated pest management programmes/systems. In the past research teams have rarely, deliberately selected in advance the intellectual and social components that determine the particular sociocognitive framework for their collaborative programme. More often, organization evolves into a stable pattern by trial and error (Swanson, 1979; Dent, 1992). However, a more positive, decisive approach to selecting a suitable, or a combination of suitable sociocognitive frameworks will undoubtedly improve the chances of successful interdisciplinary research.

5.5 PROGRAMME PLANNING AND MONITORING

A plan is a specified means of achieving goals that requires the simultaneous or sequential undertaking of several activities (Saaty and Kearns, 1985). Plans may be: (i) formal, in which case the problem is narrowed down and considered in terms of qualitative models and optimization techniques; (ii) incremental, which relies on qualitative reasoning and modification of existing approaches; or (iii) systemic, which attempts to combine positive elements of both formal and incremental planning. It involves an attempt to structure problems in terms of the whole range of influencing factors and their interactions, seeking convergence between an idealized optimal response and a feasible incremental one. The relative emphasis on the mix of planning philosophies used will depend on the particular system and pest management problems being addressed, but which ever planning approach is used it should provide a positive contribution to the progress of the programme. Planning is generally considered valuable because it provides an accurate picture of what is happening, enabling a manager to better anticipate problems and events. It can enhance coordination and commitment as well as providing the framework for programme monitoring, thus, improving the opportunity for completion on time according to set standards and budgets (Randolph and Posner, 1988).

5.5.1 Components of planning

A programme plan requires both short-term and long-term check points that both act as targets for achievement and as a measure of progress throughout the duration of the programme. The short-term checkpoints, referred to as events, are useful at the operation level to provide regular feedback on progress. By contrast milestones, longer-term checkpoints, provide more of an overview and are used to compare actual versus planned progress in a programme. They act as tangible measures of achievement such as the completed development of a monitoring device or a control measure. A milestone usually represents the culmination of activities which consisted of a number of events. For instance, development of a pheromone-monitoring device would have necessitated the identification of the pheromone, its synthesis, bioassay, trap design and evaluation. Each stage culminates in an event. Activities are the tasks that are undertaken that lead to an event and subsequently, milestones (Randolph and Posner, 1988).

In devising a plan it will be necessary to define the relationships between various activities, since some will need to be carried out in sequence to minimize delay, while others can be effectively worked on simultaneously. It is useful to ask questions about the most appropriate order for activities, and to look for different ways of approaching a problem, always bearing in mind your goals and the resources at your disposal.

5.5.2 Planning/monitoring tools and techniques

The need for planning in pest management has always been apparent but with the advent of IPM and the complexity of interdisciplinary programmes it now requires that the need should be met rather than just recognized. A commonly used tool in business and management planning that has been used in a similar role in pest management programmes (Dent, 1992) is the Gantt chart (or deliverables chart; FAMESA, 1984). This has three basic components: (i) a horizontal time line; (ii) a vertical list of activities; and (iii) a bar for the estimated length of time it will take to complete each activity (Figure 5.4). Gantt charts provide an easy-to-read visual picture of activities and events. They also can be used to depict the use of all forms of programme resources.

Although a valuable tool the Gantt chart is difficult to construct and cannot adequately deal with the inter-relationships between a number of activities. For instance, indicating that one activity cannot start until another is complete may be done by joining both bars by a dotted line. In complex programmes however, such an arrangement would quickly become unreadable. The charts are difficult to construct because locating an activity on a chart requires three simultaneous decisions: (i) the logical sequence of activities; (ii) the duration of each activity; and (iii) locating an activity in a position implies that resources are available to carry out the activity (Lockyer, 1984). To require a manager to make decisions on these three features at one time is to set a very difficult task, yet this is what is required when a Gantt chart is drawn.

Another method which does address the limitations of a Gantt chart is the network diagram, also known as the project network technique, the critical path method (CPM) or the programme evaluation and review technique (PERT). This technique was developed in the 1950s and although

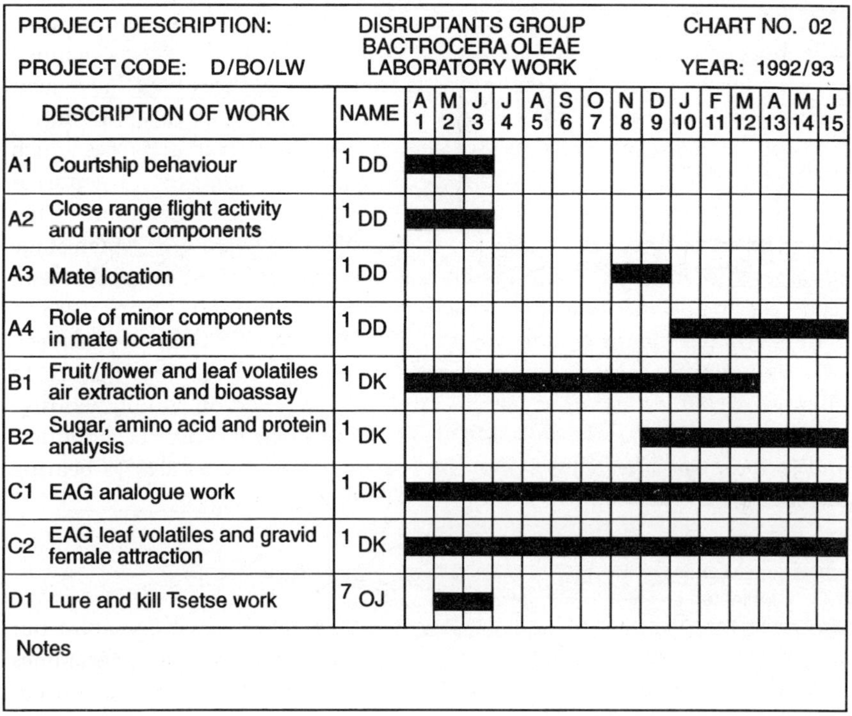

Figure 5.4 An example of a Gantt chart used in an IPM programme. (After Dent, 1991).

network diagrams do not provide a simple representation of a programme in the same way as a Gantt chart they are extremely useful in identifying and managing the sequence and flow of critical activities in a large or complex programme. The construction and use of these techniques has been covered in a number of textbooks (Lockyer, 1984; Lester, 1991; Lockyer and Gordon, 1991). The network analysis essentially consists of two basic steps: (i) drawing the network and estimating the individual activity times; and (ii) analysing these times in order to find the critical activities and the amount of slack in the non-critical ones.

The programme is represented by a diagram built up from a series of arrows and nodes (boxes or circles). Once the arrow diagram shows an acceptable logic, times are then given to the various activities of the diagram. A calculation is then carried out to discover the total time for the programme. If the total time exceeds that available then the activities that dictate this time (the control activities) are examined to determine whether they can be adequately shortened by using a different method or by changing the logic of the network itself (Lockyer, 1984). Computer software for these analyses is now commonly available.

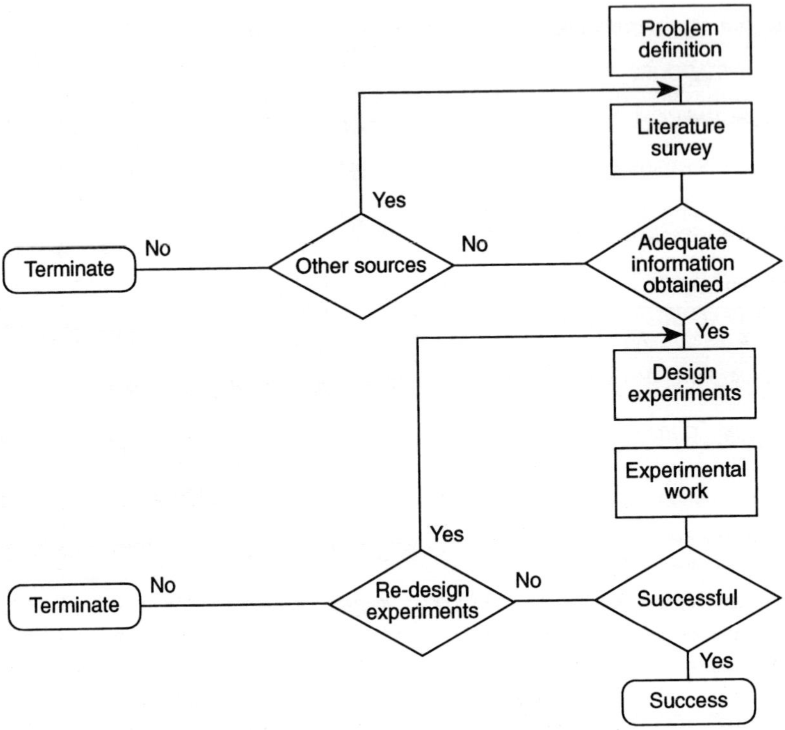

Figure 5.5 An example of a generalized research planning diagram. (After Pearson, 1988).

A further development of the network diagram, which uses commonly accepted symbols of relational diagrams (Forrester, 1961; Leffelaar and Ferrari, 1989) is the research planning diagrams (Pearson, 1988; Twiss, 1992) (Figure 5.5). Such diagrams are a valuable addition to R&D and the approach has been found most useful.

5.6 MANAGEMENT AND LEADERSHIP

The fact that leaders can have an immense effect on the performance of those under them, whether they are employees, soldiers or sports teams, has always been recognized. A leader serves to direct, coordinate, inspire and motivate their team towards achieving their goals, whatever they happen to be. A leader, coordinator or manager of an IPM programme takes on the same role, and requires the skills and abilities that will enable them to carry out this responsibility effectively.

While the power of good leadership to produce extraordinary results is a fact, it is difficult to produce facts about what it actually consists of (Dale, 1978). Karlöf (1987) provided a list of characteristics of good leaders cited by people with management experience (Table 5.3). However, throughout

Table 5.3 Characteristics cited by experienced managers as important qualities of leadership. (From Karlöf, 1987)

Open and extrovert	Calm
Inquisitive	Quick to understand
Result-orientated	Warm and empathetic
Decisive	Unbounded by prestige
Critical	Courageous
Ready to experiment and tolerate mistakes	Capable of bringing the best out in others
Confidence-inspiring	A good listener
Charismatic and enthusiasm inspiring	Unflappable
	Flexible

history recognized leaders have had only some of these characteristics and not others, or will have characteristics not included in this or similar lists. Defining the characteristics of successful leaders need not be helpful in identifying the skills and abilities required for successful management. An approach that has been more commonly considered useful is the categorization of leadership into a number of different styles. The classification varies according to which author you consider, but there are basically three types which will be referred to here as autocratic, democratic and laissez-faire (Hill, 1970). The autocrat tends to be authoritarian and self-centred, maintaining a close control over activities and allowing members of the team few opportunities for using their initiative or for independent action. The democratic leader's behaviour tends to be group-centred, providing opportunities for all group members and encouraging inter-relationships and cooperative activities. The laissez-faire style of leadership tends to be permissive, non-directive, with responsibilities and decision-making abdicated to the group as a whole. Common sense would inform us that the democratic leadership style is most suitable to the management of interdisciplinary programmes and the findings of the work by Rossini and colleagues confirm this. The groups that had democratic/facilitating leaders produced the most integrative reports followed by authoritarian and then laissez-faire leaders (Rossini *et al.*, 1978). The management style correlated with overall substantive integration, systemic integration and depth of analysis of the output. A number of other interesting points also emerged. For instance, it was easier for programme leaders who were respected by the team to manage democratically but the possession of stature within the organization also appeared to make the leaders' task easier. Having a solid intellectual grasp of the main features of each part of the overall task appears, quite understandably, to make the democratic leader's role easier to play. Social scientists and economists also came out as representing the most successful discipline for integrating the work in interdisciplinary studies.

The ability of the team leader to manage the interdisciplinary programme is obviously vital to its success. The skills required to do this competently are more often learned than intrinsic. There is a need to ensure that the best person for the job is found and where individuals with the natural flair are unavailable then training programmes may be considered appropriate.

5.6.1 Team dynamics

There will be many situations where a programme leader has a choice over the mix of expertise involved in an interdisciplinary team but not the actual individuals that are selected. This will be especially true where the programme is a collaborative exercise between a number of organizations. However, all individuals will not work equally well in an interdisciplinary situation, and where a leader has a choice then careful selection of members is required. In the first place the person must be secure and competent in their own disciplinary endeavours; secondly they must have a taste for adventure in the unknown and unfamiliar; and thirdly their interest must be fairly broad, at least in terms of what they feel is important (Petrie, 1976). Disciplinary competence and security are sometimes at odds with broad interests and imaginative speculation, largely because of the narrow disciplinary focus of graduate education. A useful blend of competence and broad interest is rare. The most suitable mix of competence, broad interests and adventurous spirit for an individual joining an interdisciplinary team will be difficult to define, but certainly no one characteristic should predominate, or it will eventually act to the detriment of the programme. For instance, an extremely competent specialist that is not also extremely adventurous and interested in the programme will inevitably be lured away to their own disciplinary interests, simply because they will perceive that the benefits accruing from the interdisciplinary work are insufficient to justify their commitment.

Behaviour tends to be directed towards satisfaction of individual needs. Hence, if specialists are to be encouraged to participate in interdisciplinary activities then they must be convinced that their essentially 'disciplinary' aspirations, desires, needs and goals will be met within the programme. Or at the least, that their disciplinary goals, etc. will not be jeopardized by involvement in the programme. In practice this means selecting team members whose goals, etc. can be met by the programme.

Scientists working in universities need to be able to publish papers and to have students carrying out research leading to higher degrees. Scientists in government research institutes and extension workers want tangible solutions to farmers' problems that are evidence of their productivity and value, while industrial partners want to see the development of marketable products for manufacture and sale (Dent, 1992). All these objectives will need to be met if the interdisciplinary effort is to be maintained. The participants must feel they are contributing something and that indicators and rewards for their achievement can be provided by the programme.

Members of the team will not only bring with them their expertise but

Table 5.4 The behaviour types required for effective group functioning. (From Belbin, 1981)

Type	*Role*
Chairman	The chairman acts as coordinator, and needs to be disciplined and balanced. He needs to work primarily *through* others.
Plant	The plant comes up with original ideas, is imaginative, and usually very intelligent. But plants can be careless of detail, and may resent criticism.
Shaper	The shaper stimulates others to act.
Monitor–evaluator	The monitor–evaluator assesses the quality of ideas or proposals.
Resource investigator	This person is very good at bringing in resources and ideas from outside. Usually extroverted and relaxed, but not original, or a driver. The team will usually have to pick up this person's contributions and run with them.
Team worker	Such a person is very important on the process side, and works to hold the team together.
Company worker	This is where the strength in practical organization lies. The company worker turns ideas into manageable tasks, and is a good administrator.
Finisher	Finishers are essential for group performance. They check details, and chase when deadlines look like being missed. For this reason they may be unpopular, despite their importance.

also a particular style of group interactive behaviour. People tend to prefer to 'play' a limited number of roles within a group (Cameron, 1991). Belbin (1981) suggested that for groups to work effectively eight behaviour types need to be present (Table 5.4). While it may not be possible to achieve the balance proposed by Belbin, it is important that the group works well together and hence, individuals should be selected not only on the basis of their expertise but also on the way in which they will interact in a group situation. It will be no good for instance, having a group composed of 'chairmen' types or 'plant' types – nothing would ever be achieved.

The dynamics of an interdisciplinary group should improve with time. It is not easy to collaborate across disciplines and to integrate effort and expertise; to do so requires a commitment to a common aim and more importantly the development of trust between the individuals concerned. Experience in IPM programmes shows that this can take as long as 2 years (Barfield *et al.*, 1987; Dent, 1992, 1994).

5.6.2 Communication

The effective management of a programme requires good communication so that everyone is kept informed and has the opportunity to influence

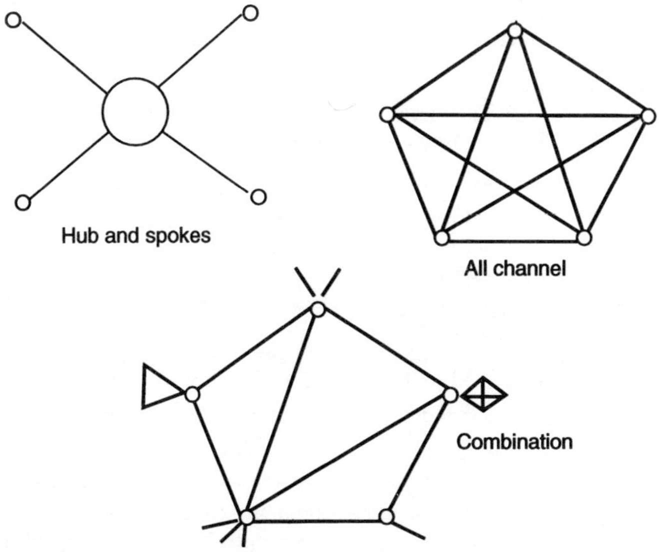

Figure 5.6 The two basic extreme communication patterns, hub and spokes and all channel, with an example of a combination pattern.

developments and decision making. The actual communication patterns adopted in a programme will be strongly influenced by the means by which management decisions are made. They will also inevitably change as the programme develops.

There are two basic extreme communication patterns, the 'all channel' pattern and the 'hub and spokes' pattern (Figure 5.6). The hub and spokes pattern is generally considered effective under conditions where simple factual information and simple tasks need to be communicated (Bavelas, 1950; Guetzkow and Simon, 1955). As the problem becomes more complex then the all channel pattern provides greater opportunities for effective communication. However, if the number of participants in an interdisciplinary programme is large, then maintenance of the all channel pattern of communication becomes prohibitive, simply because of the time and effort that needs to be expended to maintain each of the links; time and effort that could have a detrimental effect on research output. One of the key factors in developing a successful interdisciplinary programme is the need to have a core team of reasonably small size, just to ensure effective communication can be maintained. Where larger teams are considered necessary then support arrangements for core team members may be considered which will decrease the number of links when contrasted with a large all channel pattern (Figure 5.6).

In the analysis of interdisciplinary programmes carried out by Rossini

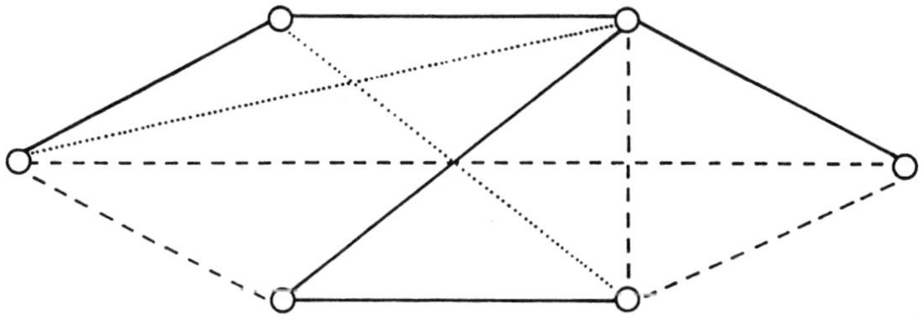

Figure 5.7 A combination communication pattern having channel links of various strengths; strong links (———); intermediate (- -); and weak links (······). (After Rossini *et al.*, 1978).

et al. (1978), one successful programme adopted a communication pattern similar to an all channel pattern but with links of varying strengths (Figure 5.7). A similar pattern emerged in a large programme involving 12 core participants developing an IPM programme for European olives (Haskell, 1992; Dent, 1994). In general, however, the study of Rossini and colleagues indicated that teams showed a preference for all channel patterns followed by an intermediate and then hub and spokes communication pattern. Certainly if the interdisciplinary nature of IPM programmes is to be established and maintained then there would be a need for all channel communication patterns aided by keeping the core team as small as technically possible (no more than five).

The establishment of effective channels of communication will greatly contribute to maintaining interaction among team members. The effectiveness with which they inter-relate and understand each other's subject will be more dependent however, on their recognition of discipline differences in terms of basic concepts, modes of enquiry, significance of problems, observational categories, representation techniques and standards of proof, i.e. disciplinary cognitive maps (Petrie, 1976). Quite plainly if team members do not share and understand their different cognitive maps then they will be quite unable to see the relevance of their colleagues' point of view to the problem at hand. Learning at least a part of others' disciplinary maps is a necessary condition for turning multidisciplinary research into interdisciplinary research. Petrie (1976) suggested that a failure to undertake such learning partly explains the relatively naive nature of so much interdisciplinary work. Failing to recognize the significant differences in cognitive maps and yet faced with the necessity for communication with each other on some level, the members retreat to the level of common knowledge and sense shared by all (common group learning; Rossini *et al.*, 1978).

The least that an interdisciplinary team member will need to learn is the observational categories of the other disciplines and the meanings of their key terms/concepts. The importance of learning the observational categories has been illustrated with the use of the ambiguous figure in Figure

Figure 5.8 Perception: a duck or a rabbit?

5.8 (Dent, 1991, 1992). Consider this figure, is it a rabbit or a duck or both? What happens if the observational skills developed in one discipline allowed an individual to see a rabbit and the observational skills developed in another discipline allow them to see a duck? What if they did not realize they were perceiving the same problem in completely different ways? The chances are that in such circumstances the view of others may be considered foolish, or further misunderstandings could arise when each blames the other for lack of intelligence, deceptiveness or insincerity (Swanson, 1979). Thus, it is important for individuals involved in interdisciplinary teams to learn and understand each others' observational categories.

Communication is only really possible where everyone involved can use the same language, i.e. the vocabulary and terms used in conversation and discussion relating to IPM and its component disciplines. Each discipline has its own, often very specific and hence unfamiliar terminology. Hence in order to understand what specialists from another discipline are saying it will be necessary to learn the important key terms that are used (Apple and Smith, 1976). Dent (1992, 1994) considered that it would be worthwhile for each new programme to compile a compendium of basic concepts and key terms used by each of the disciplines involved in the programme, as a precursor to easy initial communication. Such conscious attention to the problem of communication represents a necessary and important means of bridging the gap between disciplines and promoting an interdisciplinary perspective.

5.7 RUNNING THE PROGRAMME

The ease with which an IPM programme can be managed will be largely dependent on how well it has been established. The setting up and organization of a multidisciplinary research programme, especially its first 6 months, are the most difficult but also the most crucial to its future development and likelihood of success. If the research manager can put in place appropriate organizational structures, sociocognitive frameworks, communication patterns, monitoring techniques, and has been able to select a good team of open-minded scientists, with sufficient resources for completing the required tasks, then the job of running the programme will be largely dependent on the team's ability to carry through the science. However, if the manager has been unable to do these things then there is every likelihood that the science, although multidisciplinary, will be carried out by isolated individuals or groups and it will be difficult to coordinate any interdisciplinary effort or produce an integrated solution to the pest management problem. If IPM programmes are to produce integrated interdisciplinary answers to pest management problems then equal attention must be given to the research management of the programme as to the science. Good management will not compensate for poor research but a really successful R&D programme is dependent on both good science and good research management (Dent, 1994).

Where appropriate management structures are in place then the manager's role is largely one of maintaining communication between the different participants, reviewing progress and offering direction where appropriate. This may be achieved through the circulation of reports of research results, newsletters, E-mail, video conferences, letters, etc. Collaborations are however, based upon personal relations between partners and the establishment of working relations, and trust would be impossible without meetings (Barker, 1994). Furthermore, if collaboration is to extend beyond mere exchange of results to genuine interdisciplinary research then meetings and face-to-face discussions are essential. The arrangement for meetings, their form and frequency will depend on the size of the programme and hence the number of participants involved and the distance separating the participant organizations. Large programmes involving a number of organizations in different countries may find it difficult and costly to have frequent meetings of all participants and hence restrict meetings of the full programme to an annual plenary meeting for a detailed technical review. However, smaller groups would need to meet more frequently to promote collaboration on specific projects. Two programmes involving collaborative R&D studied by Barker (1994) held well-prepared meetings with pre-circulated material every 3–4 months and considered such events as the hub of the collaboration.

It is the face-to-face nature of such meetings which is important. To collaborate effectively there needs to be a good bonding and trust between the participants and this is best achieved through regular small group meetings (Clarke, 1994). A social element is also an essential part of these meetings for participants to be able to understand and to get to know each other. The manager also needs to develop good rapport with individual partners and the appropriate administrative and managerial staff within their participating organizations. In this way it becomes apparent whether different organizations have hidden agendas of which the manager should be aware.

Collaborative IPM programmes which aim to develop interdisciplinary solutions to pest problems require good research management. The complex nature of the problems, the multidisciplinary teams required to solve them and the collaborative nature of the work, all make research management necessary. Utilization of the approaches and techniques outlined in this chapter will assist in developing IPM programmes which can meet the demands of integration and hopefully in the process produce better targeted and appropriate IPM systems.

REFERENCES

Adams, M.E. (1982) *Agricultural Extension in Developing Countries*, Longman Scientific & Technical, London.

Andrews, K.L., Bentley, J.W. and Cave, R.D. (1992) Enhancing biological control's contributions to integrated pest management through appropriate levels of farmer participation. *Florida Entomologist*, 75(4), 429–39.

Apple, J.L. and Smith, R.F. (1976) *Integrated Pest Management*, Plenum Press, New York.

Baldwin, B. (1986) Commercialisation of microbially produced pesticides. The World Biotech Report 1986: Proceedings of Biotech '86 held in London, May 1986. Vol. 1, Applied Biotechnology. Online Publications, London, New York, pp. 39–49.

Barfield, C.S., Cardelli, D.J. and Boggess, W.G. (1987) Major problems with evaluating multiple stress factors in agriculture. *Tropical Pest Management*, **33**(2), 109–18.

Barker, K. (1994) Management of collaboration in ECR & D projects, in *Management of Collaborative European Programmes and Projects in Research, Education and Training University of Oxford – An International Conference*, Oxford, UK, 10–13 April 1994 (in press).

Bavelas, A. (1950) Communications patterns of task oriented groups. *Journal of the Acoustical Society of America*, **22**, 725–30.

Belbin, R.M. (1981) *Management Teams*, Heinemann, Oxford.

Cameron, S. (1991) *The MBA Handbook. An Essential Guide to Effective Study*, Pitman Publishing, London.

Carlson, G. and Rodriguez, R. (1984) Farmer adjustments to mandatory pest control, in *Pest and Pathogen Control Strategic, Tactical, and Policy Models*, (ed. G.R. Conway), John Wiley & Sons, New York, pp. 429–40.

Chambers, R. (1983) *Rural Development: Putting the Last First*, Longman Scientific & Technical, Harlow, England.

Chambers, R., Pacey, A. and Thrupp, L.A. (1989) *Farmer First Farmer Innovation and Agricultural Research*, Intermediate Technology Publications, London.

Clarke, A.M. (1994) The human factors of project management, in *Management of Collaborative European Programmes and Projects in Research, Education and Training University of Oxford – An International Conference*, Oxford, UK, 10–13 April, 1994 (in press).

Dale, E. (1978) *Management: Theory and Practice*, 4th edn, McGraw-Hill, Singapore.

Dent, D.R. (1991) *Insect Pest Management*, CAB International, Wallingford.

Dent, D.R. (1992) Scientific programme management in collaborative research, in *Research Collaboration in European IPM Systems*, BCPC Monograph No. 52, (ed. P.T. Haskell), BCPC, Brighton, pp. 69–76.
Dent, D.R. (1994) Strategic management of an olive IPM programme, in *International Symposium on Crop Protection*, Gent University, 3 May 1994, p. 57.
FAMESA (1984) *Management Manual for Productive R & D*, ICIPE, Nairobi.
Forrester, W.O. (1961) *Industrial Dynamics*, MIT Press, Massachusetts.
Goodell, G. (1984) Challenges to international pest management research and extension in the Third World: do we really want IPM to work? *Bulletin of the Entomological Society of America*, (Fall), 18–26.
Goodell, G.E. (1989) Social science input into IPM. *Tropical Pest Management*, **35**(3), 252–3.
Graham-Bryce, I.J. (1981) The current status and future potential of chemical approaches to crop protection. *Phil. Transactions of the Royal Society of London*, **295**, 5–16.
Guetzkow, H. and Simon, H.R. (1955) The impact of certain communication sets upon organisation and performance of task oriented groups. *Management Science*, **1**, 233–50.
Haskell, P.T. (1992) *Research Collaboration in European IPM Systems*, BCPC Monograph 52 edn, BCPC, Farnham, Surrey.
Headley, J.C. (1985) Cost-benefit analysis: defining research needs, in *Biological Control in Agricultural IPM Systems*, (eds, M.A. Hoy and D.C. Herzog), Academic Press Inc., Orlando, pp. 53–63.
Heap, J.P. (1989) *The Management of Innovation and Design*, Cassell Educational Ltd, Guildford.
Hill, S.C. (1970) A natural experiment on the influence of leadership behavioural patterns on scientific productivity. *IEEE Transactions on Engineering Management*, **17**, 10–20.
Karlöf, B. (1987) Business Strategy in Practice, John Wiley & Sons, New York.
Kolodny-Hirsch, D.M. and Schwalbe, C.P. (1990) Use of disparlure in the management of the gypsy moth, in *Behavior-Modifying Chemicals for Insect Management*, (eds R.L. Ridgway, R.M. Silverstein and M.N. Inscoe), Marcel Dekker Inc., New York, pp. 363–85.
Koyama, J., Teruya, T. and Tanaka, K. (1982) Eradication of the oriental fruit fly (Diptera: Tephritidae) from the Okinawa Islands by male annihilation. *Journal of Economic Entomology*, **77**, 468–72.
Leffelaar, P.A. and Ferrari, T.J. (1989) Some elements of dynamic simulation, in *Simulation and Systems Management in Crop Protection*, (eds R. Rabbinge, S.A. Ward and H.H. van Laar), Pudoc, Wageningen, pp. 19–45.
Lester, A. (1991) *Project Planning and Control*, 2nd edn, Butterworth-Heinemann Ltd, Oxford.
Lockyer, K. (1984) *Critical Path Analysis*, Pitman Publishing, London.
Lockyer, K. and Gordon, J. (1991) *Critical Path Analysis and Other Project Network Techniques*, Pitman Publishing, London.
Matthews, G.A. (1984) *Pest Management*, Longman Inc., London.
Neuenschwander, P. and Herren, H.R. (1988) Biological control of the cassava mealybug, *Phenacoccus manihoti*, by the exotic parasitoid *Epidinocarsis lopezi* in Africa. *Phil. Transactions of the Royal Society of London*, **318**, 319–33.
Norton, G. and Mumford, J.D. (1982) Information gaps in pest management, in *Proceedings of the International Conference on Plant Protection in the Tropics, Kuala Lumpur, Malaysia*, 1–4 March, 1982, pp. 589–97.
Norton, G.A. (1982) Crop protection decision making – an overview, in *Decision*

Making in the Practice of Crop Protection, (ed R.B. Austin). BCPC Publications, pp. 3–11.

Norton, G.A. (1990) Decision tools for pest management: their role in IPM design and delivery. (Paper presented to the FAO/UNEP/USSR Workshop on Integrated Pest Management).

Norton, G.A. and Mumford, J.D. (1993) *Decision Tools for Pest Management*, CAB International, Oxford.

Pearson, A.W. (1988) Managing research and development, in *The Gower Handbook of Management*, 2nd edn, (eds D. Lock and N. Farrow), Gower Publishing Co Ltd, London, pp. 387–403.

Petrie, H.G. (1976) Do you see what I see? The epistemology of interdisciplinary inquiry. *Journal of Aesthetic Education*, **10**, 29–43.

Randolph, W.A. and Posner, B.Z. (1988) (eds) *Effective Project Planning and Management*, Prentice Hall, Englewood Cliffs, p. 163.

Reed, W., Davies, J.C. and Green, S. (1985) Field experimentation, in *Pesticide Application: Principles and Practice*, (ed. P.T. Haskell). Clarendon Press, Oxford, pp. 153–74.

Reichelderfer, K.H., Carlson, G.A. and Norton, G.A. (1984) Economic guidelines for crop pest control. FAO Plant Production and Protection Paper **58**.

Ridgway, R.L., Silverstein, R.M. and Inscoe, M.N. (1990) *Behavior-Modifying Chemicals for Insect Management*, Marcel Dekker Inc., New York.

Rossini, F.A., Porter, A.L., Kelly, P., Chubin, D.E., Carpenter, S.R. and Lipscomb, M.A. (1978) *Frameworks and Factors Affecting Integration within Technology Assessments, Reports and Appendices*, National Technical Information Service.

Ruthenberg, H. and Jahnke, H.E. (1985) *Innovation Policy for Small Farmers in the Tropics*, Oxford University Press, Oxford.

Saaty, T.L. and Kearns, K.P. (1985) (eds) *Analytical Planning*. Pergamon Press, Oxford, New York, p. 208.

Saul, S.H. (1989) Genetics of the Mediterranean fruit fly (*Ceratitis capitata*) (Wiedemann), in *Genetical and Biochemical Aspects of Invertebrate Crop Pests*, (ed. G.E. Russell), Incercept, Hampshire, pp. 1–36.

Scopes, N.E.A. and Biggerstaff, S.M. (1971) The production, handling and distribution of the whitefly *Trialeurodes vaporariorum* and its parasite *Encarsia formosa* for use in biological control programmes in glasshouses. *Plant Pathology*, **20**, 111–16.

Southwood, T.R.E. and Norton, G.A. (1973) Economic aspects of pest management strategies and decisions. *Mem. Ecol. Soc. of Australia*, **1**, 168–84.

Steiner, L.F., Mitchell, W.C., Harris, E.J., Koyama, T.T. and Fumimoto, M.S. (1965) Oriental fruit fly eradication by male annihilation. *Journal of Economic Entomology*, **58**, 961–4.

Sumberg, J. and Okali, C. (1989) Farmers, on-farm research and new technology, in *Farmer First. Farmer Innovation and Agricultural Research*, (eds R. Chambers, A. Pacey and L.A. Thrupp). Intermediate Technology Publications, London, pp. 109–14.

Swanson, E.R. (1979) Working with other disciplines. *American Journal of Agricultural Economics*, **61**(5), 849–59.

Tait, E.J. (1987) Planning an integrated pest management system, in *Integrated Pest Management*, (eds A.J. Burn, T.H. Coaker and P.C. Jepson), Academic Press, London, pp. 189–207.

Twiss, B. (1992) *Managing Technological Innovation*, 4th edn, Pitman, London.

Ushio, S., Yoshioka, K., Nakasu, K. and Waki, K. (1984) Eradication of the oriental fruit fly from Amami Islands by male annihilation (Diptera: Tephritidae). *Japanese Journal of Applied Entomological Zoology*, **26**, 1–9.

Vickers, R.A., Rothschild, G.H.L. and Jones, E.L. (1985) Control of the oriental fruit moth, *Cydia molesta* (Busck) (Lepidoptera: Tortricidae), at a district level by mating disruption with synthetic female pheromone. *Bulletin of Entomological Research*, 75, 625–34.

CHAPTER 6

Techniques in systems analysis

D.R. Dent

6.1 INTRODUCTION

Originally developed during World War II as part of operations research to deal with long-range military problems, systems analysis is used to identify appropriate courses of action by systematically examining costs, effectiveness and risks of alternative strategies and designing additional ones if those examined are found wanting (Heong, 1985). Since World War II, the techniques of systems analysis have been widely used by a range of disciplines including business managers, engineers, sociologists and scientists. However, despite this diverse use, the definition of systems analysis provided by Quade and Boucher (1968) is still generally applicable, 'a systematic approach to helping a decision maker choose a course of action by investigating [the] full problem, searching out objectives and alternatives and comparing them in the light of their consequences, using an appropriate framework – in so far as possible analytic – to bring expert judgement and intuition to bear on the problem'. It is a definition that is considered applicable here and is highly relevant to the problems of renewable resource management, which includes integrated pest management.

The application of systems analysis to the problems of pest management has been viewed as a merging of four disciplines: ecological theory, quantitative population biology, economic theory and the computer techniques of operations research (Getz and Gutierrez, 1982). Ecologists realized the relevance of systems analysis to the problems of pest management as early as the 1960s (Watt, 1961, 1962, 1966) and there has been a growing increase in the use of these techniques since that time. Those involved in approaching pest management from an economic perspective started to apply the

Integrated Pest Management Edited by David Dent. Published in 1995 by Chapman & Hall, London. ISBN 0 412 57370 9

techniques in the 1970s (Conway, 1976; Regev *et al.*, 1976; Shoemaker, 1976). Both the ecologists and the economists approached systems analysis in a quantitative way, making use of mathematical models, and, as they became more commonplace, computers. Computer models have become a key component of most (but not all) systems analyses.

6.1.1 Models

A model is a representation of a system. It attempts to mimic the essential features of a particular system where a system is taken as a limited part of reality (see Chapter 1). Models may be classified in a number of different ways, according to the type of problem to which they are being applied, to the mathematical or computer techniques employed, to the spatial scale on which they operate or according to the disciplinary perspective of the 'scientist' developing them. No one classification is better than any other, they merely reflect the various ways in which models can be used and categorized according to particular needs and perspectives.

One of the major divisions in the categorization of models are those that involve an algorithm and those that do not, where an algorithm may be considered as a fixed rule for solving a mathematical problem. Algorithms are the formulae used in statistical procedures, e.g. a regression equation, and in linear and goal programming techniques. They provide a prescribed solution structure, often the convergence of solutions to an optimal situation. Pest management simulation models do not involve a fixed algorithm, but are based on use of a numerical search method which is highly flexible (Rossing, 1989). However, with simulation models an optimal solution is not derived from the final end point of the analysis, and some regard this as a serious limitation in their use. Simulation models also may be divided according to whether they utilize the multi-trophic – supply-demand approach (Getz and Gutierrez, 1982; Chapter 10) or the state variable-relational diagram approach (Rabbinge *et al.*, 1989). The choice will greatly affect the form and type of output from the model.

Models may also be categorized according to their intended function, so that a model may be described as a population phenology model (Yencho *et al.*, 1986; Feng *et al.*, 1988), an insect dispersion model (Rudd and Gandour 1985), a damage model (Barnard *et al.*, 1986), a yield-loss model (Yencho *et al.*, 1986) or a management model (Southwood, 1978; Rabbinge and de Wit, 1989) to name but a few. Many of the models that are designed to aid scientist or extension/farmer decision making are referred to as management models. These may involve simple rules for decisions on whether or not to apply an insecticide, or more complex models used to evaluate the impact on a pest population of a range of control measures used singly or in combination (e.g. Kidd and Gazziano, 1992). Such models cover a wide range of circumstances. Classification of models according to their function tends to provide rather an arbitrary assemblage of models often with little in common since the categories are developed which best describe the role of each specific model for a particular situation or publication. Any formal categorization of models along functional lines would be purely dependent on the perspective of, and the emphasis given by, the individual devising the scheme; what one individual would call a damage model could be a yield loss or management model to another.

Conway (1984) described a scheme based on the geographical scale and time span over which pest management models operated. He described three categories of models in pest management; tactical, strategic and policy models (Table 6.1). The tactical models are those that deal with problems applicable to individual crops and/or fields on a particular farm. They aim to offer advice on the day-to-day management of a particular crop in a specific field. Typically they are used as an aid to farmer decision making, usually to advise on whether or not to apply a pesticide (Dent, 1991). Tactical models tend to require extensive research and rely on farmers collecting pest, crop and environmental data as inputs for the model. Their use at present is restricted to only intensively studied systems. However, as more extensive data become available then sufficient will be known to develop tactical models for a greater number of systems.

Strategic models are used to elucidate more general guidelines and principles which are applicable over a wider range of circumstances. They may be divided into three types, deductive, inductive and mixed deductive–inductive (Southwood, 1978). A deductive model utilizes accepted general knowledge and theory and from this, and the use of additional inputs that appear relevant, attempts to develop a model that accords with reality. In this way the model is used to determine how a system functions and the key components involved in this (e.g. Hassell, 1978). Inductive models differ in that they use available experimental data and infer from this the way in which the system is structured and functions. Although inductive models can only deal with experimental data, which by its very nature is limited in time and space, they have been used to try to elucidate more general principles (e.g. Gilbert and Gutierrez, 1973). Mixed deductive–inductive models utilize, as the name suggests, a combination of general theory and experimental data (e.g. Hartstack and Hollingsworth, 1974; Southwood and Comins, 1976). These three types of strategic model all attempt to predict an outcome from a given set of initial conditions and in doing so may identify links which characterize the system and hence, the interactions central to an underlying strategy. Strategic models have been widely used in studies of predator–prey interactions (analytical models) (sections 6.3.1 and 7.5) and to determine the relative reliance that can be placed on different control measures in different circumstances.

Policy models deal with medium- to long-term questions in pest management at the regional, national and international levels (Conway, 1984). Few have been developed despite the potentially important contribution they can make to science policy. They are required for the evaluation of

Table 6.1 Hierarchical arrangement of decision-making models in pest and pathogen control (after Conway, 1984)

Level	Field or individual	Farm or subpopulation	Region or settlement	National	International
Decision maker	Farmer, medical or veterinary practitioner (or patient)	Farmer, medical or veterinary practitioner	Regional agricultural service or health board	Department/ Ministry/agency of Agriculture or Health	FAO/WHO
			Policy	Policy	Policy
Appropriate models		Strategy	Strategy		
	Tactics	Tactics			

important current issues in IPM such as the utilization and impact of genetic manipulation techniques and the effects of global warming on pest problems, to name but a few. Examples include Regev (1984) and Comins (1984).

Models can also be categorized in a more general way, for instance, as ecological or economic models. The perspective of the discipline of the scientists designing the model will have a major influence on its form and output. This is particularly evident when scientists from different disciplines seemingly develop models to address the same problem. Mumford and Norton (1984) provide an excellent illustration of the difficulties with the economic threshold concept. The entomologist Stern and his colleagues (Stern *et al.*, 1959) defined the economic threshold as the level to which a given pest population should be reduced to achieve the point where marginal revenue should just exceed costs. Economists addressing the same concept focused on a completely different aspect of the problem. They have taken the pest population as 'given' and asked 'what level of control is most profitable for that particular pest density, i.e. the optimal level of control'. This difference in focus has tended to confuse matters regarding the economic threshold concept (Mumford and Norton, 1984), largely caused by difference in disciplinary perspectives of economists and entomologists.

One of the most commonly acknowledged classifications of models is that based on their division into statistical, analytical, simulation and optimization models. This classification has been slightly modified by Norton *et al.* (1993) combining analytical and simulation models to provide three categories, statistical models, mechanistic models and optimization models. This classification is used here in a consideration of each of the different types of model relevant to systems analysis in pest management.

6.2 STATISTICAL MODELS

Statistics deals with techniques for collecting, analysing and drawing conclusions from data. Statistical models serve to facilitate the handling of data for statistical procedures used in the analysis of these data. Although utilized in all manner of statistical procedures, such as analysis of variance and covariance and factorial analysis, the most explicit use of statistical models in pest management and systems analysis is the application of regression and multiple regression models.

The regression model is specified by the equation

$$Y = \alpha + \beta x + \epsilon \tag{6.1}$$

where α = intercept; β the slope; Y and x the variables under study and ϵ an estimate of error.

Sokal and Rohlf (1981) cite six different applications for regression analyses: study of causation, description of scientific laws, prediction, comparison of independent variates, statistical control and substitution of variables. Among these, study of causation, description of scientific laws (or at least relationships) and prediction, constitute the most common uses of regression models in systems analysis.

A great deal of scientific effort is concerned with the relations between pairs of variables, the development and evaluation of hypotheses of cause and effect. The effect of environmental variables on pest development is one common example. Insect development rates are highly correlated with temperature and in many cases of linear regression can be used to define an equation describing the relationship (e.g. Stinner *et al.*, 1974; Reissig *et al.*, 1979).

Where more than one independent variable is thought to be involved then multiple regression may be used to establish the subset that provides the best linear prediction equation or to rate the variables in order of their importance (Snedecor and Cochran, 1978).

Regression analysis also permits the definition of scientific laws, such as Taylor's power law which addresses the distribution of individuals in natural populations. Taylor (1961, 1965, 1971) demonstrated that the distribution of individuals in natural populations is such that the variance is not independent of the mean, i.e . if the mean and variance of a series of samples are plotted, then they increase together. This relationship is described by

$$S^2 = \mathrm{a}x^{\mathrm{b}} \tag{6.2}$$

where S^2 = variance, x = mean, and a and b are constants, with 'a' largely a sampling factor and 'b' an index of aggregation characteristic of the species. Since Taylor's power law holds so widely the value 'b' can be used to determine precise transformations of distribution data to permit use of a parametric statistical method (Healy and Taylor, 1962). In contrast to Taylor's power law most regression lines are empirically fitted curves and the constants necessary to fit these curves do not possess any clear inherent meaning. The functions simply represent the best mathematical fit (usually by the criterion of least squares) to an observed set of data. However, if they do adequately describe the relations between variables then they may provide some useful insight into important phenomena. A good example of the use of regression equations in defining the relationships is found with pest infestation and crop damage or yield loss data where regressions are used in the development of economic thresholds (e.g. Gutierrez and Daxl, 1984; Barnard *et al.*, 1986). Such yield-loss functions have been sought for insects (Ingram, 1980; Yencho *et al.*, 1986), pathogens (Vogel *et al.*, 1993) and weeds (e.g. Fischer *et al.*, 1993).

One of the most common uses of regression models in pest management has been in the prediction of pest outbreaks. They have been used successfully in entomology, with the degree of infestation in spring of *Aphis fabae* predicted from the size of the spring migration (Way *et al.*, 1981) and the peak numbers of the cereal aphid *Sitobion avenae* predicted from the rate of increase of the aphid per tiller per day (Entwistle and Dixon, 1986). However, plant pathology has perhaps an even longer history of successful forecasting procedures based on regression analysis (Eversmyer and Burleigh, 1970; Royle, 1973) and multiple regression (Butt and Royle, 1974). Despite a large number of such successes it is clear however, that regression models have a number of potentially serious defects (Conway, 1984). The most basic is that the equation produced from the regression analysis can be used predictively, provided the conditions under which the predictions are made are within the range of data sets from which the relationships were originally derived. The model will work well until a combination of circumstances arise that are outside the original data sets, and then the model will fail. Also, where important causal relationships tend towards non-linearity and the time-scale of prediction is sufficient to allow their expression then again the models will eventually fail. Within these limitations however, regression and multiple regression models are of great practical utility in pest management and will undoubtedly continue in the future to play a significant role in the quantitative techniques used in systems analysis.

6.3 MECHANISTIC MODELS

According to the classification of Norton *et al.* (1993) mechanistic models include analytical, rule-based and simulation models. They are characterized by their representation of underlying, fundamental biological and ecological mechanisms or processes and their attempt to include biological realism. They are particularly useful for identifying and evaluating the mechanisms that underlie pest population dynamics, damage and/or the effects of different control measures.

6.3.1 Analytical models

Analytical models are those models for which a closed-form solution can be obtained; the results of the model can be written as an algebraic expression involving parameter values. Shoemaker (1984) gives the logistic model as an example, where

$dN/dt = rN\ (1 - N/k)$ has the closed form solution (6.3)

$$N_{(t)} = \frac{K}{[1 + c\ \exp\ (-rt)] \times 100}.$$

The model in this form provides an immediate understanding of the impact of the changes in the parameter K (carrying capacity) on the population density ($N_{(t)}$). Models such as the logistic equation tend to utilize only small numbers of relationships and parameters but despite this they are used to obtain general information about a system.

Most approaches to the use of analytical models have been based on differential or difference equations (Jones, 1993). Differential equations make use of calculus which assumes that a process is continuous. Hence, differential models are used in situations where organisms have overlapping generations or more or less continuous births and deaths, e.g. in tropical climates or heated glasshouses. However, for many pest populations, especially those in temperate areas, processes such as reproduction or development may occur at discrete time intervals. The effects of these time delays are best illustrated by difference equations (Hassell, 1978). A greater emphasis has been placed on developing analytical models based on difference equations. One of the major applications of these models in pest management has been the development of predator–prey, host–parasitoid theory (e.g. Hassell, 1978) (section 6.5), but they have also been used to explore interactions of competing species and to address many questions involved in epidemiology (Getz and Gutierrez, 1982), particularly human disease epidemiology (Anderson and May, 1991).

Major criticisms of analytical models are the inability to relate their output to any real system and the dearth of experimentation to verify the models (Getz and Gutierrez, 1982; Waage, 1986). Analytical models are also seriously limited in their application to many situations because it is usually not possible to find closed-form solutions for a series of non-linear equations with a large number of variables (Shoemaker, 1984). Despite these drawbacks however, analytical models fulfil an important role in producing a conceptual framework for discussion of broad classes of phenomena (May, 1974). Such frameworks are essential for indicating key areas or relevant questions and contentious issues for the scientists of all disciplines involved in pest management.

6.3.2 Simulation models

In agricultural pest management the most widely applied modelling techniques have been simulation models (Shoemaker, 1984). One of the reasons for this is probably that the simulation approach does not require a high degree of mathematical sophistication. Hence, simulation models are a mathematical tool accessible to most biologists when they need to address quantitative problems in systems analysis (Getz and Gutierrez, 1982). To this end biologists have made good and effective use of the techniques over the last 20 years. Although the majority of models are used to simulate the population dynamics of crops, pests (Holt *et al.*, 1987; Kendall *et al.*, 1992), natural enemies of pests (Vorley and Wratten, 1985) or combinations of these (Holt and Norton, 1993) a number have been developed to evaluate the impact of various control measures either to produce practical pest management tools (Allen, 1981; Sawyer and Fick, 1987; Feng *et al.*, 1988;

Wall *et al.*, 1993) or as a means of more effectively directing research effort (Carruthers *et al.*, 1986; Kidd and Gazziano, 1992). In these different ways simulation models have now become an integral part of the investigative and problem-solving procedures of pest management.

Simulation techniques are highly versatile. They can be developed and run in real-time, utilizing up-to-date inputs from the field, or they can utilize historical data and evaluate situations retrospectively. Deterministic models (where parameter values included in the model are input as constants) are useful in retrospective analysis of decisions (Watt *et al.*, 1984). Alternatively parameter values can be based on stochastic processes, where parameter values such as fecundity, mortality or immigration are drawn from a distribution (ideally reflecting the extent of variation found in nature). Thus, in direct contrast with deterministic models, where the model always yields the same answer for a given set of inputs, stochastic models will always provide a different answer reflecting the levels of introduced variation. Since stochastic models tend to become mathematically intractable (Maynard-Smith, 1974) most simulation models are deterministic in nature.

A further dichotomy of approach is seen in the construction of the model either using the state variable relational diagram approach or the multitrophic supply–demand concept. The state variable approach is probably more widely used and is based on the assumption that the state of a system at any point in time can be quantified and that changes in state can be described by a mathematical equation (Rabbinge and de Wit, 1989). Models are constructed using state, rate and driving variables all of which can be depicted in relational diagrams (Figure 6.1). State variables are quantities such as number of eggs, larvae and adults that characterize a system at a particular time. Each state variable is associated with a rate variable which characterizes its rate of change at a certain instant, e.g. the proportion of eggs that hatch. Rate variables may be constant or be functions of other variables, e.g. a density-dependent function. Driving variables characterize the effect of the environment on the system at its boundaries and include such effects as weather, and pest immigration and emigration. The simulation works by the movement of individuals for instance, between different states (e.g. developmental stages) according to a rate set by the rate variable determined over a number of time steps, i.e. the state variable at time $t + \Delta t$ equals the state variable at time t plus the rate at time t multiplied by Δt (Rabbinge and de Wit, 1989). Thus, over an appropriate number of time steps individuals (with allowance for some mortality) will move between stages, e.g. eggs to larvae to pupae to adults. For the purposes of computation the relational diagram is represented by a series of equations.

The alternative to the relational diagram representation is the supply–demand concept used in multitrophic models (Gutierrez *et al.*, 1991a,b; Graf *et al.*, 1992) (Chapter 11). Most models ignore trophic level effects that include the influence of weather on the dynamics of herbivores and

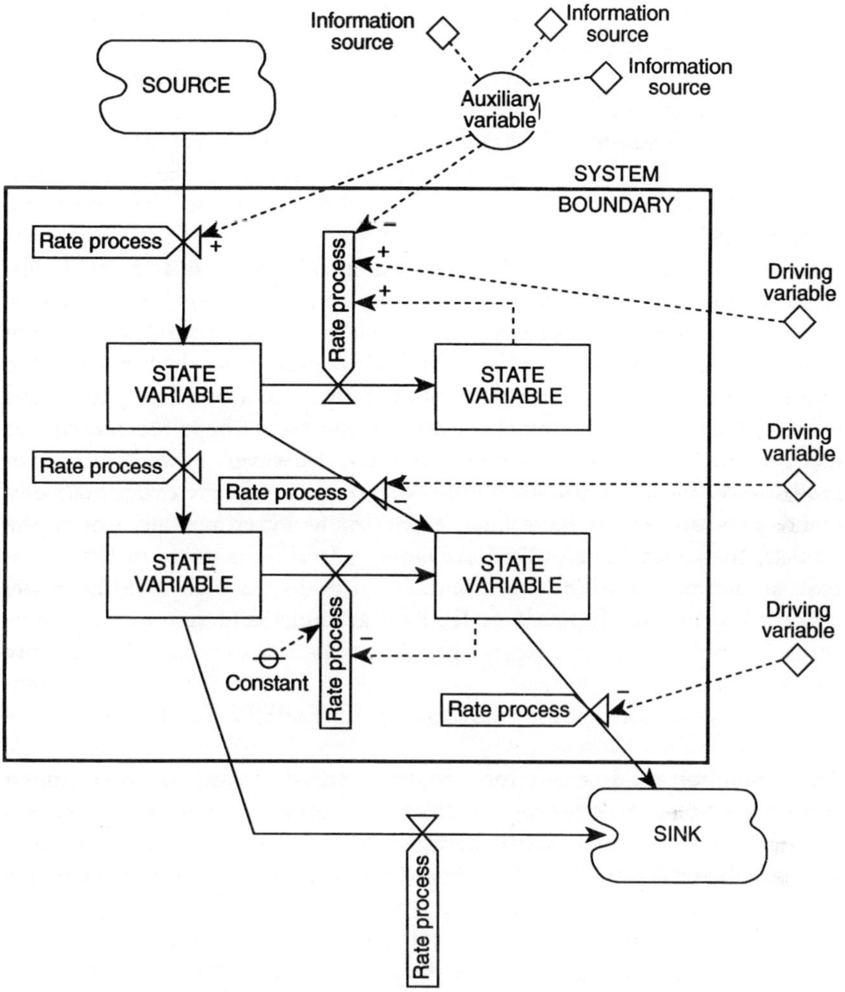

Figure 6.1 A hypothetical system showing some of the characteristic components. State variables: quantities (such as population numbers) that can be measured at any point in time. Rate processes: processes that cause changes in the levels of state variables through time (for example, mortality, reproduction, immigration, and emigration). Driving variables: factors that affect the rate of operation of such processes, but that operate from outside the system as defined. Constants: fixed values that may affect also the rate of operation of driving variables. Sources and Sinks: pools of material that feed into or receive material from state variables. Material flows (solid lines): the actual movement of material within the system and across the System Boundary. Information flows (dashed lines): influences on the rate of operation of rate processes that sometimes are grouped conveniently into named Auxiliary variables. (After Kitching, 1991).

their natural enemies, whereas the supply–demand concept makes greater attempts towards this biological realism rather than mathematical elegance and high level theory (Gutierrez *et al.*, 1993). This approach involves building simulation models with a modular structure that consist of assemblages of similar sub-models that describe the mass and energy flow through the system (Godfray and Waage, 1991). However, such models are difficult to parameterize and to validate and are expensive on resources (Holt and Norton, 1993).

In general, the drawbacks of simulation models are those associated with constructing models that are too large and complex so that the output is difficult to interpret (models must be kept as simple as needed to do the job), or lack of awareness of key biological processes which renders the model useless or worse misrepresents the real situation. The power and versatility of modern computers lend themselves, perhaps too readily, to the construction of more complex models. However, where simulation models have been maintained within reasonable bounds of complexity and where time and effort have been expended to experimentally verify the models, then they have proved extremely useful. It is likely in the future that simulation modelling will continue to be increasingly used by those involved in the development of IPM programmes and systems.

6.4 RULE-BASED MODELS AND EXPERT SYSTEMS

The simulation models described above are used to simulate how system components respond over time to changes in other components. The components are represented by real number variables and changes in variables are described by equations. Rule-based models are the qualitative equivalent of a simulation model in which components are represented by a small set of discrete states and changes are described by 'IF–THEN' type rules (Holt and Day, 1993). Quantitative data for simulation models are replaced in a rule-based approach, with subjective knowledge about a system. This has the advantage that pest managers can themselves structure and evaluate the models, and hence they tend to be readily comprehensible. Using basic information the components of the system are identified and the range of states considered feasible are assigned to each component. For instance, a pest population size (the component) attacking a crop may be considered to be small, intermediate, large or extremely large (the different states). On the basis of these possibilities a number of options are then presented for the user to choose from. This basic building block of an 'IF–THEN' type rule is also fundamental to the construction of expert systems. In this way rule-based models and expert systems are very similar (Holt and Day, 1993).

The use of expert systems in pest management has recently been reviewed by Mumford and Norton (1993). They consider that an expert system mimics the processes employed by a human expert in diagnosing a problem and in giving advice. The intention is to quantify and make explicit the insight and knowledge of the experienced decision maker in a form that is accessible to the farmer/decision maker. Expert systems are usually computer-based but they can be produced as charts, matrices or decision trees, for situations where computers are unavailable or inappropriate.

The key to a successful expert system is the quality of the experts' advice. A good expert must be fully conversant with the technology to be employed, understand the benefits of developing such a knowledge-based system and be able to communicate their ideas and thought processes well (Mumford and Norton, 1993). Good decision makers are not just those with long experience since long experience of incorrectly made decisions is no use to anyone. A good decision maker makes use of the most up-to-date quantitative information and evaluates and interprets this in the light of the success or otherwise of decisions they have made in similar circumstances in the past. Obviously the older the person the greater the range of experience they will have of similar 'decision circumstances'. However, this experience is only of value if the decision maker has taken an objective view of their decisions, i.e. they recognize the positive and negative effects of their choice. It is in this way that a decision maker learns by experience and hence over time is successful at making decisions and is regarded as an expert. Since few experts are actually objective concerning their own decision making it is always advisable to base an expert system on the opinions of a number of experts and where there is disagreement they should be encouraged to justify their decision-making process to each other to test the validity of their thought processes.

One of the major features of expert systems is that they provide the most appropriate advice on a specific pest management situation by asking the user a minimum number of questions (Norton, 1990). The system is made up from a number of rules (there are 70 rules for the COMAX cotton management expert system; McKinnon, 1989) often based on the IF–THEN–ELSE sequence. Qualifiers form the conditions which must be met so that choices can be selected. A qualifier may be a statement such as the 'crop growth stage is . . .' with a number of stages, e.g. 'booting', 'grain filling', 'ripening', while a choice may be a decision as to whether or not to apply a control measure.

Expert systems can be devised to assist decision making in a number of areas relevant to pest management (Table 6.2) but in practice there have been few published attempts to produce them. Among those that have been published are expert systems for pest control in tropical grain stores (Compton *et al.*, 1992), the management of wheat bulb fly in the UK (Jones *et al.*, 1990) and cotton crop management (McKinnon, 1989).

Table 6.2 Areas where expert systems could be employed to assist decision makers in IPM. (From Mumford and Norton, 1993)

Expert systems
Pest identification
Assessment of levels of pest attack
Assessment of crop loss levels
Determination of available control options and their effectiveness
Cost/benefit assessment
Determination of other objectives and constraints

6.5 OPTIMIZATION MODELS

Optimization models embody a range of techniques that are designed to choose the best solution to a problem from a range of alternatives. Very often these solutions are obtained from searching for a strategy that optimizes some mathematically defined expression which is thought to characterize the utility of the management procedure (Getz and Gutierrez, 1982). Typically a strategy would seek to optimize crop yield or maximize profit while, for instance, constraining pesticide use. However, the objectives need not be economic, they may be environmental, such as devising an optimal strategy for control where environmental factors govern criteria for achieving an optimal solution. With most of the techniques however, the objectives must be quantifiable. Of the techniques that may be classified under the heading of optimization models, dynamic programming has received most attention in pest management, while the techniques of linear and goal programming have potential for use but their application to pest management problems to date has been very limited.

6.5.1 Dynamic programming

Dynamic programming (Bellman, 1957) is a technique which determines the optimal solution in problems in which separate but related decisions occur in a set of sequential time periods. Dynamic programming is generally appropriate to pest management decisions which are made sequentially (Rossing, 1989), and where the decision at one point in time can affect the decision and outcome at another. Essentially, the dynamic programming approach begins at the final stage of the decision process and then works backwards along an optimal pathway to the beginning (Conway, 1984). This approach is quite efficient because not all combinations of a 'route' have to be checked. However, there are no standard dynamic programming procedures. Typically the most difficult step in dynamic programming is the development of a mathematical model which can adequately describe the impact of variables such as weather and the management decisions on the dynamics of the system (Shoemaker, 1982). This is especially difficult because of what is known as the 'curse of dimensionality', a constraint that reflects the fact that as the number of variables in the model increases, the number of calculations required by the computation to solve a problem increases very rapidly. This means that most dynamic programming models are limited to four state variables (maximum of six) which tends to limit its usefulness, although the inclusion of random events, the number of discrete values of a state component, and the complexity of the transformation function may alter this (Rossing, 1989). Despite this problem, the models may be dynamic, non-linear or stochastic, and have been found to be of value in various pest management situations, but particularly for defining optimal use of pesticides (Conway *et al.*, 1975; El-Shishiny, 1984; Shoemaker, 1979, 1982, 1984).

6.5.2 Linear and goal programming

One optimization technique which has found wide application including in the scheduling of farm operations is linear programming (Rossing, 1989). This is a general purpose technique used by economists for determining the best allocation of scarce resources. However, it is a technique that can be readily applied to any pest management problem that needs to optimize the use of two or more resources (Dent, 1991). Linear programming problems cannot be analytical, but use an algorithm called the simplex method. The problems are characterized by three things: (i) an objective function; (ii) a set of decision variables; and (iii) a set of resource constraints. The objective function, a measure of how good an allocation is, must be linear for the decision variables (the way in which scarce resources can be allocated) which means that a change in one unit in the decision variables results in a constant change of the objective function and the resource contraints (limitations placed on the decision variables to reflect the scarcity of the resource). Linear programming can be solved graphically where there are only two decision variables (see Dent, 1991 for a hypothetical example relevant to pest management) but for more complicated situations the simplex method is required (Walsh, 1985; Markland and Sweigart, 1987).

Goal programming is a technique that aims to provide an optimal solution in situations where there may be several different objectives, some of which may be conflicting. The method requires that the objectives are placed in order of priority and then starting with the highest priority objective the method attempts to satisfy each goal, or at least to minimize undesirable deviations. A solution is eventually found which minimizes the amount of under-achievement for any goal that cannot be met, without worsening the achievement of any higher priority goal (Rossing, 1989). The technique is applicable to pest management problems and particularly to research management where choices between a number of conflicting options need to be made. Dent (1991) provides a hypothetical example using a graphical analysis (the simplex method is also applicable to goal programming) but like linear programming, goal programming has not been used in quantitative analysis of pest management systems. A general description of goal programming can be found in Markland and Sweigart (1987).

6.6 MODELS: RELATIVE ADVANTAGES AND DISADVANTAGES

The main classes of model considered above, statistical, mechanistic (analytical and simulation) and optimization models all have a role in the qualifying relationships in pest management systems. However, each has a number of advantages and disadvantages which constrain its use to particular situations. Table 6.3 summarizes some of the advantages and disadvantages of analytical, simulation and optimization models with regard to three criteria: (i) response to changes in parameters; (ii) a large number of variables; and (iii) a large number of management alternatives.

Simulation models are by far the most widely used models and have the advantage of being highly flexible and able to describe a system in greater detail than either analytical or optimization models because they can include such a larger number of variables. They have the disadvantage that there is a need to recompute to evaluate the impact of each management option, normally a very large number of parameters have to be estimated in order to construct the model (Godfray and Waage, 1991), and they tend to be site-specific (Horn, 1988). They also do not have a predescribed solution structure; hence, convergence to an optimal solution is not guaranteed as it is for dynamic and linear programming (Rossing, 1989).

Optimization methods such as dynamic programming are much more efficient than simulation models in evaluating a large number of management options. However, for non-linear systems, optimization models are not able to incorporate as many dynamic variables as are commonly used in simulation models (Shoemaker, 1984). Most optimization models are also unsuitable for dealing with the combined effects of uncertainty, dynamic interaction between decisions and subsequent events and complex interdependencies among variables in the system (Rossing, 1989). They can

Table 6.3 Advantages and disadvantages of different types of model used to analyse the behaviour and management of dynamic ecosystems

Type of model	*Suitability for analysis*		
	Response to changes in	*Large number of variables*	*Large number of management alternatives*
Analytical	Excellent	Poor	Fair
Simulation	Fair	Excellent	Poor
Optimization	Fair	Fair	Excellent

however, be used to analyse systems with many more variables than is typically possible with analytical models.

Analytical models are more efficient than simulation models in addressing specific questions relating to the underlying mechanisms that influence the behaviour of the system (Getz and Gutierrez, 1982) but are limited in their application because it is not always possible to find closed-form solutions for a problem (Shoemaker, 1984).

Ultimately the models used to address a problem are not always mutually exclusive with approaches combining models of different types to find solutions to particular problems. For instance, optimization models sometimes benefit from being combined with simulation models (Dudley *et al.*, 1989; Rossing, 1989); regression models are also commonly used in both analytical and simulation models. More recently Godfray and Waage (1991) have introduced the concept of an intermediate complexity of models which are generally more complex than simple analytical models but less complex than many detailed simulation models. Such advances clearly indicate the dynamic nature of modelling and promise a continued development in this important area. However, despite this, a general claim of dissatisfaction with the modelling approach can be levelled, since there are currently many more models than there are studies to test them (Waage, 1986). A greater emphasis in future must be placed on the verification and validation of models to ensure the solutions they offer are appropriate and will meet the needs of those for whom they are developed.

REFERENCES

Allen, J.C. (1981) The citrus mite game: a simulation model of pest losses. *Environmental Entomology*, **10**, 171–6.

Anderson, R.M. and May, R.M. (1991) *The Population Biology of Infectious Diseases*. Oxford University Press, Oxford.

Barnard, D.R., Ervin, R.T. and Epplin, F.M. (1986) Production system-based model for defining economic thresholds in preweaner beef cattle, *Bos taurus*, infested with the Lone Star tick, *Amblyomma americanum* (Acari: Ixodidae). *Journal of Economic Entomology*, **79**(1), 141–3.

Bellman, R.E. (1957) *Dynamic Programming*, Princeton University Press, Princeton, New Jersey.

Butt, D.J. and Royle, D.J. (1974) Multiple regression analysis in the epidemiology of plant diseases, in *Epidemics and Plant Disease*, (ed. J. Kranz), Springer-Verlag, Berlin, pp. 158–75.

Carruthers, R.I., Whitfield, G.H., Tummala, R.L. and Haynes, D.L. (1986) A systems approach to research and simulation of insect pest dynamics in the onion agro-ecosystem. *Ecological Modelling*, **33**, 101–21.

Comins, H.N. (1984) The mathematical evaluation of options for managing pesticide resistance, in *Pest & Pathogen Control: Strategic, Tactical and Policy Models*, (ed. G.R. Conway), John Wiley & Sons, New York, Chichester, pp. 454–69.

Compton, J.A.F., Tyler P.S., Mumford, J.D., Norton, G.A., Jones, T.H. and Hindmarsh, P.S. (1992) Potential for an expert system for pest control in tropical grain stores. *Tropical Science*, **32**, 295–303.

Conway, G. (1976) Man versus pests, in *Theoretical Ecology Principles and Applications*, (ed. R.M. May), Blackwell Sci. Publ., Oxford, pp. 257–81.

Conway, G.R. (1984) Strategic models, in *Pest and Pathogen Control: Strategic, Tactical and Policy Models*, (ed. G.R. Conway), John Wiley & Sons, New York, Chichester, pp. 15–28.

Conway, G.R., Norton, G.A., King, A.B.S. and Small, N.J. (1975) A systems approach to the control of the sugar cane froghopper, in *Study of Agricultural Systems*, (ed. G.E. Dalton), Applied Science Publishers, London, pp. 193–229.

Dent, D.R. (1991) *Insect Pest Management*, CAB International, Wallingford.

Dudley, N.J., Mueller, R.A.E. and Wightman, J.A. (1989) Application of dynamic programming for guiding IPM on groundnut leafminer in India. *Crop Protection*, **8**, 349–57.

El-Shishiny, H. (1984) Optimal chemical control of the greenhouse whitefly, in *Pest and Pathogen Control: Strategic, Tactical and Policy Models*, (ed. G.R. Conway), John Wiley & Sons, New York, Chichester, pp. 310–18.

Entwistle, J.C. and Dixon, A.F.G. (1986) Short-term forecasting of peak population density of the grain aphid ("*Sitobion avenae*") on wheat. *Annals of Applied Biology*, **109**, 215–22.

Eversmyer, M.G. and Burleigh, J.R. (1970) A method of predicting epidemic development of wheat leaf rust. *Phytopathology*, **60**, 805–11.

Feng, Z., Carruthers, R.I., Larkin, T.S. and Roberts, D.W. (1988) A phenology model and field evaluation of *Beauveria bassiana* (Bals.) Vuillemin (Deuteromycotina: Hyphomycetes) mycosis of the European corn borer, *Ostrinia nubilalis* (Hbn.) (Lepidoptera: Pyralidae). *The Canadian Entomologist*, **120**, 133–44.

Fischer, A.J., Lozano, J., Ramirez, A. and Sanint, L.R. (1993) Yield loss prediction for integrated weed management in direct-seeded rice. *International Journal of Pest Management*, **39**(2), 175–80.

Getz, W.M. and Gutierrez, A.P. (1982) A perspective on systems analysis in crop production and insect pest management. *Annual Review of Entomology*, **27**, 447–66.

Gilbert, N. and Gutierrez, A.P. (1973) An applied plant-aphid–parasite relationship. *Journal of Animal Ecology*, **42**, 323–40.

Godfray, H.C.J. and Waage, J.K. (1991) Predictive modelling in biological control: the mango mealy bug ("*Rastrococcus invadens*") and its parasitoids. *Journal of Applied Ecology*, **28**, 434–53.

Graf, B., Lamb, R., Heong, K.L. and Fabellar, L. (1992) A simulation model for the population dynamics of rice leaf-folders (Lepidoptera: Pyralidae) and their interactions with rice. *Journal of Applied Ecology*, **29**, 558–70.

Gutierrez, A.P. and Daxl, R. (1984) Economic thresholds for cotton pests in Nicaragua: ecological and evolutionary perspectives, in *Pest and Pathogen Control: Strategic, Tactical and Policy Models*, (ed. G.R. Conway), John Wiley & Sons, New York, Chichester, pp. 184–205.

Gutierrez, A.P., Dos Santos, W.J., Pizzamiglio, M.A., Villacorta, A.M., Ellis, C.K., Fernandes, C.A.P. and Tutida, I. (1991a) Modelling the interaction of cotton and the cotton boll weevil. II. Bollweevil ("*Anthonomus grandis*") in Brazil. *Journal of Applied Ecology*, **28**, 398–418.

Gutierrez, A.P., Dos Santos, W.J., Villacorta, A., Pizzamiglio, M.A., Ellis, C.K., Carvalho, L.H. and Stone, N.D. (1991b) Modelling the interaction of cotton and the cotton boll weevil. I. A comparison of growth and development of cotton varieties. *Journal of Applied Ecology*, **28**, 371–97.

Gutierrez, A.P., Neuenschwander, P. and van Alphen, J.J.M. (1993) Factors affecting biological control of cassava mealybug by exotic parasitoids: a ratio-dependent supply-demand driven model. *Journal of Applied Ecology*, **30**, 706–21.

Hartstack, A.W. and Hollingsworth, J.P. (1974) A computer model for predicting *Heliothis* populations. *Transactions of the ASAE*, **17**, 112–15.

Hassell, M.P. (1978) *The Dynamics of Arthropod Predator–Prey Systems*, Princeton University Press, Princeton, New Jersey.

Healy, M.J.R. and Taylor, L.R. (1962) Tables for power-law transformations. *Biometrika*, **49**, 557–9.

Heong, K.L. (1985) Systems analysis in solving pest management problems, in *Integrated Pest Management in Malaysia*, (eds B.S. Lee, W.H. Loke and K.L. Heong), Malaysian Plant Protection Society, Kuala Lumpur, pp. 133–49.

Holt, J. and Day, R.K. (1993) Rule-based models, in *Decision Tools for Pest Management*, (eds G.A. Norton and J.D. Mumford), CAB International, Oxford, pp. 147–58.

Holt, J. and Norton, G.A. (1993) Simulation models, in *Decision Tools for Pest Management*, (eds G.A. Norton and J.D. Mumford), CAB International, Wallingford, pp. 119–46.

Holt, J., Cook, A.G., Perfect, T.J. and Norton, G.A. (1987) Simulation analysis of brown planthopper (*Nilaparvata lugens*) population dynamics on rice in the Philippines. *Journal of Applied Ecology*, **24**, 87–102.

Horn, D.J. (1988) *Ecological Approach to Pest Management*, Elsevier Applied Science Publishers, London.

Ingram, W.R. (1980) Studies of the pink bollworm *Pectinophora gossypiella*, on sea island cotton in Barbados. *Tropical Pest Management*, **26**(2), 118–37.

Jones, T.H. (1993) Analytical models, in *Decision Tools for Pest Management*, (eds G.A. Norton and J.D. Mumford), CAB International, Oxford, pp. 101–18.

Jones, T.H., Young, J.E.B., Norton, G.A. and Mumford, J.D. (1990) An expert system for the management of wheat bulb fly *Delia coarctata* (Diptera: Anthomyiidae) in the United Kingdom. *Journal of Economic Entomology*, **83**, 2063–72.

Kendall, D.A., Brain, P. and Chinn, N.E. (1992) A simulation model of the epidemiology of barley yellow dwarf virus in winter sown cereals and its application to forecasting. *Journal of Applied Ecology*, **29**, 414–26.

Kidd, N.A.C. and Gazziano, S. (1992) Development of population models for olive pest management, in *Research Collaboration in European IPM Systems*, BCPC Monograph No. 52, (ed. P.T. Haskell), BCPC, Brighton, pp. 41–6.

Kitching, R.L. (1991), in *Heliothis: Research Methods and Prospects*, (ed. M.P. Zalucki), Springer-Verlag, Berlin, p. 174.

Markland, R.E. and Sweigart, J.R. (1987) *Quantitative Methods: Applications to Managerial Decision Making*, John Wiley & Sons, Chichester.

May, R.M. (1974) *Stability and Complexity in Model Ecosystems. Monographs in Population Biology 6*, 2nd edn, Princeton University Press, New York.

Maynard-Smith, J. (1974) *Models in Ecology*, Cambridge University Press, Cambridge.

McKinnon, J.M. (1989) Modelling and economics, in *Progress and Prospects in Insect Control. BCPC Monograph No. 43*, (ed. N.R. McFarlane), BCPC, Farnham, pp. 205–15.

Mumford, J.D. and Norton, G.A. (1982) Economics of decision making in pest management. *Annual Review of Entomology*, **29**, 157–74.

Mumford, J.D. and Norton, G.A. (1984) Economics of decision making in pest management. *Annual Review of Entomology*, **29**, 157–74.

Mumford, J.D. and Norton, G.A. (1993) Survey and knowledge acquisition techniques, in *Decision Tools for Pest Management*, (eds G.A. Norton and J.D. Mumford), CAB International, Wallingford, pp. 79–88.

Norton, G.A. (1990) Decision tools for pest management: their role in IPM design and delivery. (Paper presented to the FAO/UNEP/USSR Workshop on Integrated Pest Management).

Norton, G.A., Holt, J. and Mumford, J.D. (1993) Introduction to pest models, in *Decision Tools for Pest Management*, (eds G.A. Norton and J.D. Mumford), CAB International, Wallingford, pp. 89–100.

Quade, E.S. and Boucher, W.I. (1968) *Systems Analysis and Policy Planning-Applications in Defense*, Elsevier, New York.

Rabbinge, R. and de Wit, C.T. (1989) Systems, models and simulation, in *Simulation and Systems Management in Crop Protection*, (eds R. Rabbinge, S.A. Ward and H.H. van Laar), Pudoc, Wageningen, pp. 3–15.

Rabbinge, R., Ward, S.A. and van Laar, H.H. (1989) *Simulation and Systems Management in Crop Protection*, Pudoc, Wageningen.

Regev, U. (1984) An economic analysis of man's addiction to pesticides, in *Pest and Pathogen Control: Strategic, Tactical and Policy Models*, (ed. G.R. Conway), John Wiley & Sons, Chichester, pp. 441–53.

Regev, U., Gutierrez, A.P. and Feder, G. (1976) Pests as a common property resource: a case study of alfalfa weevil control. *American Journal of Agricultural Economics*, **58**, 185–95.

Reissig, W.H., Barnard, J., Weires, R.W., Glass, E.H. and Dean, R.W. (1979) Prediction of apple maggot fly emergence from thermal unit accumulation. *Environmental Entomology*, **8**(1), 51–4.

Rossing, W.A.H. (1989) Application of operations research techniques in crop protection, in *Simulation and Systems Management in Crop Protection*, (eds R. Rabbinge, S.A. Ward and H.H. van Laar), Pudoc, Wageningen, pp. 279–98.

Royle, D.J. (1973) Quantitative relationships between infection by the hop downy mildew pathogen, *Pseudoperonospora humuli*, and weather and inoculum factors. *Annals of Applied Biology*, **73**, 19–30.

Rudd, W.G. and Gandour, R.W. (1985) Diffusion model for insect dispersal. *Journal of Economic Entomology*, **78**, 295–301.

Sawyer, A.J. and Fick, G.W. (1987) Potential for injury to alfalfa by alfalfa blotch leafminer (Diptera: Agromyzidae): simulations with a plant model. *Environmental Entomology*, **16**, 575–85.

Shoemaker, C.A. (1976) Optimal management of an alfalfa ecosystem, in *Pest Management International Institute for Applied Systems Analysis Proc. Ser*, (eds G.A. Norton and C.S. Holking), Pergamon, Oxford, pp. 301–16.

Shoemaker, C.A. (1979) Optimal timing of multiple applications of pesticides with residual toxicity. *Biometrics*, **35**, 803–12.

Shoemaker, C.A. (1982) Optimal integrated control of univoltine pest populations with age structure. *Operations Research*, **30**(1), 40–61.

Shoemaker, C.A. (1984) The optimal timing of multiple applications of residual pesticides: deterministic and stochastic analyses, in *Pest and Pathogen Control: Strategic, Tactical and Policy Models*, (ed. G.R. Conway), John Wiley & Sons, Chichester, pp. 290–310.

Snedecor, G.W. and Cochran, W.G. (1978) *Statistical Methods*, Iowa State University, Ames, Iowa.

Sokal, R.R. and Rohlf, F.J. (1981) *Biometry*, 2nd edn, W.H. Freeman & Co., New York.

Southwood, T.R.E. (1978) *Ecological Methods with Particular Reference to the Study of Insect Populations*, 2nd edn, Chapman & Hall, London, pp. 407–19.

Southwood, T.R.E. and Comins, H.N. (1976) A synoptic population model. *Journal of Animal Ecology*, **45**, 949–65.

Stern, V.M., Smith, R.F., van den Bosch, R. and Hagen, K.S. (1959) The integrated control concept. *Hilgardia*, **29**, 81-101.

Stinner, R.E., Gutierrez, A.P. and Butler, G.D. (1974) An algorithm for temperature-dependent growth rate simulation. *The Canadian Entomologist*, **106**, 519–24.

Taylor, L.R. (1961) Aggregation, variance and the mean. *Nature*, **189**, 732–5.

Taylor, L.R. (1965) A natural law for the spatial disposition of insects, in *Proceedings XII International Congress of Entomology*, pp. 396–7.

Taylor, L.R. (1971) Aggregation as a species characteristic, in *Statistical Ecology 1*, (eds G.P. Patil, E.C. Pielou and W.E. Waters), Pennsylvania State University, pp. 357–77.

Vogel, W.O., Hennessey, R.D., Berhes, T. and Matungulu, K.M. (1993) Yield losses to maize streak disease and *Busseola fusca* (Lepidoptera: Noctuidae), and economic benefits of streak-resistant maize to small farmers in Zaïre. *International Journal of Pest Management*, **39**(2), 229–38.

Vorley, W.T. and Wratten, S.D. (1985) A simulation model of the role of parasitoids in the population development of *Sitobion avenae* (Hemiptera: Aphididae) on cereals. *Journal of Applied Ecology*, **22**, 813–23.

Waage, J.K. (1986) Family planning in parasitoids: adaptive patterns of progeny and sex allocation, in *Insect Parasitoids. 13th Symposium of the Royal Entomological Society of London*, (eds J.K. Waage and D. Greathead), Academic Press, Orlando, pp. 63–96.

Wall, R., French, N.P. and Morgan, K.L. (1993) Sheep blowfly population control: development of a simulation model and analysis of management strategies. *Journal of Applied Ecology*, **30**, 743–51.

Walsh, G.R. (1985) *An Introduction to Linear Programming*, 2nd edn, John Wiley & Sons, Chichester.

Watt, A.D., Vickerman, G.P. and Wratten, S.D. (1984) The effect of the grain aphid, *Sitobion avenae* on wheat in England: an analysis of the economics of control practice and forecasting systems. *Crop Protection*, **3**(2), 209–222.

Watt, K.E.F. (1961) Mathematical models for use in insect pest control. *The Canadian Entomologist*, **19**(Suppl), 62.

Watt, K.E.F. (1962) Use of mathematics in population ecology. *Annual Review of Entomology*, **7**, 243–60.

Watt, K.E.F. (1966) *Systems Analysis in Ecology*, Academic Press, New York.

Way, M.J., Cammell, M.E., Taylor, L.R. and Woiwod, I.P. (1981) The use of egg counts and suction trap samples to forecast the infestation of spring-grown field beans, *Vicia faba*, by the black bean aphid, *Aphis fabae*. *Annals of Applied Biology*, **98**, 21–34.

Yencho, G.C., Getzin, L.W. and Long, G.E. (1986) Economic injury level, action threshold and a yield-loss model for the pea aphid *Acyrthosiphon pisum* (Homoptera: Aphididae), on green peas, *Pisum sativum*. *Journal of Economic Entomology*, **79**(6), 1681–7.

CHAPTER 7

Experimental paradigms

D.R. Dent

7.1 INTRODUCTION

Pest management research involves scientists from a wide range of disciplines, for instance, chemists (studying pesticides, semiochemicals, hostplant allelochemicals), ecologists, social scientists, biological control practitioners, systems analysts, plant breeders, to name but a few. Specialization also exists within a single discipline, in terms of the approaches used, the theories and hypotheses adhered to, the instrumentation or techniques utilized. In the terminology of Kuhn (1970, 1977) the scientists who are practitioners of a scientific speciality make up a discrete 'scientific community' and the speciality they adhere to is known as a 'paradigm'. Although the concept of the 'scientific community' and particularly the 'paradigm' has met with some criticism (Musgrave, 1980; Shapere, 1980) it has begrudgingly gained acceptance and credence as part of the general philosophy of science (Medawar, 1981, 1986).

The term paradigm is used here to describe the common way of thinking, approach, goals and established scientific opinion that are shared by particular scientific communities. The paradigm presents a limited framework within which each 'specialist subject' is explored; at least until its limitations and constraints become overwhelmingly 'apparent' and cannot be ignored. Then an extraordinary scientist or extraordinary scientific phenomenon supplants the prevailing paradigm with a new orthodoxy (Medawar, 1981)

Integrated Pest Management (IPM) is of course itself a paradigm. IPM arose from the deficiencies of the paradigm of 'chemical control' (Perkins, 1982). Within the disciplines which make up IPM, different paradigms also exist. Scientists working in IPM, especially those managing programme or system development, need to be aware of the existence of these different paradigms, because of their relevance to programme decision making. Established paradigms can provide limitations for programme development

Integrated Pest Management Edited by David Dent. Published in 1995 by Chapman & Hall, London. ISBN 0 412 57370 9

by: (i) creating problems that do not exist by adherence to particular divisions, polarizations, conceptualizations; (ii) acting as conceptual traps or prisons which prevent a more useful arrangement of information; and (iii) through blocking by adequacy (de Bono, 1970). Hence, from a programme manager's perspective paradigms may limit scope for tackling problems, or the type of research that is undertaken, simply because the scientists involved with the programme subscribe to a particular set of assumptions and ways of thinking. An understanding of the different scientific communities and their associated paradigms permits a more informed choice of the type of research that will be carried out in the development of an IPM programme. However, the situation is dynamic, paradigms will continue to change, and scientific managers will need to keep abreast of these changes. A number of examples of the paradigms in pest management are provided to illustrate the concept and to expand on a few important areas considered earlier in the text.

7.2 PESTICIDES

The introduction of the notorious DDT as a commercial insecticide in 1941 (Mellanby, 1989) heralded an era of pest control dominated by chemical pesticides that persists to this day. However, the study of pesticides, their development, efficacy and application has changed over that time to meet the various demands of the different generations of scientists, their science, consumers and the general public. The agrochemical industry has succeeded

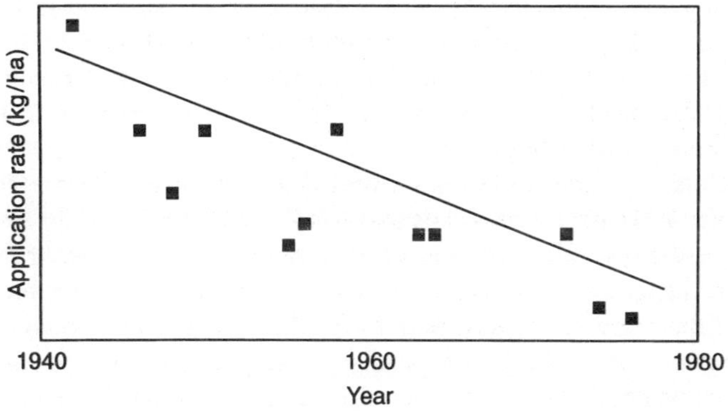

Figure 7.1 The reduction in application rates of insecticides since 1940. The graph is based on the following insecticides: DDT (1942), parathion (1946), dieldrin (1948), diazinon (1951), azinphos-methyl (1955), dimethoate (1956), carbaryl (1958), chlordimeform (1963), monocrotophos (1965), diflubenzuron (1972), permethrin (1973), deltamethrin (1975). (After Geissbühler, 1981).

in increasing the biological activity of pesticides which has meant a steady decline in application rates (Figure 7.1), improved chemical selectivity, increased safety evaluations and risk assessments, and the development of application techniques to improve efficiency and targeting of pesticides (Geissbühler, 1981). In achieving this, emphasis in research has been placed on three approaches, referred to here as: (i) the laboratory bioassay; (ii) field experimentation; and (iii) pesticide transfer/behaviour studies. The first two of these areas of research (laboratory bioassay and field experimentation) have formed the backbone of pesticide research since the 1940s; however, the latter (pesticide transfer/behaviour) is a more recent philosophy which has largely developed from a difficulty faced by scientists of predicting field performance of pesticides from laboratory data. Although some of the differences in approach are due to different objectives, the emergence of the transfer/behaviour studies represents a major shift in emphasis and thinking and has subsequently spawned a community of scientists advocating this approach – a new 'paradigm'.

7.2.1 Laboratory bioassays

Laboratory-based pesticide bioassays are characterized by their use of highly controlled laboratory conditions, standardized biological material and subsequent determination of dose–effect relations. Although carried out to evaluate either the potency of a pesticide (intrinsic toxicity, formulations, pesticide mixtures) or the susceptibility of biological material (pesticide resistance studies) (Busvine, 1971), the rationale is essentially the same; emphasis is placed on the development of precise, artificial techniques that permit comparisons over time under standard, controlled conditions. The most extreme example is perhaps reflected in toxicological tests which use topical applications. A micro-syringe is used to apply technical-grade insecticide (in an appropriate solvent) to individual laboratory-reared insects, on a specified part of their anatomy for the purpose of measuring, for instance, levels of physiological insecticide resistance (FAO, 1980; ffrench-Constant and Roush, 1990).

Entomologists, plant pathologists and weed scientists share a common problem of standardizing materials and conditions for use in their respective pesticide bioassays. The problem of maintaining constant environmental conditions, especially temperature, is not difficult for insect bioassays. Because of their size, thousands of individual insects can be treated and kept in temperature-controlled cabinets or rooms. Temperature (probably the main environmental factor considered in insect bioassays) is usually kept constant to permit comparisons between occasions and different laboratories (Matthews, 1984), although some attempts have been made to use cyclical temperatures under laboratory and greenhouse conditions (Dent, 1990). The situation for plant pathologists and weed scientists is complicated by the greater range of environmental variables that have to be taken into account, especially lighting conditions. Growth chambers are ideal where available, but glasshouses need to have provision for supplementary lighting and temperature control to overcome seasonal changes (Frans *et al.*, 1986). Since plant morphology and physiology can differentially affect pesticide or plant susceptibility (Blair and Martin, 1989; Davies and Blackman, 1989), it is important for comparison between bioassays that conditions remain constant or are standardized as much as possible.

The conditions required for standardizing insect bioassays have been well documented (Busvine, 1971; Matthews, 1984) and include the requirement for control of insect size, sex, age, and physiological and behavioural condition. To ensure insect material is suitable and available for laboratory bioassays a great deal of effort is expended on the culture and rearing of suitable insect populations. Standard techniques now exist for many insect species (Singh and Moore, 1985). However, the need for large numbers of uniform disease-free insects (Felton *et al.*, 1987) can cause some problems, particularly through genetic uniformity (McKauer, 1972; Berlocher and

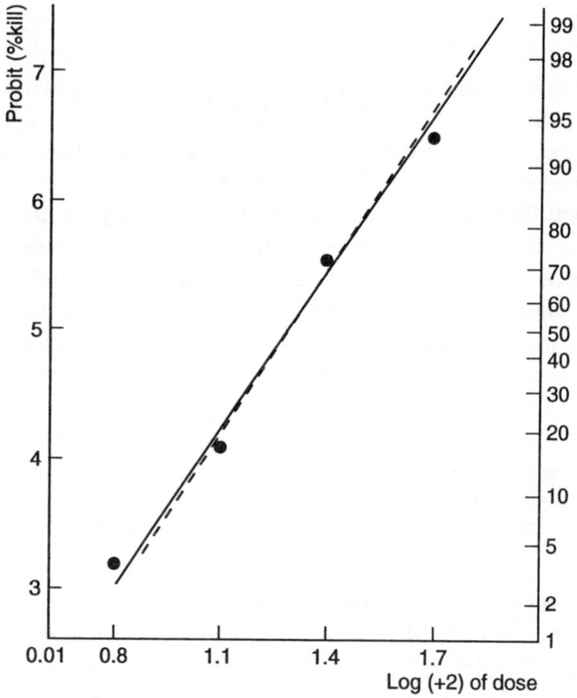

Figure 7.2 A hypothetical example of a regression line (solid line) used to calculate an LD_{50} through probit analysis. The dotted line is a provisional line from which the expected probits are determined. Response % is the percentage of insects killed; probits are a transformed estimate of the % response. (After Matthews, 1984.)

Friedman, 1981) and insects which adapt to culture conditions to produce characteristics which are unrepresentative of field populations (Cannon, 1989; Dent, 1990).

The results of a bioassay (the mortality recorded over a range of pesticide concentrations) are plotted as dosage mortality curves and analysed by probit analysis (Finney, 1971; Gunning, 1991) (Figure 7.2). The toxicity of a pesticide is usually quoted in milligrams of active ingredient for each kilogram of body weight (i.e. parts per million of the test organism) (Matthews, 1992). It is most commonly measured in such toxicological studies as the dose at which 50% of the test organisms are killed, in a specified time (often 24 hours), and is referred to as the LD_{50} (LD = lethal dose) (Busvine, 1971; Finney, 1971). Several workers have expressed plant sensitivity to herbicides in terms corresponding to LD_{50}s (Lavy and Santelmann, 1986), variously called ID_{50} (Inhibition Dosage) GR_{50} (Growth Reduction) and ED_{50} (Effective Dosage) for 50% inhibition of growth. The latter term is also used in studies of fungicide efficacy (Shephard *et al.*, 1986; Shephard, 1989). Unfortunately, the LD_{50} obtained in the laboratory bioassay does not usually relate to the performance of pesticides in field trials (Copping *et al.*, 1989).

7.2.2 Field experimentation

The systematic evaluation of different agricultural inputs became a science in the 1930s when R.A. Fisher, F. Yates and others gave their attention to field experimentation and developed replicated testing in block designs (Reed *et al.*, 1985). Trial design and analysis have been developed and expounded since, by among others, Cochran and Cox (1957), Steel and Torrie (1960), Bailey (1981) and Püntener (1981).

Pesticide trials may be undertaken to determine an optimum dosage, timing of application, spectrum of action, tolerance of different crops, to name

Table 7.1 Possible objectives of pesticide trials. (From Püntener, 1981)

Optimum dosage
Optimum timing of application
Suitability of various application methods
Selectivity properties
Comparison of different formulations
Influence of environment on properties of pesticide
Economic comparison between test and standard
Spectrum of action
Determination of residues
Difference in tolerance levels of crop
Effects on succeeding crops
Specific yield assessments

but a few (Table 7.1). The criterion used to judge success of a trial is usually based on a measure of yield in terms of a cost/benefit comparison of untreated and treated crops. However, measures of plant tolerance, degree of infection (fungi), insect mortality or a decrease in natural number of plants per weed species may be used in addition (Püntener, 1981). In the process of acquiring such information a great deal of planning, sampling and basic research on the crop and the target pest are required. Pesticide trials are costly affairs and may take many months or years to complete. Hence it is essential to keep avoidable mistakes to a minimum (Dent, 1991). This requires a great deal of thought and careful planning, supervision and monitoring of the experiment. Unterstenhöfen (1976), in a review of field trials techniques warned that it is necessary to acquire exact knowledge of the biology, ecology and epidemiology of the target pest before embarking on any pesticide trial, for it is these factors in conjunction with mechanism of action and other properties which should determine the planning and implementation of the experiment.

Normally, the field experimentation will progress from small plot trials to large-scale testing on research station fields. Then, successful trials will be carried out as multilocational experiments on fields of cooperating farmers as a prelude to recommendation for use (Reed *et al.*, 1985).

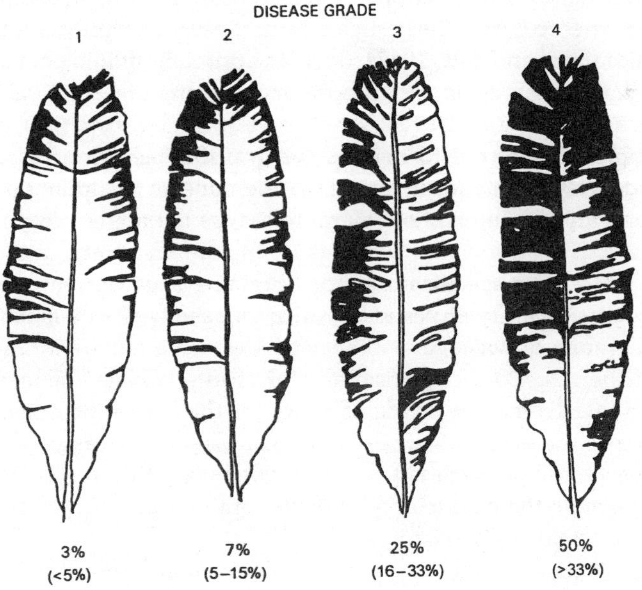

Figure 7.3 Banana leaf spot assessment key; black areas represent the percentage of destroyed tissue as if all spots (all spotted areas) were fitted into one leaf position. (After Stower, 1971).

7.2.3 Pesticide transfer/behaviour studies

The progression outlined above for a series of pesticide field trials, ultimately provides the empirical data on which recommendations for application are largely based. This empirically based approach can be costly and time consuming. It also suffers from the problem that any recommendations inevitably reflect the preliminary ideas about the appropriate set of spray properties available to the investigator at the design stage of the trial (Ford and Salt, 1987). The traditional laboratory bioassay approach however, does not provide wholly appropriate kinds of data either. There is general agreement, whether dealing with fungicides (Shephard, 1989), soil or foliar-applied insecticides (Gordon *et al.*, 1989), insect sterilants (Smet *et al.*, 1989) or herbicides (Hance *et al.*, 1968) that laboratory data poorly relates to the performance of pesticides in field experiments (Copping *et al.*, 1989). There are numerous reasons cited for this, such as the difference between laboratory and field biological material (Tadros, 1987; Kudsk, 1989), the simplistic application methods used in the laboratory (Hall *et al.*, 1989; Shephard, 1989; Merritt, 1989) and the environmental conditions in the laboratory being totally unrepresentative of the field (Hislop, 1989; Dent, 1990). If however, laboratory conditions can be made to be more realistic then an alternative approach does exist. This approach emphasizes the need to determine the pattern and properties of a pesticide deposit on the 'target' surface necessary for its maximum effectiveness, and the subsequent use of these properties to define the pesticide formulation and spray characteristics for field use (Ford and Salt, 1987). Such an approach should provide a more realistic way of carrying out laboratory research relevant to the field situation.

The approach referred to here as the 'transfer/behaviour studies' have developed to meet this ideal, based on the concept of optimum pesticide deposition. This is defined, in general terms, as the application of a biologically effective dose on a target with maximum safety and economy (Hislop, 1987). The specifications for obtaining such optimum placement with a specific quantity of pesticide which will safely exert a maximal effect with a minimum of wastage, will be dependent on a number of factors: the nature of the target, the organism and/or crop involved, the method used for pesticide delivery, the mode of action of the active ingredient, its formulation and the environmental conditions. Given this range of variables, provided each can be modified in a prescribed way, then it should be possible to determine the characteristics of the optimal pesticide deposit necessary to exert maximal control.

The problems of conducting pesticide performance research in the laboratory are now well acknowledged (Copping *et al.*, 1989), and because of this there has been a general shift of emphasis away from maximizing biological response towards a compromise between maximizing response and the simulation of 'real' application conditions. Spray chambers are now much more sophisticated (in some cases even incorporating control of wind speed; Hislop, 1987), and a range of laboratory devices capable of producing appropriate droplet deposition characteristics is now available (Munthali and Scopes, 1982; Young, 1986; Reichard *et al.*, 1987; Reynolds *et al.*, 1987). Studies which incorporate both controlled deposition of pesticides and observations of pest behaviour have been used to identify optimum forms of deposits for acaricides (Munthali, 1981; Munthali and Scopes, 1982; Munthali and Wyatt, 1986) and insecticides (Adams *et al.*, 1987; Omar and Matthews, 1987; Adams and Hall, 1989; Omar *et al.*, 1991). Computer models are also proving useful in predicting the pick-up and transfer efficiency of insecticides to the extent that the choice of formulation based on the laboratory assays are in close agreement with subsequent field studies (Crease *et al.*, 1987; Ford *et al.*, 1987). Such approaches bode well for the future use of pesticide transfer/behaviour studies for identifying optimum deposition characteristics.

While such laboratory studies may indicate the ideal form of insecticide deposit on a leaf surface, their use may also present some practical problems. For instance, the use of small droplet size ($< 100\mu m$), which in the laboratory may be shown to provide the optimum deposition characteristics, under field conditions is actually quite difficult to apply because such droplets tend to be vulnerable to downwind drift (Matthews, 1992). A different approach that incorporates the use of small outdoor miniplots (Grayson and McCarthy, 1987) provides a great deal of scope for further development of methodology in this area (Merritt, 1989).

7.3 INTERCROPPING

Intercropping is a traditional method of crop production, particularly in the tropics, which in addition can be used as a cultural control method. Intercropping has been shown to successfully suppress a range of pests including nematodes (Atwal and Mangar, 1969; Khan *et al.*, 1971; Castillo *et al.*, 1976; Egunjobi, 1984), pathogens (van Rheenen *et al.*, 1981; Moreno and Mora, 1984), insects (Perrin, 1977; Perrin and Phillips, 1978) and weeds (Beltrao, 1994). The method certainly also has some application and potential for use in low input farming of temperate regions. Intercropping is unlikely however, to find a place in most modern agriculture until the research technology for intercrops is as well developed as it is today for monoculture and sole crops (Coaker, 1990).

Research carried out on the role of intercrops in reducing pest abundance falls into two distinct categories: an approach based on: (i) the study of ecological mechanisms; and (ii) pest control experimentation. The former approach is concerned with the underlying principles contributing to differences in pest abundance between intercrops and sole crops, while the latter seeks to identify crop combinations and spatial patterns that will reduce levels of pest infestation for a specific primary crop, in a particular farming system.

7.3.1 Pest control experimentation

Research work on intercrops dates back to 1934 (Okigbo and Greenland, 1976). Since that time the vast majority of work with intercrops has concentrated on improving yield through various crop combinations, densities and spatial designs and a whole range of agronomic factors (Vandermeer, 1989). Among these have been experiments to determine the effects of intercrops in reducing levels of pest abundance. Comparisons have been made between the size of pest populations in sole crops and various types of intercrop combination to find the most suitable intercrop for suppressing pest populations (Amoaka-Atta and Omolo, 1983; Power, 1987; Ram *et al.*, 1989; Gold and Wightman, 1991). However, it is often the case that such studies reveal inconsistent results, with certain crop combinations reducing pest abundance of some species but not others. For instance, in studies of field beans (*Phaseolus vulgaris*) in a relay intercrop, with winter wheat, the intercrops reduced populations of *Empoasca fabae* and *Aphis fabae* but increased densities of *Lygus lineotaris* and *Systema frontalis* (Tingey and Lamont, 1988). In an intercrop of Brussels sprouts (*Brassica oleracea gemnifera*) with spurry (*Spergula arvensis*) the number of *Mamestra brassicae* and *Evergestis forficalis* were reduced but the intercrop had no effect on levels of infestation of *Pieris rapae* (Theunissen and den Ouden, 1980). Studies of some crop combinations have produced conflicting results. In the cowpea maize and cowpea sorghum intercrop the work of Karel *et al.* (1982) indicated a reduction of insect population size while-others reported increases (Kayumbo, 1976; Ochieng, 1977; Ezueh and Taylor, 1983; Ezeuh, 1991). These examples illustrate the location-specific nature of the effects of intercrops on pest abundance and the need for extensive research over a number of years before various intercrop combinations can be recommended (Omolo *et al.*, 1993).

A drawback of the majority of studies carried out to determine the influence of intercrops on pest infestation has been their emphasis on measuring pest abundance without any assessment of crop yield (Coaker, 1990). In an extensive study of the literature only 12.6% of studies reported on crop yield (Risch *et al.*, 1983) but interpretation of these was difficult because none used the conventional method for comparing yield of sole and intercrops; land equivalent ratios (a measure of yield per unit of land) (Mead and Willey, 1980). The evaluation of yield effects is absolutely essential, since there is no assurance that greater levels of infestation in monocrops will necessarily reduce crop yield (Bach, 1980a,b; Dempster, 1969). Coaker (1990) considered that yield effects could be best studied by comparing intercrops and monocultures treated with or without pesticides to deduce whether the yield differences are caused by differences in pest populations, an approach not used to date in intercrop pest control experiments.

7.3.2 Ecological mechanisms

The experimental approaches discussed above have identified a tendency for some intercrops to reduce levels of pest abundance. However, in order to develop a predictive theory of how diversity affects pest populations and to allow finer tuning of the agronomics of crop mixtures, it is necessary to gain a greater understanding of the ecological mechanisms involved that are responsible for the differences in pest abundances (Coaker, 1990).

The earliest proposed mechanisms were those of Aiyer (1949). Aiyer hypothesized that: (i) hostplants are more widely spread in intercrops and hence are harder to find; (ii) one species serves as a trap-crop to prevent the pest from finding the other crop; and (iii) one species serves as a repellant to the pest. These hypotheses were replaced at a later date by the 'resource concentration' and 'natural enemy' hypotheses of Root (1973). A number of studies have been carried out to test these hypotheses.

The 'resource concentration' hypothesis considers that the presence of different plant species has direct effects on the ability of the pest to find and utilize its hostplant, and predicts lower pest abundance in diverse communities. The reasons for this may be the presence of confusing or masking chemical stimuli (e.g. onion volatiles masking carrot chemical stimuli; Uvah and Coaker, 1984), or physical barriers to movement (Perrin and Phillips, 1978). In general, it would be expected that a hostplant stand which is adequately camouflaged by diverse vegetation background should encounter lower levels of insect immigration (Uvah and Coaker, 1984), faster foraging movement (Tukahirwa and Coaker, 1982) and higher rates of emigration of insects (Wetzler and Risch, 1984; Bach, 1980a,b; Risch, 1981).

The 'natural enemies' hypothesis emphasizes the role of diverse habitats in attracting and maintaining higher populations of natural enemies that subsequently exert a level of control over the pest species present. An intercrop is considered to provide more favourable conditions than a monocrop by providing a greater temporal and spatial distribution of nectar and pollen sources and alternative prey when the pest species are scarce (Risch, 1981). There have been a number of studies that indicate the natural enemy abundance is increased in more diverse intercrop situations. However, in most studies undertaken there has been little evidence to suggest that they significantly contribute to reduced pest levels (Bach, 1980a,b; Tukahirwa and Coaker, 1982; Uvah and Coaker, 1984; Letourneau, 1987; Theunissen *et al.*, 1992).

The hypotheses proposed by Aiyer (1949) and Root (1973) have provided the main theoretical framework for considering mechanisms of reducing pest population size in intercrops, however, more recently Vandermeer (1989) proposed an alternative scheme involving three hypotheses: (i) the disruptive crop hypothesis; (ii) the trap-crop hypothesis; and (iii) the natural enemy hypothesis. The disruptive crop is equivalent to (i) and (iii) of Aiyer and the trap-crop of (ii). The disruptive crop and trap-crop together are equivalent to the resource concentration hypothesis of Root and the 'natural enemy' hypothesis are identical in each scheme (Vandermeer, 1989). The separation of the resource concentration into two components is done to emphasize the requirement for different theoretical treatments of the two strategies. Whether this scheme will replace that of Root (1973) remains to be seen.

7.4 HOSTPLANT RESISTANCE

The subject of hostplant resistance has always been dominated by the practical need to breed crop plants resistant to pests and diseases. Until early this century plant breeding was the domain of 'biometricians' studying quantitative genetics, i.e. the inheritance of characters which differ by degree and show continuous variation. The basis of their thinking was the normal distribution. The biometricians however, placed no particular emphasis on breeding for resistant crop plants. In 1900, Hugo de Vries in the Netherlands, Carl Correns in Germany and Erich Tschermak von Seysenegg in Austria, independently and simultaneously rediscovered Mendel's laws of inheritance, and a new school of genetics immediately came into existence based on a study of qualitative genetics, i.e. a study of characters which differ in kind and are either present or absent, with no intermediates (Robinson, 1991). Despite the obvious significance of Mendel's laws to the study of genetics it soon became apparent that few characters were of practical significance and of use in plant breeding. Then, in 1912 Biffen discovered that resistance of wheat to yellow rust was controlled by a major gene (Parry, 1990) and in 1916 Harlan demonstrated that resistance to the leaf blister mite *Eriophyes gossypii* was a heritable trait in cotton (Smith, 1989), and the 'Mendelians' (as they were known), had discovered an important practical application for their endeavours. Accordingly they pursued the topic with great vigour, to the extent that for most of this century the breeding of crops for resistance to parasites has belonged almost totally to the Mendelian school of genetics (Robinson, 1991).

The history of discovering useful resistance genes and incorporating them into agronomically acceptable cultivars has proved particularly successful for the breeding of crops resistant to pathogens. This long history of success led to examinations of the pathogen virulence and host resistance structures and the subsequent development of the gene-for-gene paradigm (Flor, 1942; van der Plank, 1963; Robinson, 1976, 1987). Over the same period successes with breeding for crop resistance to insects was far less spectacular (de Ponti, 1983). Emphasis in insect/hostplant resistance studies was placed on identifying mechanisms of resistance in terms of the three categories defined by Painter (1951); non-preference, tolerance and antibiosis (see Chapter 3). The biochemical and morphological characteristics of plants were studied and although several investigations have shown that these plant traits have a genetic basis, on the insect side, the genetic variation underlying the ability to utilize different hosts has not been related to variation in actual traits that influence such virulence (Marquis and Alexander, 1992).

The divergence of approaches has led to the development of particular experimental techniques according to the school of thought to which one subscribes. However, given the more recent advances in biotechnology and their implications for hostplant resistance it seems likely that greater emphasis will in the future be placed on the genetic interactions of all types of parasites and their crop hosts, including insects.

The experimental paradigms of hostplant resistance are considered in terms of inheritance of resistance, mechanisms of resistance and biotechnology.

7.4.1 Inheritance of resistance

The two schools of genetics, the 'Mendelians' and the 'biometricians', based on qualitative and quantitative genetics respectively, inevitably produced two entirely different methods of plant breeding (Robinson, 1991). The techniques used for qualitative characters are based on the transfer of a single gene character from one plant to another. These are the gene-transfer methods referred to as pedigree breeding and backcrossing (Mayo, 1987; Simmonds, 1979; Dent, 1991). A resistant character for a wild progenitor or primitive cultivar may be transferred to a modern high-yielding, but susceptible, cultivar, by a controlled cross-pollination. The resistance genes will show clear-cut and discrete segregation in the next generation. The most resistant progeny (they will have segregated into Mendelian ratios of susceptible and resistant depending on the dominance of resistance gene expression) will then be back-crossed to the original modern cultivar in order to restore some of its valuable high yielding and agronomically valuable characteristics. This process will then be repeated a number of times until the qualities of the original cultivar have been restored in a plant which carries the gene for pest or disease resistance (Robinson, 1991).

Quantitative genetics, working with many genes, each contributing in some degree to the required characteristics, requires a completely different approach to plant breeding. This involves population-breeding techniques, referred to as recurrent mass selection, which aim to change the gene frequencies for a particular character in an existing genetically mixed population. Large populations of crop plants are screened for the small number that have the desired characters. These plants are selected and randomly poly-crossed over many generations so that the mean value of the desired characteristics shifts away from the original population. Resistance will accumulate to useful levels after 10–15 generations which could mean only 5–7 year cycles with two cropping seasons per year. The technique has been used successfully to breed for resistance to leafhoppers and leafhopper yellowing rust and bacterial wilt in alfalfa (Hanson *et al.*, 1972) and to wilt, nematode and insect pests of sweet potato (Jones *et al.*, 1976). However, there are far more examples of resistance breeding involving qualitative techniques and these have been recently reviewed (Singh, 1986; Khush and Brar, 1991).

7.4.2 Mechanisms of resistance

Mechanistic terms such as 'hypersensitive', 'tolerance' and 'antixenosis' typically provide a shorthand way to describe a highly complex, poorly understood situation (Harris and Frederiksen, 1984). They tend to serve the purpose of providing a general direction to further investigation and of providing a point of reference with which to compare similar findings. They are however, particularly vulnerable to misunderstanding and ambiguity.

Among pathologists, mechanistic terminology has been particularly varied but the mechanisms defined by Painter (1951), antibiosis, non-preference (now antixenosis) and tolerance have gained wide acceptance among entomologists. This has subsequently produced experimental methodologies and evaluation procedures by which these categories of resistance are assessed.

Antixenosis may be measured in terms of the preference shown for particular crop cultivars in either 'choice or no-choice' experiments. Experiments which evaluate preference between cultivars by quantifying responses in a no-choice situation may provide spurious results. This is because the response of the insects to a particular cultivar may be conditioned by the presence of other cultivars (Cantelo and Sanford, 1984). In laboratory studies, the host or type of diet on which the insects have been reared will also affect their preference (see below). Preference for a host can be measured, not just in terms of the numbers of insects alighting on a particular cultivar, but also in terms of the numbers leaving cultivars. The rationale behind this approach is that insects which have located susceptible plants will be less inclined to leave them than insects on resistant plants. Studies of this kind have been carried out both in the laboratory (Dent, 1986) and in the field (Müller, 1958).

Observations of insect behaviour provide a useful technique in the study of antixenotic resistance, often because behavioural responses exhibited by the insects provide an indication of the resistance mechanism involved (Hsiao, 1969; Bernays *et al.*, 1983, 1985; Givovich *et al.*, 1988).

Resistance to insects and mites is more often assessed in terms of antibiotic resistance, i.e. the effect of resistance on pest development, fecundity and survival. Insects are usually confined to the host cultivar by cages in either the laboratory or the field. Estimates are obtained of development in terms of development rate (Dent and Wratten, 1986), difference in size (Uthamasamy, 1986; Salifu *et al.*, 1988) or weight (Perrin, 1978; Dent and Wratten, 1986; Sekhon and Sajjan, 1987), fecundity/fertility in terms of numbers per plant (Lowe, 1978) embryo number (Dewar, 1977), r_m values (Birch and Wratten, 1984; Holt and Wratten, 1986) and life-table analysis (Easwaramoorthy and Nandagopal, 1986).

Tolerance, (a mechanism potentially common to all parasite/host interactions), the ability of the host to endure the presence of the parasite without suffering significant yield loss, is difficult to assess (Parlevliet, 1981). It is evaluated by measuring yield differences, but such difference could equally well result from variation in levels of resistance (antixenotic or antibiotic). When a breeder assesses yield losses the estimate obtained is a measure of the combined effects of both tolerance and resistance. To separate these two effects, an assessment of both yield loss and pest incidence (or severity) has to be determined. Parlevliet (1981) provides an example (Table 7.2) of how this may be achieved based on the yield loss (y) equation $y = 0.3x - 3$ for the percentage of leaf area of barley affected by mildew (*Erysiphe graminis*) (x) 3 weeks after heading. The equation indicates that there is no yield loss at 10% leaf area affected and that above this level each 10% leaf area affected decreases the yield by 3%. Differences in resistance (percentage leaf area affected) between varieties are indicated in Table 7.2. The variation in yield loss caused by the variation in resistance is described by the linear regression of yield loss on the disease rating. Deviations from this regression indicate relatively low or high levels of tolerance. Varieties A, C and F (Table 7.2) do not deviate from the regression line (zero values in final column) and hence, their yield loss is caused by variation in resistance. Variety E, on the other hand, has a yield loss equal to that of C (C = 12%; E = 12.5%) although it is more resistant (it exhibits a low affected leaf area). The yield loss is therefore, caused by a lack of tolerance. Variety D is, however, relatively tolerant because it has

Table 7.2 Effect of powdery mildew, *Erysiphe graminis*, on yield reduction of six barley varieties with different levels of tolerance and/or resistance. (From Parlevliet, 1981)

Variety	*Percent of leaf area affected*	*Yield reduction (%)*		
		Total	*Due to lack of resistance*	*Due to lack of tolerance*
A	80	21.0	21.0	0.0
B	60	18.5	15.0	3.5
C	50	12.0	12.0	0.0
D	50	7.0	12.0	5.0
E	40	12.5	9.0	3.5
F	20	3.0	3.0	0.0

a yield loss considerably smaller than C which has the same level of resistance (Table 7.2). It should be noted however, that unless the estimates of pest incidence are accurate then differences in resistance may well go unnoticed and the resulting yield loss assumed to be derived from host tolerance (e.g. Parlevliet, 1981; Brönniman, 1975).

Bioassays to assess antixenotic and antibiotic resistance are often carried out, not only to assess the levels of resistance of various cultivars but also to actually identify mechanisms of resistance in terms of their morphological or biochemical basis. When a resistance trait has been identified the plant-screening process can then be based on the presence or absence of that character or the degree to which it is expressed (Dent, 1991). For biochemical traits this could mean the development of an analytical technique which could be used to screen large numbers of plants quickly and efficiently for the required chemical. Identification of readily visible morphological characters would also obviate the need for time-consuming bioassays, allowing screening of resistance on the basis of the presence or absence of the character. Hence, the potential benefits are enormous, both in terms of reduced labour, time and an increased efficiency, provided that the character can be reliably identified.

Traditionally, screening for genotypes resistant to pests starts with current agronomically acceptable cultivars, followed by evaluation of abandoned cultivars, foreign cultivars and land races and if these do not provide a useful source of resistance then breeders turn to aboriginal and related plant species (Harris and Frederiksen, 1984). Plants in these latter two categories are largely ignored, however (Gould, 1983).

Screening procedures tend to emphasize measures of the levels of pest infestation, incidence or damage. Rarely in the initial stages of a programme is yield evaluated. Since there is a need to evaluate large numbers of genotypes, techniques tend to involve readily identifiable traits that allow rapid assessment of material. These techniques are based on visual assessment and scoring systems (e.g. Figure 7.2) (Bellotti and Kawana, 1980) and are largely carried out in the field. However, despite their common use, the techniques are subjective and their reliability dependent on familiarity with the scale and the assessors' relative experience (Ismail and Valentine, 1983; Valentine and Ismail, 1983).

Experience with races and biotypes of pathogens and arthropods has shown that consideration of genetic variability within the pest species is important (Harris and Frederiksen, 1984). Plant genotypes need to be screened against pest populations of known genetic constitution in order to ensure that only parts exhibiting resistance to these known biotypes/races are allowed to progress through the breeding scheme. This means screening plants against the whole range of biotypes and races available. While it is possible to maintain and catalogue many populations, isolates and pathotypes of plant pathogens, the situation with insects is far more difficult. Insect pest populations utilized for screening tend to be field-collected populations, or those reared in a laboratory colony. Genetic diversity of laboratory colonies of insects is generally low and insects used rarely come from more than one laboratory (Gould, 1983). Although less problematic, field screening procedures are also limited because they often involve only one or a few local insect populations. These populations may never have encountered the resistant cultivars before and hence, do not provide a reliable measure of the potential range of biotypes in the insect species. Hence, insect screening procedures based on such restricted genetic material will not provide reliable results.

The problem is one of assessing the potential of pest populations to increase in fitness on a new variety. One method which could be used to circumvent this problem would be to undertake a pest breeding programme. Single pest populations or a hybrid population can be selected for higher fitness on the new resistant cultivar by being reared on that variety for as many generations as possible. In this manner the most fit genotypes would rise in frequency in the population and hence, exhibit a detectable response (Gould, 1978). If the increase in fitness were large, plant breeders would be forewarned that resistance in the new variety may be only temporary, especially if used extensively. Obviously only extensive pest breeding and/or exhaustive sampling of wild populations would reveal rare genes, but the methodology has some potential and could readily be adapted to suit a range of pest species. As Gould (1978) points out, given the investment needed to produce resistant cultivars (commonly 10 years' effort), the costs of developing a sub-programme for pest breeding are relatively low compared to the potential gain in longevity of the resultant cultivar.

7.4.3 Biotechnology

Biotechnology represents one of the fastest developing fields of science over the last decade and it has direct application to the development of resistant crop cultivars. What were seen as just promising gene manipulation techniques a few years ago, have now already produced new crop varieties (Monti, 1992). This non-sexual means of gene transfer, whether it is used for direct substitution of existing alleles in very much the same way as in conventional plant breeding or whether it is used for introducing 'alien' genes from completely unrelated species (e.g. the gene for *Bt* toxin), is becoming a very powerful tool in plant breeding (Evans, 1993).

The development of genetic manipulation through a range of somatic cell and molecular procedures, concurrent with elucidation of many of the molecular events associated with the transformation of plant cells by Gram-negative soil bacteria of the genus *Agrobacterium*, has led to the development of a technique for inserting genes into plant cells (Davey and Finch, 1991). Although other techniques, such as micro-injection of DNA into host cells, the use of micro-projectiles to convey DNA to its target (particle gun), or the direct uptake of DNA by plant protoplasts are available, it has been the use of the vector *Agrobacterium* Ti plasmid that has proved the most efficient non-sexual method of acheiving complete integration of genes into dicotyledonous plants (Evans, 1993). Basically transferring genes from one organism to another requires the availability of: (i) a DNA vector, which can replicate in living cells after foreign DNA has been inserted into it; (ii) a DNA donor molecule to be transferred; (iii) a method of joining the vector and the donor DNA; (iv) a means of introducing the joined DNA molecule into the recipient organism in which it will replicate; and (v) a means of screening for recombinant lines that have replicated the desired recombinant molecule (Lindquist and Busch-Petersen, 1987).

The technique has proved successful for conferring plant resistance to viruses and insect pests (Vaeck *et al.*, 1987; Nejidat and Beachy, 1990; Perlak *et al.*, 1990; Hill *et al.*, 1991; Ling *et al.*, 1991). These and other studies of their kind point the way to a highly productive future for host-plant resistance (Khush and Toenniessen, 1992).

7.5 NATURAL ENEMY THEORETICAL MODELS

Classical biological control, involving the introduction of exotic natural enemies for the control of pest insects has a long and relatively successful history. Most of the successes have been with pests of fruit, forest and range crops (Greathead and Waage, 1983; Huffaker, 1985) reflecting the advantage of stable systems which permit continuous interaction between the natural enemy and its pest host (Hassell, 1978). This level of success however, has been achieved against a background of a number of failures. Many biological control agents have been introduced but for one reason or another have failed to establish or exert the required levels of control. A success rate of about 1 in 7 appears to be the general rule (Hokkanen and Pimentel, 1984). Given this probability of success it is hardly surprising that biological control practitioners have sought to identify the factors which contribute to a successful introduction and in addition define the essential features of an effective biological control agent. Mathematical theory has been considered useful in this context, to the extent that it can capture the important aspects of nature and make general statements about the useful characteristics possessed by successful biological control agents (Murdoch *et al.*, 1985).

Population modelling has tended to follow two very different approaches in the study of pest–natural enemy interactions: (i) simulation models; and (ii) simple analytical models (Godfray and Waage, 1991). It is the use of simple analytical models incorporating only the minimal biological details necessary to describe the system, that are considered here in the context of defining the characteristics of biological control agents. These models are, in most cases, based on the Lotka–Voltera equations or the Nicholson–

Bailey (1935) models which make a very important assumption; they assume that a low stable equilibrium pest population is the most desirable outcome from introducing a natural enemy for the purposes of control. This is based on the premise that a natural enemy which drives its pest host to extinction will itself become extinct. Hence, to maintain stable populations of the pest and the natural enemy a proportion of the pest population should always avoid attack by the natural enemy. Control will be exerted by maintaining the pest population at a lower stable equilibrium level. This standard assumption however, has been challenged (Murdoch *et al.*, 1985). An alternative paradigm has been put forward that considers the role of spatial scale as a key factor in maintaining stable population levels and that pest extinction, at least at a local population level, is an acceptable goal in pest control.

Superficially it may not seem particularly important which of the two approaches represents and explains successful biological control, except perhaps to the scientists embroiled in the debate; however, there are practical implications to these findings. This is because the different modelling approaches lead one to look for completely different properties of biological control agents to explain successful control. If biological control methods are to be improved then there is a need to establish the essential characteristics of a good biological control agent. The arguments are considered under the headings of deterministic–stable equilibrium models and stochastic–non-equilibrium models.

7.5.1 Deterministic–stable equilibrium models

Deterministic models assume that populations of the pest and the natural enemy are of infinite size and that any reproductive individual gives rise to a set number of offspring within a defined period of time (Maynard-Smith, 1974). Such models ignore environmental randomness and, since many of them have been developed using difference equations, they are mostly only applicable to temperate insect pests having discrete single generation cycles (but see Godfray and Hassell, 1987). Given these limitations the deterministic models have concentrated on the simple dynamics of host-specific parasitoids, which have a single age class and a number of progeny equal to the number of parasitoid hosts.

The basic model used to study the dynamics of parasitoid–prey interactions have largely revolved around:

$$N_{t+1} = Fg(N_t)\ N_t\ f(N_t, P_t)$$

$$P_{t+1} = csN_t\ [1 - f(N_t, P_t)$$

where N_t is the initial host population size and N_{t+1} the number in the next generation. Similarly P_t and P_{t+1} represent the initial and subsequent parasitoid population size and subsequent population size in the next generation, respectively. $Fg(N_t)$ is the per capita net rate of increase of the host population while $g(N_t)$ is a density-dependent function. c and s are respectively the average number of progeny produced per host individual attacked and the proportion of these progeny that are female. The function $f(N_t, P_t)$ defines host survival (see Hassell and Waage, 1984; May *et al.*, 1988).

Over a period of about 20 years (1969–1989) a methodical theoretical appraisal was carried out to determine how various factors would affect the equilibrium levels and stability of the pest and natural enemy populations. The changes in the equilibrium level were considered in relation to: (i) changes in the rate of increase of the host population (Waage and Hassell, 1982; Hassell and Waage, 1984); (ii) parasitoid mortality (Waage and Hassell, 1982); (iii) density-dependent sex ratios (Hassell *et al.*, 1983); and (iv) parasitoid feeding habit (Kidd and Jervis, 1989). The stabilizing mechanisms of the host–parasitoid interactions were similarly considered, particularly in terms of: (i) density dependence of the prey population (Beddington *et al.*, 1975); (ii) the type of functional response (Hassell and Rogers, 1972; Hassell, 1984); (iii) mutual interference of attacking parasitoids (Hassell, 1984); (iv) host susceptibility (Turnock, 1972; Hassell and Anderson, 1984); (v) variable parasitoid sex ratios (Hassell *et al.*, 1983; Comins and Wellings, 1985); (vi) parasitoid synovigeny (Kidd and Jervis, 1989); and most importantly (vii) non-random search by parasitoids in a patchy prey environment (Beddington *et al.*, 1978; Hassell, 1982, 1984). Ironically it is the aggregation of parasitoids in patches of high host density (leaving other patches as partial refuges from parasitism) that seems to provide the major mechanism for stabilising the population at low equilibrium levels and thus ensuring control of the pest (Beddington *et al.*, 1978; Hassell, 1978, 1980, 1982; Waage, 1983).

The results of these theoretical studies implied that the following characteristics were important for successful biological control agents: (i) host specificity; (ii) capability for rapid increase in response to the pest population growth; (iii) the need for only one pest individual to reproduce itself; (iv) a high search rate; and (v) an ability to aggregate in areas of high pest density. Not surprisingly these attributes are more characteristic of parasitoids than of predators. Hence, host-specific parasitoids have been accorded priority status in the search for suitable biological control agents, a status based on their ability to depress pest populations to new, low and stable equilibrium levels.

7.5.2 Stochastic–non-equilibrium models

The depression of a pest population to a new but lower stable equilibrium level is an unnecessary prerequisite for achieving effective control. Provided a pest population can be maintained below its economic threshold it doesn't matter whether or not the pest population is stable. It doesn't even matter if the pest population becomes extinct. Low stable equilibria have always been sought because it has been assumed that pest extinction implies natural enemy extinction. This would be considered unsatisfactory from a biological control practitioner's point of view. However, if the need for such low stable equilibrium is not assumed and the type of natural enemy effective at control is not considered to be limited to host-specific parasitoids then other options immediately become evident.

Murdoch *et al.* (1985) challenged the deterministic–low stable equilibrium paradigm by considering the implications of utilizing polyphagous natural enemies as control agents and by viewing the problem of stability in terms of metapopulation dynamics. Polyphagous natural enemies are characterized by their ability to survive and reproduce in the absence of the pest (by attacking alternative prey). Hence, with polyphagous natural enemies there is little justification for requiring a stable equilibrium which maintains viable populations of the natural enemy and its pest host. Removing this limitation then permits a greater range of model parameter values that are compatible with successful control.

Metapopulation theory describes the dynamics of a number of local populations with limited interpopulation migration, in which local populations become extinct but may be reestablished by occasional successful reinvasion, i.e. the local populations all display unstable dynamics (Crowley, 1981; Reeve, 1988). Probably all species persist as metapopulations at an appropriate spatial scale (Harvey *et al.*, 1992). Understanding stability in metapopulation dynamics is dependent on spatial scale. If we define our universe to be small enough (e.g. an individual) then in the short term extinction will be a likely event. However, as the size of the universe is increased the probability of extinction decreases (Figure 7.4). This decrease

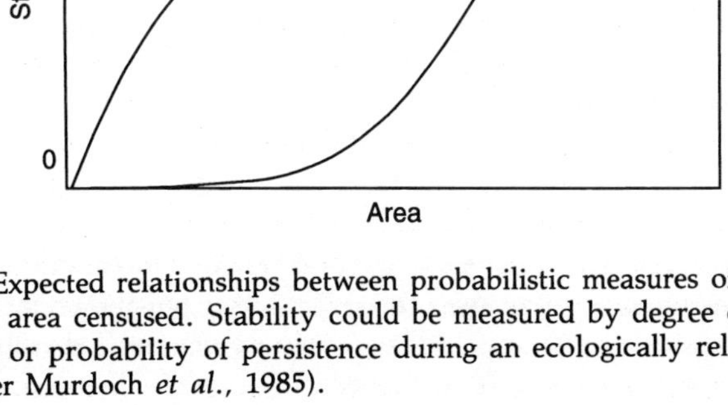

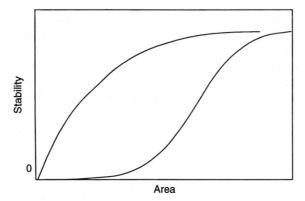

Figure 7.4 Expected relationships between probabilistic measures of population stability and area censused. Stability could be measured by degree of stochastic boundedness or probability of persistence during an ecologically relevant time period. (After Murdoch *et al.*, 1985).

results from a degree of asynchrony of the local populations in space. If the system is large enough it should tend to fragment into two or more local populations with low exchange rates that permit these populations to remain out of phase with each other (Crowley, 1981). Such a feature creates a mosaic of fluctuating local populations that can yield remarkably constant population densities for the system as a whole.

Population density is viewed differently in deterministic models. In the first instance they do not include the possibility of extinction and are insensitive to questions of scale and secondly, their stabilizing mechanisms are considered to occur uniformly throughout the environment. Further, there seem to be few field studies which actually support the standard view that control is achieved through the establishment of a low stable pest–enemy equilibrium (Murdoch *et al.*, 1985). The field data also fail to confirm that the characteristics exemplified by host-specific parasitoids are actually essential features of biological control agents. Given all of this, one would assume that a change of paradigm is imminent. However, for the most part the paradigm of local pest population dynamics and low stable equilibrium will continue, simply because it suits the scientists that it should do so.

7.6 IPM RESEARCH AND DEVELOPMENT

If an holistic view is taken of IPM research and development it becomes obvious that political, social, economic as well as scientific factors influence the emphasis and direction that has been taken. These, often subtle, factors can have a profound effect on, and shape the type of, research that is considered appropriate and acceptable. IPM programme managers need to be able to place their programme in a political, social, economic and scientific context. This is particularly important with regard to the relative emphasis placed on development and use of pest control products and techniques, on environmentally friendly control measures, and the increasing prominence given to 'major gene' control technologies. It is important that pest management programmes are designed to solve pest problems rather than meet the demands of trendy or politically correct ideology.

7.6.1 Products versus techniques

Scientific research gives rise to inventions such as new devices, materials or processes, ideas and know-how. Such results can be exploited. In terms of pest management exploitable research of value to farmers may be categorized as either products or techniques (Dent, 1993). A product may be defined as a device, material or substance (which may be a living organism, chemical or plant material) which is usually manufactured, produced, formulated or packaged for the purpose of sale. They are essentially purchased as off-farm inputs and are characterized by their commercial

Table 7.3 Pest control products and techniques that are potentially available for insect pest management

Products ←	*Continuum*	→ *Techniques*
Pesticides Chemical insecticides Microbial insecticides Insect growth regulators		
Semiochemicals		
High-yielding hybrids ——	Hostplant resistance ——	Traditional farming Selection of seeds from crop
Augmentation of natural enemies ——	Biological control ——	Conservation of natural enemies
		Cultural control Intercropping Plant spacing Planting date Tillage practices Habitat modification

value, their general application and for many, their broad-spectrum effects. A technique, by contrast, is a form of procedure, skill or method that may be utilized by the farmer from available on-farm measures. They are often based in principle on traditional agronomic and husbandry practices and tend to be specific at least to the level of the cropping system (Dent, 1993). The main types of control measures used in pest management can be divided on a continuum between products and techniques (Table 7.3).

The archetypal pest control products are chemical pesticides. From the sales of approximately 100 major chemical companies, pesticides currently have a world market worth \$19 000 m (Worthing and Hance, 1991; Whittaker, 1993). However, chemical pesticides are not the only pest control 'products'; there are companies which produce and sell microbial pesticides (based on bacteria, viruses and nematodes), insect growth regulators, semiochemicals and natural enemies (parasitoids and predators). More recently, seed companies have identified a new niche in the pest control market with the potential sale of genetically manipulated crop plants having engineered resistance to pests.

The extent and the diversity of these pest control products is the result of extensive investment in R&D both in the private and the public sector. This is reflected in the relative number of scientific papers published in the various subject disciplines (Table 7.4). It is interesting to note that even in these modern times with an awareness and interest in IPM, there is still a very large amount of research carried out on the different aspects of

Table 7.4 The number of publications (1979–1990) dealing with different control options in pest management. Obtained from on-line computer search of CAB Abstracts Database. (From Dent, 1993)

Subject (keywords)	*Number of publications*
Pesticides	30 679
Herbicides	8736
Insecticides	7500
Fungicides	5186
Biological control or biological control agents	10 434
Pest and disease resistance	9013
Pheromones	2568
Cultural control or cultural methods	1728
Sterile insect technique or sterile male technique	56

pesticides and their use (Dent, 1993). By comparison however, there have been relatively few studies on cultural control methods. It would not be sensible to place too much emphasis on the absolute numbers involved in each case, given the obvious problems with such data. However, they do serve to highlight the relative emphasis given to research on products (particularly pesticides), than to research on techniques (largely encompassed in cultural control) (Tables 7.3 and 7.4). Such an emphasis matters little provided that the farmer benefits from their use, i.e. use of products provides safe, more reliable, economic and sustainable means of pest control. However, while it may be true that many products are reliable and economic, it can also be said of many others that they are hazardous to apply and their use is unsustainable (see below). A great deal of pest management research seems to be locked into a rationale based largely on the use of control products. There are a number of influences which seem to have contributed to this, including not only the agro-industry but also scientists, and their funding agencies (Dent, 1993). Despite this however, control measures based on control products do not have to dominate rationale and pest management strategy. Pest control techniques can have an important role to play in the development of sustainable pest management, but for this to happen programme managers must begin to place greater emphasis on research into cultural control techniques (El Titi and Landes, 1990). The growing interest in integrated farming systems provides a testament to their value.

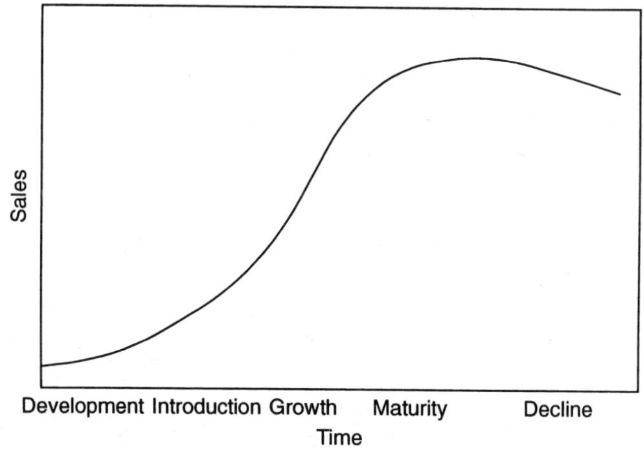

Figure 7.5 A product life-cycle. (After Heap, 1989).

7.6.2 The major gene phenomenon

R&D has to be viewed as an investment in terms of producing benefits which exceed its costs (Pearson, 1988). Economic returns from a product must more than cover its R&D costs, as well as its production, marketing and sales costs, in order for a company to obtain a profit. Hence, the life-cycle and performance of a product are critical issues in assessing the success of a product. If the product performs poorly and has a short life then the company may fail to gain a sufficient return on its investment. If however, a product is successful and has a long life, then a company can potentially make a large profit. A hypothetical product's life-cycle is presented in Figure 7.5.

The costs of developing chemical pesticides are increasing, mainly because it is becoming increasingly difficult to identify new compounds and those that are identified tend to be complex structures requiring a greater number of steps in their synthesis. There are also increases associated with more stringent registration requirements. This means that agrochemical companies are even more aware of the need to ensure their products are 'durable' to enable them to recoup their costs and make a profit. Hence, the interest and assistance provided by agrochemical companies to programmes involved in pesticide resistance management, prolonging the useful life of their pesticides.

In the public sector increasing attention is being given by plant breeders to maintaining the viability of pest-resistant cultivars. Although not assessed in economic terms, the shamefully short lives of some resistant crop cultivars due to the development of virulent pest races or biotypes is a cause for concern. When the resistance of the cultivars breaks down within just a few years it becomes questionable whether the exercise is tenable.

In both situations, the development of pest resistance to pesticides and the breakdown of pest-resistant cultivars, the products become inoperative because genetic changes in the pest result in adaptations enabling survival on or in, the presence of the resistant cultivar or pesticide, respectively. To a very large extent these adaptations appear to be under the control of single or a few major genes. Where resistance has developed (in pesticides) or broken down (hostplant cultivars) then major genes have very often been associated with the control of the adaptive mechanism. This is true of each of the classes of insecticides (organochlorines, organophosphates, carbamates and pyrethroids) (Roush and McKenzie, 1987; Roush and Daly, 1990) for systemic fungicides (Ragsdale and Siegel, 1985; Parry, 1990), herbicides (Caseley *et al.*, 1991) and is implicated in the resistance of insects to insect growth regulators (Cheng *et al.*, 1990; Pree, 1990). In studies of hostplant resistance, gene-for-gene relationships have been identified between crop plants and fungi (van der Plank, 1984; Parry, 1990), viruses (Walkey, 1985), bacteria (Parry, 1990), nematodes (Bingefors, 1982), insects (Khush and Brar, 1991) and weeds (Singh and Emechebe, 1992). In addition, a whole range of resistance characters have been identified that are under the control of major genes but for which gene-for-gene relationships are as yet not known (Wilcox, 1982; Khush and Brar, 1991). Major genes have also been shown to be responsible for control of the protein toxin (Kronstad *et al.*, 1983) and breakdown of resistance in *B. thuringiensis* (McGaughey, 1985) and in the control and production of insect sex pheromones (Löfstedt, 1990). Given that this range of products represent the main pest control measures currently available and that the next generation of pest control measures based on biotechnology will also utilize major genes, the potential for pest adaptation, breakdown and loss of valuable control agents is quite high. Obviously the possibility of such widespread problems must be tempered by the spectrum of differences in pest biology, ecology and operational factors. However, by looking at the problem of development of pest control products from a larger, more holistic perspective there is presently a predominance of measures that are dependent on major gene-based phenomena for their viability, and this should be a cause for concern.

There are of course alternative approaches not based on major gene phenomena but based on polygenic controlled characteristics. These, at least in theory, should provide greater durability. The most obvious is the use of plant breeding techniques for horizontal resistance (see section 3.3) based on the additive effects of many minor genes. Also, a number of pesticides that have not developed resistant pest populations after decades of use (Bordeaux mixture) deserve further study to elucidate mechanisms and modes of inheritance. It could be possible to develop pesticides with mechanisms controlled by many minor genes, thus decreasing the likelihood of resistance developing. Even crop cultivars based on gene-manipulated resistances may in the future be based on polygenic characters In the short term however, incorporation of different control measures in IPM or PRM (pesticide resistance management) systems should reduce the likelihood of resistance 'development' or 'breakdown' occurring. In the long term, and particularly as far as agro-business is concerned, the life of pest control products will be best extended through development of measures based on many minor genes.

7.6.3 Environmentally friendly control measures

The concern for the environment is now widespread and has even become an important political issue, particularly among Western nations. When an issue gains such predominance it becomes difficult to make decisions that run counter to the prevailing view.

The use of chemical pesticides is now generally held as being detrimental to the environment, and there is no doubt that the environmental and social costs associated with their use may be high (Table 7.5). However, these costs should not be considered in isolation from the benefits that are obtained from their use. Programme managers must be very aware of the dangers of joining a 'bandwagon' when addressing the problems of devising pest management strategies. Control measures such as pesticides should not be dismissed out of hand but viewed against the relative costs and benefits associated with their use in each particular situation. One of the most poignant examples of this problem in the history of pest control has been the debate that has raged over the chemical insecticide DDT.

DDT was most successfully used in campaigns to control the spread of mosquito-borne malaria. The mosquitoes were controlled by treating their roosting sites inside buildings with an annual application of 1 or 2g DDT per square metre. The disease was eliminated from Mauritius and Cyprus, the Netherlands and Spain, and in 1952 it was estimated that 100 000 000 people in various countries were protected by DDT (Mellanby, 1989). Rachel Carson's *Silent Spring* did much to encourage various developing

Table 7.5 Estimated total environmental and social costs associated with pesticide use in the USA. (From Pimentel *et al.*, 1991)

Cause	*Cost ($ million)*
Human pesticide poisoning	250
Animal pesticide poisoning and contaminated livestock products	15
Reduction of natural enemies of pests	150
Resistance of pests to pesticide	150
Honeybee poisoning and reduced pollination	150
Losses of crops and trees	75
Losses of fish and wildlife	15
Government pesticide pollution regulations	150
Total	955

countries to stop using DDT at a time when malaria was still being controlled. Mellanby (1989) contends that those responsible for the premature bans on DDT must bear the responsibility for thousands, perhaps millions, of deaths which might have been avoided. It is perhaps ironic that despite legitimate concerns for the environment DDT was, and is, probably one of the safest and least hazardous insecticides for handling and application (Spindler, 1983). From studies of production workers over 20 years it has been shown that DDT causes no adverse effects, and is not carcinogenic or teratogenic. Spindler (1983) summed up his studies by stating 'if the extremely low health risk of DDT for man is weighed against the enormous benefits resulting from its global use, more particularly in control of malaria, but also in agriculture, all the criticism alleging that DDT had an alarming effect on the environment, including man is completely unrealistic'.

Environmentalism, for all its virtues, is only a paradigm - an approach based on currently acceptable assumptions and premises. The paradigm will inevitably be replaced by another, perhaps even towards greater concern, for instance, 'globalism', but decision makers in IPM must be sufficiently objective to recognize and understand the nature of existing paradigms. Only in this way can they make rational, objective decisions concerning the development of strategies that will meet the real needs of the farmers and the communities they serve.

REFERENCES

Adams, A.J. and Hall, F.R. (1989) Influence of bifenthrin spray deposit quality on the mortality of *Trichoplusia ni* (Lepidoptera: Noctuidae) on cabbage. *Crop Protection*, **8**, 206–1.

Adams, A.J., Abdalla, M.R., Wyatt, I.J. and Palmer, A. (1987) The relative influence of the factors which determine the spray droplet density required to control the glasshouse whitefly, *Trialeurodes vaporariorum*. *Aspects of Applied Biology*, **14**, 257–66.

Aiyer, A.K.Y.N. (1949) Mixed cropping in India. *Indian Journal of Agricultural Science*, **19**, 439–543.

Amoaka-Atta, B. and Omolo, E. (1983) Yield losses caused by the stem/pod borer complex within maize–cowpea–sorghum intercropping system. *Insect Science and its Application*, **4**, 39–46.

Atwal, A.S. and Mangar, A. (1969) Repellent action of root exudates of *Sesamum orientale* against the root-knot nematode *Meloidogyne incognita*. *Indian Journal of Entomology*, **31**, 286.

Bach, C.E. (1980a) Effects of plant density and diversity on the population dynamics of a specialist herbivore, the striped cucumber beetle, *Acalymma vittata* (Fab.). *Ecology*, **61**(6), 1515–30.

Bach, C.E. (1980b) Effects of plant diversity and time of colonization on an herbivore–plant interaction. *Oecologia (Berl.)*, **44**, 319–26.

Bailey, N.T.J. (1981) *Statistical Models in Biology*, 2nd edn, Hodder & Stoughton, London.

Beddington, J.R., Free, C.A. and Lawton, J.H. (1975) Dynamic complexity in predator-prey models framed in difference equations. *Nature*, **225**, 58–60.

Beddington, J.R., Free, C.A. and Lawton, J.H. (1978) Characteristics of successful natural enemies in models of biological control of insect pests. *Nature*, **273**, 513–19.

Bellotti, A. and Kawana, K. (1980) Breeding approaches in cassava, in *Breeding Plants Resistant to Insects*, (eds F.G. Maxwell and P.R. Jennings), John Wiley & Sons, Chichester, pp. 313–16.

Beltrao, N.E. de M. (1994) Weed management in cotton, in *Weed Management for Developing Countries*, (eds R. Labarda, J.C. Caseley and C. Parker), FAO, pp. 340–5.

Berlocher, S.H. and Friedman, S. (1981) Loss of genetic variation in laboratory colonies of *Pharmia regina*. *Entomologia Experimentalis et Applicata*, **30**, 205–8.

Bernays, E., Woodhead, S. and Haines, L. (1985) Climbing by newly hatched larvae of the spotted stalk borer *Chilo partellus* to the top of sorghum plants. *Entomologia Experimentalis et Applicata*, **39**, 73–9.

Bernays, E.A., Chapman, R.F. and Woodhead, S. (1983) Behaviour of newly hatched larvae of *Chilo partellus* (Lepidoptera: Pyralidae) associated with their establishment in the host-plant, sorghum. *Bulletin of Entomological Research*, **73**, 75–83.

Bingefors, S. (1982) Nature of inherited nematode resistance in plants, in *Pathogens, Vectors, and Plant Diseases: Approaches to Control*, (eds K.F. Harris and K. Maramorosch), Academic Press, New York, pp. 187–219.

Birch, N. and Wratten, S.D. (1984) Patterns of aphid resistance in the genus *Vicia*. *Annals of Applied Biology*, **104**, 327–38.

Blair, A.M. and Martin, T.D. (1989) Effects of laboratory and field environments on plant root growth and consequences upon pesticide uptake by roots, in *Comparing Laboratory and Field Pesticide Performance*, Aspects of Applied Biology 21, (eds L.G. Copping, C.R. Merritt, B.T. Grayson, S.B. Wakerley and R.C. Reay), BCPC, Brighton, pp. 25–38.

Brönniman, A. (1975) Beitrag der Toleranz auf *Septoria nodorum* Berk. bie Weizen (*Triticum aestivum*). *Zeitschrift feur Pflanzenzüchtung*, **75**, 138–60.

Busvine, J.R. (1971) *A Critical Review of the Techniques for Testing Insecticides*, CAB International, Oxford.

Cannon, R.J.C. (1989) Differences between field-controlled and laboratory-reared insects and mites: implications for foliar insecticide/acaricide evaluation programmes, in *Comparing Laboratory and Field Pesticide Performance*, Aspects of Applied Biology 21, (eds L.G. Copping, C.R. Merritt, B.T. Grayson, S.B. Wakerley and R.C. Reay), BCPC, Brighton, pp. 65–80.

Cantelo, W.W. and Sanford, L.L. (1984) Insect population response to mixed and uniform plantings of resistant and susceptible plant material. *Environmental Entomology*, **13**, 1443–5.

Caseley, J.C., Cussons, G.W. and Atkin, R.K. (1991) *Herbicide Resistance in Weeds and Crops*, Butterworth-Heinemann Ltd, Oxford.

Castillo, M.B., Alejar, M.S. and Harwood, R.R. (1976) Nematodes in cropping patterns. II. Control of *Meliodogyne incognita* through cropping patterns and cultural practices. *Philippine Agriculturist*, **59**, 295–312.

Cheng, J.A., Norton, G.A. and Holt, J. (1990) A systems analysis approach to brown planthopper control on rice in Zhejiang Province, China. II. Investigation of control strategies. *Journal of Applied Ecology*, **27**, 100–12.

Coaker, T.H. (1990) Intercropping for pest control, in *1990 BCPC Monograph No. 45, Organic and Low Input Agriculture*, BCPC, Brighton, pp. 71–6.

Cochran, W.G. and Cox, G.M. (1957) *Experimental Designs*, 2nd edn, Wiley, New York.

Comins, H.N. and Wellings, P.W. (1985) Density-related parasitoid sex-ratio: influence on host–parasitoid dynamics. *Journal of Animal Ecology*, **54**, 583–94.

Copping, L.G., Merritt, C.R., Grayson, B.T., Wakerley, S.B. and Reay, R.C. (1989) *Comparing Laboratory and Field Persistence Performance*, 2nd edn, The Association of Applied Biologists, Warwick, 245 pp.

Crease, G.J., Ford, M.F. and Salt, D.W. (1987) The use of high viscosity carrier oils to enhance the insecticidal efficacy of ULV formulations of cypermethrin. *Aspects of Applied Biology*, **14**, 307–22.

Crowley, P.H. (1981) Dispersal and the stability of predator–prey interactions. *The American Naturalist*, **118**(5), 673–701.

Davey, M.R. and Finch, R.P. (1991) Genetic manipulation for crop protection: application of somatic cell approaches and recombinant DNA technology, in *Molecular Biology: Its Practice and Role in Crop Protection*, BCPC Monograph No. 48, (eds G. Marshall and D. Atkinson), BCPC, Brighton, pp. 33–58.

Davies, W.J. and Blackman, P.G. (1989) Growth and development of plants in controlled environments and in the field, in *Comparing Laboratory and Field Pesticide Performance*, Aspects of Applied Biology 21, (eds L.G. Copping, C.R. Merritt, B.T. Grayson, S.B. Wakerley and R.C. Reay), BCPC, Brighton, pp. 1–12.

de Bono, E. (1970) *Lateral Thinking: A Textbook of Creativity*, Penguin Books, Aylesbury, Bucks.

de Ponti, O.M.B. (1983) Resistance to insects promotes the stability of integrated pest control, in *Durable Resistance in Crops*, (eds F. Lamberti, J.M. Waller and N.A. Van der Graff), Plenum Press, New York, pp. 211–25.

Dempster, J.P. (1969) Some effects of weed control on the numbers of the small cabbage white (*Pieris rapae* L.) on brussels sprouts, *Journal of Applied Ecology*, **6** 339–45.

Dent, D.R. (1986) Resistance to the aphid *Meopolophium festucae cerealium*: effects of the host plant on flight and reproduction. *Annals of Applied Biology*, **108**, 577–83.

Dent, D.R. (1990) *Bacillus thuringiensis* for the control of *Heliothis armigera*: bridging the gap between the laboratory and the field. *Aspects of Applied Biology*, **24**, 179–85.

Dent D.R. (1991) *Insect Pest Management*, CAB International, Wallingford.

Dent, D.R. (1993) Products versus techniques in community-based pest management, in *Community-Based and Environmentally Safe Pest Management*, (eds R.K. Saini and P.T. Haskell), ICIPE Science Press, Nairobi, pp. 169–80.

Dent, D.R. and Wratten, S.D. (1986) The host-plant relationships of apterous virginoparae of the grass aphid *Metopolophium festucae cerealium*. *Annals of Applied Biology*, **108**, 1–10.

Dewar, A.M. (1977) Assessment of methods for testing varietal resistance to aphids in cereals. *Annals of Applied Biology*, **87**, 183–90.

Easwaramoorthy, S. and Nandagopal, V. (1986) Life tables of internode borer, *Chilo sacchariphagus indicus* (K.), on resistant and susceptible varieties of sugarcane. *Tropical Pest Management*, **32**(3), 221–8.

Egunjobi, O.A. (1984) Effects of intercropping maize with grain legumes and fertilizer treatment on populations of *Pratylenchus brachyurus* Todfrey (Nematoda) on the yield of maize (*Zea mays* L.). *Protection Ecology*, **6**, 153–67.

El Titi, A. and Landes, H. (1990) Integrated farming system of Lautenbach: a practical contribution towards sustainable agriculture in Europe, in *Sustainable*

Agricultural Systems, Soil and Water Conservation Society, Ankeny, Iowa, pp. 265–86.
Evans, G.M. (1993) The use of microorganisms in plant breeding, in *Exploitation of Microorganisms*, (ed. D.G. Jones), Chapman & Hall, London, pp. 225–48.
Ezueh, M.I. (1991) Prospects for cultural and biological control of cowpea pests. *Insect Science Application*, **12**(5/6), 585–92.
Ezueh, M.I. and Taylor, T.A. (1983) Effects of time of intercropping with maize on cowpea susceptibility to three major pests. *Tropical Agriculture*, **61**, 82–6.
FAO (1980) *Recommended Methods for Measurement of Pest Resistance to Pesticides*, FAO, Rome.
Felton, J.C., Watkinson, I.A. and Whitbread, S.E. (1987) The provision of test insects for continuous screening programmes, in *Crop Protection Agents–The Biological Evaluation*, (ed. N. McFarlane), Academic Press, Orlando, pp. 531–40.
ffrench-Constant, R.H. and Roush, R.T. (1990) Resistance detection and documentation: the relative roles of pesticidal and biochemical assays, in *Pesticide Resistance in Arthropods*, (eds R.T. Roush and B.E. Tabashnik), Chapman & Hall, London, pp. 4–38.
Finney, D.J. (1971) *Probit Analysis*, 3rd edn., Cambridge University Press, Cambridge.
Flor, H.H. (1942) Inheritance of pathogenicity in *Malampsora lini*. *Phytopathology*, **32**, 653–69.
Ford, M.G. and Salt, D.W. (1987) Behaviour of insecticide deposits and their transfer from plant to insect surfaces, in *Critical Reports on Applied Chemistry 'Pesticides on Plant Surfaces'*, (ed. H. Cottrell), John Wiley, Chichester, pp. 26–81.
Ford, M.G., Reay, R.C., Lane, P. and El Jadd, L. (1987) Factors affecting the performance of application of cypermethrin for the control of *Spodoptera littoralis* (Boisd.). *Aspects of Applied Biology*, **14**, 217–32.
Frans, R., Talbert, R., Marx, D. and Crowley, H. (1986) Experimental design and techniques for measuring and analysing plant responses to weed control practices, in *Research Methods in Weed Science*, (ed. N.D. Camper), Southern Weed Science Society, Champaign, USA, pp. 30–46.
Geissbühler, H. (1981) The agrochemical industry's approach to integrated pest control. *Phil. Transactions of the Royal Society of London*, **295**, 111–23.
Givovich, A., Weibull, J. and Pettersson, J. (1988) Cowpea aphid performance and behaviour on two resistant cowpea lines. *Entomologia Experimentalis et Applicata*, **49**, 259–64.
Godfray, H.C.J. and Hassell, M.P. (1987) Natural enemies may be a cause of discrete generations in tropical insects. *Nature*, **327**, 144–7.
Godfray, H.C.J. and Waage, J.K. (1991) Predictive modelling in biological control: the mango mealy bug (*"Rastrococcus invadens"*) and its parasitoids. *Journal of Applied Ecology*, **28**, 434–53.
Gold, C.S. and Wightman, J.A. (1991) Effects of intercropping groundnut with sunnhemp on termite incidence and damage in India. *Insect Science Application*, **12**(1/2/3), 177–82.
Gordon, R.F.S., Alcock, K.T. and Jutsum, A.R. (1989) Soil-applied insecticides: comparability of performance between laboratory and field, in *Comparing Laboratory and Field Persistence Performance*, (eds L.G. Copping, C.R. Merritt, B.T. Grayson, S.B. Wakerley and R.C. Reay), The Association of Applied Biologists, Warwick, pp. 107–17.
Gould, F. (1978) Predicting the future resistance of crop varieties to pest populations: a case study of mites and cucumbers. *Environmental Entomology*, **7**, 622–6.

Gould, F. (1983) Genetics of plant–herbivore systems: interactions between applied and basic study, in *Variable Plants and Herbivores in Natural and Managed Systems*, (eds R.F. Denno and M.S. McClure), Academic Press, New York, pp. 599–653.

Grayson, B.T. and McCarthy, W.V. (1987) Spray deposition studies in field miniplots of wheat at the emerged ear growth stage. *Aspects of Applied Biology*, **14**, 193–216.

Greathead, D.J. and Waage, J.K. (1983) *Opportunities for Biological Control of Agricultural Pests in Developing Countries*, The World Bank, Washington.

Gunning, R.V. (1991) Measuring insecticide resistance, in *Heliothis: Research Methods and Prospects*, (ed. M.P. Zalucki), Springer-Verlag, Berlin, pp. 151–6.

Hall, F.R., Adams, A.J. and Hoy, C.W. (1989) Correlation of precisely defined spray deposit parameters with biological responses of resistant diamond moth (DBM) field populations, in *Comparing Laboratory and Field Pesticide Performance*, Aspects of Applied Biology 21, (eds L.G. Copping, C.R. Merritt, B.T. Grayson, S.B. Wakerley and R.C. Reay), BCPC, Brighton, pp. 125–7.

Hance, R.J., Hocombe, S.D. and Holroyd, J. (1968) The phytotoxicity of some herbicides in field and pot experiments in relation to soil properties. *Weed Research*, **8**, 136–44.

Hanson, C.H., Busbice, T.H., Hill, R.R., Hunt, O.J. and Oakes, A.J. (1972) Directed mass selection for developing multiple pest resistance and conserving germplasm in alfalfa. *Journal of Environmental Quality*, **1**(1), 106–11.

Harris, M.K. and Frederiksen, R.A. (1984) Concepts and methods regarding host plant resistance to arthropods and pathogens. *Annual Review of Phytopathology*, **22**, 247–72.

Harvey, P.H., Nee, S., Mooers, A.O. and Partridge, L. (1992) These hierarchical views of life: phylogenies and metapopulations, in *Genes in Ecology*, (eds R.J. Berry, T.J. Crawford and G.M. Hewitt), Blackwell Scientific Publications, Oxford, pp. 123–37.

Hassell, M.P. (1978) *The Dynamics of Arthropod Predatory–Prey Systems*, Princeton University Press, Princeton, New Jersey.

Hassell, M.P. (1980) Foraging strategies, population models and biological control: a case study. *Journal of Animal Ecology*, **49**, 603–28.

Hassell, M.P. (1982) Patterns of parasitism by insect parasitoids in patchy environments. *Ecological Entomology*, **7**, 365–77.

Hassell, M.P. (1984) Host–parasitoid models and biological control, in *Pest and Pathogen Control: Strategic, Tactical and Policy Models*, (ed. G.R. Conway), John Wiley & Sons, Chichester, pp. 73–92.

Hassell, M.P. and Anderson, R.M. (1984) Host susceptibility as a component in host–parasitoid systems. *Journal of Animal Ecology*, **53**, 611–21.

Hassell, M.P. and Rogers, D.J. (1972) Insect parasite responses in the development of population models. *Journal of Animal Ecology*, **41**, 661–76.

Hassell, M.P. and Waage, J.K. (1984) Host–parasitoid population interactions. *Annual Review of Entomology*, **29**, 89–114.

Hassell, M.P., Lawton, J.H. and May, R.M. (1983) Patterns of dynamical behaviour in single-species populations. *Journal of Animal Ecology*, **45**, 471–86.

Heap, J.P. (1989) *The Management of Innovation and Design*, Cassell Educational Ltd, Guildford.

Hill, K.K., Jarvis-Eagan, N., Halk, E.L. *et al.* (1991) The development of virus-resistant alfalfa, *Medicago sativa L. Bio/Technology*, **9**, 373–7.

Hislop, E.C. (1987) Can we define and achieve optimum pesticide deposits? *Aspects of Applied Biology*, **14**, 153–72.

Hislop, E.C. (1989) Crop spraying under controlled conditions, in *Comparing*

Laboratory and Field Persistence Performance, (eds L.G. Copping, C.R. Merritt, B.T. Grayson, S.B. Wakerley and R.C. Reay), The Association of Applied Biologists, Warwick, pp. 119–22.

Hokkanen, H. and Pimentel, D. (1984) New approach for selecting biological control agents. *The Canadian Entomologist*, **116**, 1109–21.

Holt, J. and Wratten, S.D. (1986) Components of resistance to *Aphis fabae* in faba bean cultivars. *Entomologia Experimentalis et Applicata*, **40**, 35–40.

Hsiao, T.H. (1969) Chemical basis of host selection and plant resistance in oliphagous insects. *Entomologia Experimentalis et Applicata*, **12**, 777–88.

Huffaker, C.B. (1985) Biological control in integrated pest management: an entomological perspective, in *Biological Control in Agricultural IPM Systems*, (eds M.A. Hoy and D.C. Herzog), Academic Press, Orlando, pp. 13–24.

Ismail, A.B. and Valentine, J. (1983) The efficiency of visual assessment of grain yield and its components in spring barley rows. *Annals of Applied Biology*, **102**, 539–49.

Jones, A., Dukes, P.D. and Cuthbert, F.P. (1976) Mass selection in sweet potato: breeding for resistance to insects and diseases and for horticultural characteristics. *Journal of American Society Horticultural Science*, **101**(6), 701–4.

Karel, A.K., Lakhani, D.A. and Ndunguru, B.N. (1982) Intercropping of maize and cowpea: effect of plant populations on insect pests and seed yield, in *Proceedings of the Second Symposium on Intercropping in Semi Arid Areas, Morogoro, Tanzania*, 4–7 August 1980, Ottawa, Canada, (eds F.L. Keswami and B.J. Ndunguru), IDRC, Ottawa, Canada.

Kayumbo, H.Y. (1976) Crop protection in mixed ecosystem. Paper presented at the Symposium on Intercropping in Semi-Arid Areas. Morogoro, Tanzania, 10–12 May 1976.

Khan, A.M., Saxena, S.K. and Siddiqui, Z.A. (1971) Efficacy of *Tagetes erecta* in reducing root infesting nematodes of tomato and okra. *Indian Phytopathology*, **24**, 166–9.

Khush, G.S. and Brar, D.S. (1991) Genetics of resistance to insects in crop plants, in *Advances in Agronomy*, Academic Press Inc, Orlando, pp. 223–74.

Khush, G.S. and Toenniessen, G.H. (1992) Biotechnology and pest management in 2000 AD, in *Pest Management and the Environment in 2000*, (eds A.A.S.A. Kadir and H.S. Barlow), CAB International, Oxford, pp. 348–59.

Kidd, N.A.C. and Jervis, M.A. (1989) The effects of host-feeding behaviour on the dynamics of parasitoid–host interactions. *Research in Population Ecology*, **31**, 211–50.

Kronstad, J.W., Schnepf, H.E. and Whiteley, H.R. (1983) Diversity of locations for *Bacillus thuringiensis* crystal protein genes. *Journal of Bacteriology*, **154**, 419–28.

Kudsk, P. (1989) Herbicide rainfastness on indoor and outdoor pot-grown plants, in *Comparing Laboratory and Field Persistence Performance*, (eds L.G. Copping, C.R. Merritt, B.T. Grayson, S.B. Wakerley and R.C. Reay), The Association of Applied Biologists, Warwick, pp. 133–4.

Kuhn, T.S. (1970) *The Structure of Scientific Revolutions*, 2nd edn, University of Chicago Press, Chicago.

Kuhn, T.S. (1977) The essential tension: tradition and innovation in scientific research, in *Essential Tension*, (ed. T.S. Kuhn), The University of Chicago Press, Chicago, pp. 225–39.

Lavy, T.L. and Santelmann, P.W. (1986) Herbicide bioassay as a research tool, in *Research Methods in Weed Science*, (ed. N.D. Camper), Southern Weed Society, Champaign, USA, pp. 201–17.

Letourneau, D.K. (1987) The enemies hypothesis: tritrophic interactions and vegetational diversity in tropical agroecosystems. *Ecology*, **68**(6), 1616–22.

Lindquist, D.A. and Busch-Petersen, E. (1987) Applied insect genetics and IPM, in *Integrated Pest Management, Protection Integrée, Quo Vadis? An International Perspective*, Parasitis 86, (ed. V. Delucchi), Geneva, pp. 237–55.

Ling, K., Namba, S., Gonsalves, C., Slightom, J.L. and Gonsalves, D. (1991) Protection against detrimental effects of polyvirus infection in transgenic tobacco plant expressing the papaya ringspot virus coat protein gene. *Bio/Technology*, **9**, 752–8.

Löfstedt, C. (1990) Population variation and genetic control of pheromone communication systems in moths. *Entomologia Experimentalis et Applicata*, **54**, 199–218.

Lowe, H.J.B. (1978) Detection of resistance to aphids in cereals. *Annals of Applied Biology*, **88**, 401–6.

Marquis, R.J. and Alexander, H.M. (1992) Evolution of resistance and virulence in plant–herbivore and plant–pathogen interactions. *TREE*, **7**(4), 126–9.

Matthews, G.A. (1984) *Pest Management*, Longman, London, New York.

Matthews, G.A. (1992) *Pesticide Application Methods*, 2nd edn, Longman Scientific & Technical, London.

May, R.M., Hassell, F.R.S. and Hassell, M.P. (1988) Population dynamics and biological control. *Phil. Transactions of the Royal Society of London*, **318**, 129–69.

Maynard-Smith, J. (1974) *Models in Ecology*, Cambridge University Press, Cambridge.

Mayo, O. (1987) *The Theory of Plant Breeding*, Clarendon Press, Oxford.

McGaughey, W.H. (1985) Insect resistance to the biological insecticide *Bacillus thuringiensis*. *Science*, **229**, 193–5.

McKauer, M. (1972) Genetic aspects of insect production. *Entomophaga*, **17**, 27–48.

Mead, R. and Willey, R.W. (1980) The concept of a 'Land Equivalent Ratio' and advantages in yields from intercropping. *Experimental Africulture* **16**, 217–18.

Medawar, P.B. (1981) *Advice to a Young Scientist*, Pan Books, London, Sydney.

Medawar, P.B. (1986) *The Limits of Science*, Oxford University Press, Oxford.

Mellanby, K. (1989) DDT in Perspective, in *Progress and Prospects in Insect Control*, BCPC Monograph No. 43, (ed. N.R. McFarlane), BCPC, Brighton, pp. 3–20.

Merritt, C.R. (1989) Comparison of spray losses in laboratory and field situations, in *Comparing Laboratory and Field Persistence Performance*, (eds L.G. Copping, C.R. Merritt, B.T. Grayson, S.B. Wakerley and R.C. Reay), The Association of Applied Biologists, Warwick, pp. 137–47.

Monti, L.W. (1992) The role of biotechnology in agricultural research, in *Biotechnology: Enhancing Research on Tropical Crops in Africa*, (eds G. Thottappilly, L.W. Monti, D.R. Mohan Raj and A.W. Moore), CTA/IITA, Nigeria, pp. 1–10.

Moreno, R.A. and Mora, L.E. (1984) Cropping pattern and soil management influence on plant disease. II Bean rust epidemiology. *Turrialba*, **34**, 41–5.

Müller, H.J. (1958) The behaviour of *Aphis fabae* in selecting its host plants, especially different varieties of *Vicia faba*. *Entomologia Experimentalis et Applicata*, **1**, 66–72.

Munthali, C.D. and Wyatt, I.J. (1986) Factors affecting the biological efficiency of small pesticide droplets against *Tetranychus urticae* eggs. *Pesticide Science*, **17**, 155–64.

Munthali, D.C. (1981) Biological efficacy of small pesticide droplets. PhD Dissertation, University of London.

Munthali, D.C. and Scopes, N.E.A. (1982) A technique for studying the biological efficiency of small droplets of pesticide solutions and a consideration of the implications. *Pesticide Science*, **13**, 60–2.

Murdoch, W.W., Chesson, J. and Chesson, P.L. (1985) Biological control in theory and practice. *The American Naturalist*, **125**(3), 344–66.

Musgrave, A.E. (1980) Kuhn's second thoughts, in *Paradigms and Revolutions*, (ed. G. Cutting), University of Notre Dame Press, Notre Dame, pp. 39–53.

Nejidat, A. and Beachy, R.N. (1990) Transgenic tobacco plants expressing a tobacco virus coat protein gene are resistant to some tobamoviruses. *Molecular Plant Microbial Interactions*, **3**, 247–51.

Nicholson, A.J. and Bailey, V.A. (1935) The balance of animal populations. *Proceedings Zoological Society of London*, **Part 1**, 551–98.

Ochieng, R.S. (1977) Studies on the bionomics of two major pests of cowpea (*Vigna unguiculata*): *Ootheca mutabilis* and *Anaplocnemis carvipes*. PhD Dissertation, University of Ibadan, Ibadan, Nigeria.

Okigbo, B.N. and Greenland, D.J. (1976) Intercropping systems in tropical Africa, in *Multiple Cropping*, (ed. M. Stelly), American Society of Agronomy, Wisconsin, pp. 63–101.

Omar, D. and Matthews, G.A. (1987) Biological efficiency of spray droplets of permethrin ULV against the diamondback moth. *Aspects of Applied Biology*, **14**, 173–9.

Omar, D., Matthews, G.A., Ford, M.G. and Salt, D.W. (1991) The influence of spray droplet characteristics on the efficacy of permethrin against the diamondback moth *Plutella xylostella*: the effect of drop size and concentration on the potency of ULV- and EC-based residual deposits. *Pesticide Science*, **32**, 439–50.

Omolo, E.O., Nyambo, B., Simbi, C.O. and Ollimo, P. (1993) The role of host plant resistance and intercropping in integrated pest management (IPM) with specific reference to the Oyugis project. *International Journal of Pest Management*, **39**(3), 265–72.

Painter, R.H. (1951) *Insect Resistance in Crop Plants*, The MacMillan Co., New York.

Parlevliet, J.E. (1981) Durable resistance in self-fertilizing annuals, in *Durable Resistance in Crops*, (eds F. Lamberti, J.M. Waller and N.A. Van der Graaff), Plenum Press, New York, pp. 347–62.

Parlevliet, R.H. (1981) Crop loss assessment as an aid in the screening for resistance and tolerance, in *Crop Loss Assessment Methods – Supplement 3*, (ed. L. Chiarappa, FAO/CAB), Farnham, pp. 111–114.

Parry, D.W. (1990) How do we control disease?, in *Plant Pathology in Agriculture*, (ed. D.W. Parry), Cambridge University Press, Cambridge, pp. 86–158.

Pearson, A.W. (1988) Managing research and development, in *The Gower Handbook of Management*, 2nd edn, (eds D. Lock and N. Farrow), Gower Publishing Co Ltd, London, pp. 387–403.

Perkins, J.H. (1982) *Insects, Experts and the Insecticide Crisis: the Quest for New Pest Management Strategies*, Plenum, New York.

Perlak, F.J., Deaton, R.W., Armstrong, T.A., Fuchs, R.L., Sims, S.S., Greenplate, J.T. and Fischoff, D.A. (1990) Insect resistant cotton plants. *Bio/Technology*, **8**, 939–43.

Perrin, R.M. (1977) Pest management in multiple cropping systems. *Agro-Ecosystems*, **3**, 93–118.

Perrin, R.M. (1978) The effect of some cowpea varieties on the development and survival of larvae of the seed moth, *Cydia ptychora* (Meyrick) (Lepidoptera: Tortricidae). *Bulletin of Entomological Research*, **68**, 57–63.

Perrin, R.M. and Phillips, M.L. (1978) Some effects of mixed cropping on the population dynamics of insect pests. *Entomologia Experimentalis et Applicata*, 24, 385–93.

Pimentel, D., McLaughlin, L. and Zepp, A. (1991) Environmental and economic impacts of reducing US agricultural pesticide use, in *CRC Handbook of Pest Management in Agriculture*, 2nd edn, (ed. D. Pimentel), CRC Press, Boca Raton, Vol. 1, pp. 679–718.

Power, A.G. (1987) Plant community diversity, herbivore movement, and an insect-transmitted disease of maize. *Ecology*, **68**(6), 1658–69.

Pree, D.J. (1990) Resistance management in multiple-pest apple orchard ecosystems in Eastern North America, in *Pesticide Resistance in Arthropods* (ed. R. Roush), Chapman & Hall, London, pp. 261–76.

Püntener, W. (1981) *Manual for Field Trials in Plant Protection*, 2nd edn, Ciba-Geigy Ltd, Switzerland.

Ragsdale, N.N. and Siegel, M.R. (1985) New approaches to chemical control of plant pathogens, in *Bioregulators for Pest Control*, (ed. P.A. Hedin), American Chemical Society, Washington, DC, pp. 35–45.

Ram, S., Gupta, M.P. and Patil, B.D. (1989) Pest management in fodder cowpea (*Vigna unguiculata* L. Walp.) through mixed and inter-cropping in India. *Tropical Pest Management*, **35**(4), 345–7.

Reed, W., Davies, J.C. and Green, S. (1985) Field experimentation, in *Pesticide Application: Principles and Practice*, (ed. P.T. Haskell), Clarendon Press, Oxford, pp. 153–74.

Reeve, J.D. (1988) Environmental variability, migration and persistence in host-parasitoid systems. *The American Naturalist*, **132**, 810–36.

Reichard, D.L., Alm, S.R. and Hall, F.R. (1987) Equipment for studying effects of spray drop size, distribution and dosage on pest control. *Journal of Economic Entomology*, **80**, 540–3.

Reynolds, K.M., Madden, L.V., Reichard, D.L. and Ellis, M.A. (1987) Methods for study of raindrop impaction on plant surfaces with application to predicting inoculum dispersal by rain. *Phytopathology*, **77**, 226–32.

Risch, S. (1981) Insect herbivore abundance in tropical monocultures and polycultures: an experimental test of two hypotheses. *Ecology*, **62**(5), 1325–40.

Risch, S.J., Andow, D. and Altieri, M.A. (1983) Agroecosystem diversity and pest control: data, tentative conclusions, and new research directions. *Environmental Entomology*, **12**, 625–9.

Robinson, R.A. (1976) *Plant Pathosystems*, Springer-Verlag, Berlin.

Robinson, R.A. (1987) *Host Management in Crop Pathosystems*, MacMillan Publishing Co., New York.

Robinson, R.A. (1991) The controversy concerning vertical and horizontal resistance. *Revista Mexicana de Fitopatologia*, **9**, 57–63.

Root, R. (1973) Organisation of a plant–arthropod association in simple and diverse habitats: the fauna of collards (*Brassica oleracea*). *Ecological Monographs*, **43**, 469–75.

Roush, R.T. and Daly, J.C. (1990) The role of population genetics in resistance research and management, in *Pesticide Resistance in Arthropods*, (eds R.T. Roush and B.E. Tabashnik), Chapman & Hall, London, pp. 97–152.

Roush, R.T. and McKenzie, J.A. (1987) Ecological genetics of insecticide and acaricide resistance. *Annual Review of Entomology*, **32**, 361–80.

Salifu, A.B., Singh, R. and Hodgson, C.J. (1988) Mechanisms of resistance in cowpea (*Vigna unguiculata* (L.) Walp.) genotype TV × 3236, to the bean flower thrips *Megalurothrips sjostedti* (Trybom) (Thysanoptera: Thripidae) 2. Non-preference and antibiosis. *Tropical Pest Management*, **34**, 185–8.

Sekhon, S.S. and Sajjan, S.S. (1987) Antibiosis in maize (*Zea mays*) (L.) to maize borer, *Chilo partellus* (Swinhoe) (Pyralidae: Lepidoptera) in India. *Tropical Pest Management*, **33**(1), 55–60.

Shapere, D. (1980) The structure of scientific revolution, in *Paradigms and Revolutions*, (ed. G. Cutting), University of Notre Dame Press, Notre Dame, pp. 27–38.

Shephard, M.C. (1989) Biological factors which affect the performance of fungicides in the laboratory and field, in *Comparing Laboratory and Field Pesticide Performance*, Aspects of Applied Biology 21, (eds L.G. Copping, C.R. Merritt, B.T. Grayson, S.B. Wakerley and R.C. Reay), BCPC, Brighton, pp. 81–93.

Shephard, M.C., Noon, R.A., Worthington, P.A., McClellan, W.D. and Lever, B.G. (1986) Hexaconazole: a novel triazole fungicide, in *British Crop Protection Conference – Pests and Diseases*, 1st edn, BCPC, Brighton, pp. 19–26.

Simmonds, N.W. (1979) *Principles of Crop Improvement*, Longman Group Limited, Harlow, Essex.

Singh, B.B. and Emechebe, A.M. (1992) Breeding for resistance to *Striga* and *Alectra* in cowpea. *IITA Research*, **4**, 5–8.

Singh, D.P. (1986) Breeding for resistance to diseases and pests, in *Breeding for Resistance to Diseases and Insect Pests*, Springer-Verlag, Berlin.

Singh, P. and Moore, R.F. (1985) *Handbook of Insect Rearing*, Elsevier, Oxford.

Smet, H., van Mellaert, H., Rans, M. and de Loof, A. (1989) Efficiency of J2710 in laboratory and field conditions for control of the Colorado potato beetle, in *Comparing Laboratory and Field Persistence Performance*, (eds L.G. Copping, C.R. Merritt, B.T. Grayson, S.B. Wakerley and R.C. Reay), The Association of Applied Biologists, Warwick, pp. 123–4.

Smith, C.M. (1989) Genetics and Inheritance of plant resistance to insects, in *Plant Resistance to Insects – A Fundamental Approach*, (ed. C.M. Smith), John Wiley & Sons, New York, pp. 189–219.

Spindler, M. (1983) DDT: Health aspects in relation to man and risk/benefit assessment based thereupon. *Residue Review*, **90**, 1–34.

Steel, R.G.D. and Torrie, J.H. (1960) *Principles and Procedures in Statistics*, McGraw Hill, New York.

Stower, R.H. (1971) Banana leaf spot caused by *Mycosphaerella musicola*: containing features of Sigatoka and black leaf streak control. *Plant Disease Reporter*, **55**, 437–9.

Tadros, Th.F. (1987) Interactions at interfaces and effects on transfer and performance. *Aspects of Applied Biology*, **14**, 1–22.

Theunissen, J. and den Ouden, H. (1980) Effects of intercropping with *"Spergula arvensis"* on pests of brussels sprouts. *Entomologia Experimentalis et Applicata*, **27**, 260–8.

Theunissen, J., Booij, C.J.H., Schelling, G. and Noorlander, J. (1992) Intercropping white cabbage with clover, in *Proceedings Working Group Meeting – 'Integrated Control in Field Vegetable Crops'*, XV/4, (eds S. Finch and J. Freuler), IOBC/WPRS, pp. 104–14.

Tingey, W.M. and Lamont, W.J. (1988) Insect abundance in field beans altered by intercropping. *Bulletin of Entomological Research*, **78**, 527–35.

Tukahirwa, E.M. and Coaker, T.H. (1982) Effect of mixed cropping on some insect pests of brassicas; reduced *Brevicoryne brassicae* infestations and influences on epigeal predators and the disturbance of oviposition behaviour in *Delia brassicae*. *Entomologia Experimentalis et Applicata*, *32*, 129–40.

Turnock, W.J. (1972) Geographical historical variability in population patterns and life systems of the larch sawfly. *The Canadian Entomologist*, **104**, 1883–900.

Unterstenhöfen, G. (1976) The basic principles of crop protection field trials Pflanzenschultz-Nachrichten, Bayer, 29, 85.

Uthamasamy, S. (1986) Studies on the resistance in okra, *Abelmoschus esculentus* (L.) Moench. to the leafhopper, *Amrasca devastans* (Dist.). *Tropical Pest Management*, **32**(2), 146–7.

Uvah, I.I.I. and Coaker, T.H. (1984) Effect of mixed cropping on some insect pests of carrots and onions. *Entomologia Experimentalis et Applicata*, **36**, 159–67.

Vaeck, M., Reynaerts, A., Hofte, H. *et al.* (1987) Transgenic plants protected from insect attack. *Nature*, **328**, 33–7.

Valentine, J. and Ismail, A.B. (1983) The efficiency of visual assessment for yield and its components in winter oat rows. *Annals of Applied Biology*, **102**, 551–6.

van der Plank, J.E. (1963) *Plant Diseases: Epidemics and Control*, Academic Press, New York.

van der Plank, J.E. (1984) *Disease Resistance in Plants*, Academic Press, New York.

van Rheenen, H.A., Hasselbach, O.E. and Muigai, S.G.S. (1981) The effect of growing beans together with maize on the incidence of bean diseases and pests. *Netherlands Journal of Plant Pathology*, **7**, 193–9.

Vandermeer, J. (1989) *The Ecology of Intercropping*, Cambridge University Press, Cambridge.

Waage, J.K. (1983) Aggregation in field parasitoid populations: foraging time allocation by a population of *Diadegma* (Hymenoptera: Ichneumonidae). *Ecological Entomology*, **8**, 447–53.

Waage, J.K. and Hassell, M.P. (1982) Parasitoids as biological control agents – a fundamental approach. *Parasitology*, **84**, 241–68.

Walkey, D.G.A. (1985) Basic Control Measures, in *Applied Plant Virology*, (ed. D.G.A. Walkey), Heinemann, London, pp. 216–33.

Wetzler, R.E. and Risch, S.J. (1984) Experimental studies of beetle diffusion in simple and complex crop habitats. *Journal of Animal Ecology*, **53**, 1–19.

Whittaker, M.J. (1993) The challenge of pesticide education and training for tropical smallholders. *International Journal of Pest Management*, **39**(2), 117–25.

Wilcox, J.R. (1982) Breeding soybeans resistant to diseases, in *Plant Breeding Reviews*, (ed. J. Janick), AVI Publishing Company, Westport, Connecticut, pp. 183–235.

Worthing, C.R. and Hance, R.L. (1991) (eds) *The Pesticide Manual: A World Compendium*, 9th edn, BCPC, Farnham.

Young, B.W. (1986) The need for a greater understanding in the application of pesticides. *Outlook on Agriculture*, **15**(2), 80–7.

CHAPTER 8

Implementation of an IPM system

D.R. Dent

8.1 INTRODUCTION

Experience in many countries over many years reveals that the availability of an effective pest management product or technique alone is no guarantee that it will be adopted and prove generally effective in the field (Swaminathan, 1993). In situations where pest outbreaks have occurred and current control measures are ineffective then farmers will be more open to adoption of new technologies (big bang implementation; Norton, 1982b). However, where farmers are satisfied with current pest control measures then the process of introducing new approaches and technologies, such as an IPM system will take time, education and sometimes even incentives (incremental implementation; Norton, 1982b). In this latter situation the closer the IPM system fits a farmer's needs then the greater is the likelihood of adoption; hence the emphasis throughout the book on the need for better targeting of R&D. An alternative and sometimes complementary approach is one involving education of the user so that through increased knowledge and understanding the benefits of the proposed new system are apparent, and lead to greater uptake and utilization by the farmer.

It is generally acknowledged that the problem of transfer of IPM technology represents the principal bottle-neck limiting progress of IPM worldwide. The bridge to transfer the results of research from the scientist to the farmers is provided in most countries by the extension or advisory services. These services provide a vital link in the implementation of IPM systems, but they are very often under resourced and over-stretched and are presented with inappropriate technologies for dissemination; their work is however, inseparable from the implementation process. A knowledge of extension, its role and methods is crucial to an IPM programme manager

Integrated Pest Management Edited by David Dent. Published in 1995 by Chapman & Hall, London. ISBN 0 412 57370 9

guiding a programme from its R&D phase towards its implementation as a fully operational, field-based IPM system.

8.2 EXTENSION

Extension may be provided by a number of different types of organization. It is usually provided by a publicly funded extension service run by ministries of agriculture but they can, and often are, also provided by private consultancy services which farmers pay for, or are services provided free of charge by agricultural companies, e.g. agrochemical pesticide advisors. These different organizations will have different objectives. The agrochemical advisors will of course, wish to promote the use of their companies' products and often have a vast array of materials and resources at their disposal to encourage farmers to listen to their advice and purchase their products. Consultants are employed by individual farmers or cooperatives and are expected to deliver good, sound and impartial advice (their continued employment may depend on it). A consultant will be very aware of the need to meet a client's needs in the type of extension that he/she provides but they may not be influenced by secondary issues such as recommending control measures that preserve the environment. The type of extension provided by government organizations is determined by government policy which affects the type of service, advice, training and inputs that the extension service provides. IPM systems may be advocated by all three types of extension organizations. However, the type of system they recommend may vary markedly between them.

In general extension involves: (i) the transfer of technology; (ii) provision of information and advice; (iii) problem solving; (iv) education and training; (v) strengthening the organization base of farmers; and (vi) supplying inputs, credit and technical services (Garforth, 1993). An extension programme devised for implementing an IPM system may involve one, all or a combination of a number of these factors. There is general acknowledgement that the complexity of IPM can create problems for extension. Take, for instance, the difficulties associated with the lack of immediately observable effects in the use of IPM relative to the use of pesticides. The immediate knockdown effects of some contact insecticides are easily recognizable indicators of the success and usefulness of the control measure. By contrast with an IPM system, after having noticed a pest problem in a crop, a farmer may be encouraged not to spray pesticide but to rely on the ability of some other insects to enter the crop and parasitize the pest. The farmer must understand the role of natural enemies and the impact they can have, a situation which requires farmer education. The complex nature of IPM also makes simple on-farm trials more difficult, and economic benefits are often more long term and intangible, i.e. build-up of natural enemy populations, reduced environmental pollution. Such drawbacks may substantially reduce the attractiveness and hence, the adoption rates of IPM, especially in circumstances where farmers have a poor education in matters relating to pest management.

Education, technical difficulties and availability of resources have all been cited as obstacles to implementation of IPM systems (Wearing, 1988). The complex nature of IPM necessitates a certain degree of knowledge and some understanding of the basic concepts of biology and ecology. Where extension workers over-estimate levels of farmer knowledge then misunderstandings and difficulties may soon arise. It is important to understand the level of the farmers' knowledge in terms of both the context and the conceptual framework within which the knowledge is available to them. Garforth (1993) cites the example where a farmer may recognize a spider or a ladybird but may not have the accompanying concept of 'predator' or 'beneficial insect' to enable him to make sense of the advice on how not to kill them. Many such examples may arise in IPM.

Technical obstacles might be expected to be greatest in the early stages of IPM implementation and decline over time (Wearing, 1988), whereas resource availability can have implications for implementation at any time during the extension programme. Extension services are costly to establish and maintain in any country, but with little foreign aid allocated for extension many developing countries find them prohibitively expensive (Goodell, 1984). Lack of funds for an appropriately equipped and educated extension staff reduces extension capability and the likelihood of farmer adoption of new technologies, such as IPM. Shortage of resources during the implementation phase of an IPM system may also limit the number of interested farmers, particularly if they are expected to pay a consultant for a new but as yet unproven system (Huss-Bruun, 1992). The implications of this are even greater when one considers that IPM systems advocated by government extension agents may be in direct competition and at odds with the promotion and sale of pesticides by agrochemical companies. The comparatively small amount of resources available for the promotion of IPM systems may pale into insignificance in comparison with the money available to these companies for massive advertising campaigns, free samples, T-shirts and other promotional gimmicks (Escalada and Heong, 1993). Excessive use of pesticides in Vietnam seems to have been associated with prominent advertising by the chemical companies (Vo *et al.*, 1993). Thus, resource availability for implementing an IPM system represents an important factor in the likelihood of its adoption and will have a major impact on the extension methods that can be used to implement the system.

8.3 EXTENSION METHODS

At the core of all extension work are processes of communication. Traditionally this communication was considered a one-way process, a top-down approach from the extension service to the farmer. More recently the need for a counter balancing bottom-up approach has been recognized where it is accepted that farmers' perceptions, experience and knowledge are also relevant to the decision-making process (Chambers, 1983; Chambers *et al.*, 1989). The combined top-down/bottom-up approaches ensure exchange of information, and discussion in which farmer and extension worker try to understand one another as a basis for developing and implementing pest management strategies and for improving decision making.

The communication process of extension may take place at a number of levels. The methods used are often classified in terms of their target audience as individual methods, group methods and mass media methods, according to the type of dialogue required and the numbers of farmers involved (Adams, 1982).

8.3.1 Individual methods

Individual methods involve one-to-one farmer–extension worker discussions held, usually, on-farm or sometimes at an extension centre or, more rarely, over the telephone. The farmer has the undivided attention of the extension worker which allows discussion of very specific problems and issues. These methods have been used by 'UK glasshouse' growers to discuss a whole range of subjects from the biology of pests and natural enemies, aspects of pesticide resistance, threats from new pests and the design of IPM systems (Wardlow, 1992). The approach has many benefits for the grower/farmer but is costly in terms of an extension worker's time. In the Iganga district in Uganda with a large number of farms and a small number of extension staff, if each extension worker can visit five farms a day and each farm was visited only once, it would take an extension worker 425 days to visit every farm (Baliddawa, 1991). Hence, each farmer would receive a visit from an extension worker once in every 14 months. Such levels of extension support are obviously insufficient. The one-to-one approach is however, feasible where there are a few large farms in an area having similar extension requirements (e.g. Percy-Smith and Philipsen, 1992). In general though, despite their value to the farmer, the use of individual methods for extension are on the decline (Wearing, 1988; Allen and Rajotte, 1990) mainly due to the need to improve levels of cost effectiveness.

8.3.2 Group methods

Group methods obviously involve interaction between the extension worker and a number of farmers at one time. They provide a compromise between depth of analysis and numbers of farmers reached, rarely allowing for the in-depth discussion of specific details in the same way as individual methods, but the process does involve dissemination to greater numbers of farmers. The group methods range from formal, carefully planned events such as field days, study tours and method demonstrations to informal discussions with no prepared agenda (Garforth, 1993).

Farmers' tours and visits are used to expose farmers to new ideas that might be appropriate to their area. The farmers are invited to the research or extension centre where the information and advice are available. Such tours may include result and method demonstrations although these are usually best carried out on the land of a cooperative farmer, thereby demonstrating the relevance of the result or method to a particular locality. The result demonstration may involve, for instance, the growing of new pest-resistant cultivars. The farmers can see with their own eyes that the cultivars are useful under conditions similar to their own, which provides a more powerful demonstration of a product's value than a similar experiment conducted at some distant centre. When new techniques are developed, on-site demonstrations of these can be used to teach farmers how to carry out the techniques for themselves, in some cases providing an opportunity for the farmers to actually try out the technique under supervision.

Farmer schools or training centres provide a thorough training for farmers in an environment which permits more sustained attendance and concentration than is normally possible in a village meeting (Adams, 1982). The short courses have to fit into the seasonal cycle of the farm work and it has to be accepted that it is the more progressive farmers who are likely to attend. However, through informal discussion, practical demonstration and participation farmers can be encouraged to try new techniques and approaches and more importantly to experiment with them on their own farms. Farmers may be encouraged to measure off an area of a field for use as an experimental plot. Comparisons can then be made between this area and the rest of the field. If farmers keep a record of their pest management practices throughout the season and are supported in this by visits from scientists and extension workers, then a means of lateral learning may be established (Escalada and Heong, 1993). At the end of the season farmers compare yields and a workshop is organized where farmers share their experiences with other farmers, extension workers and scientists. These types of approaches can reduce the work load of the extension services while empowering the farmers through 'learning by doing'. Such group methods have enormous potential and are yet to be fully exploited.

8.3.3 Mass methods

Mass methods are used to reach large numbers of farmers with the same information in a relatively short period of time. They are particularly useful for raising general awareness of a pest problem or of the need for change and to shape public attitudes toward alternative forms of control. However, mass media are not well suited to teaching detailed knowledge or skills, unless they are used to reinforce existing extension messages. Mass methods include radio and television, film, video, audio cassettes, drama, newspapers and other print media.

Radio and television provide a versatile tool for non-formal education, can relay important information without distortion, and reach a large audience (who need not be literate) simultaneously. The use of radio and television as a means of dissemination does not depend on a good network of roads, fuel supplies or serviceable transport. However, despite these advantages the use of radio and television has the disadvantage that learning may not be sustained, it is not interactive (i.e. it is one-way) or specific, and only rarely can local needs be given adequate attention. There is often also a cultural gap between the speaker and the audience (language, dialect) (Adams, 1982). The larger the audience the more difficult it is to design programme content that will be relevant and comprehensible to all (Garforth, 1993). Ultimately however, the size of the audience will be dependent on the availability and distribution of radios and televisions and accompanying broadcasting services. This may not present a problem in developed countries but in some developing countries a vast proportion of the farmers may not have access to television or radio.

Other mass methods tend also to be limited by farmer accessibility to the media, whether printed matter, exhibition, films or bulletins. When farmers attend gatherings (social events) or visit markets then printed matter may be distributed, bulletins posted, films and exhibitions put on show.

Printed media on pest management are now quite common. These may range from handbooks suitable for extension workers, e.g. insect pest identification handbooks (Reed *et al.*, 1989; Wightman and Rango Rao, 1993) to the *Farmer's Primer* series of books produced by IRRI and IITA which provide pictorial representations of agronomic practices, including pest management, using simple, concise notes to explain the illustrations. The *Farmers Primer on Growing Rice* has been published in 33 languages and is the most widely published agricultural textbook in existence. Pamphlets and leaflets may be used to explain specific control measures or to highlight certain pests and the damage they cause. Bulletins may also be used to warn of pending pest outbreaks or the need to apply control measures (Huus-Brown, 1992; Wildbolz, 1992). In some European countries technical booklets and posters are used to assist growers of glasshouse crops in the recognition of pests, their damage and their natural enemies; however, their production and distribution tend to be expensive (Wardlow, 1992).

The cost of the production and distribution of different forms of mass media will influence the selection of those utilized. It is clear that each medium has its own advantages and disadvantages and for this reason a combination of methods is often used, depending on the objectives of the extension programme and the resources available. Where farmer resources permit, there is a growing trend for use of extension delivery systems based on the use of computers and computer simulation models. It is predicted that the traditional methods of extension are likely to decline in importance while the use of computers and mass electronic media will increase in importance (Allen and Rajotte, 1990) although, in 1988 computer technology was still regarded as having little to offer on crop protection decision making (Wearing, 1988). However, even though computer and electronic communications systems hold a great deal of promise it seems likely that the extension services will need to take the leading role in introducing farmers to this technology (Allen and Rajotte, 1990).

One of the first on-line computer systems for pest management was PMEX (Pest Management Executive System), introduced by Michigan State University in 1975 and concerned with apple pest management and later expanding to include field crops and vegetables (Croft *et al.*, 1976; Horn, 1988). In, Europe, EPIPRE is the most widely known system. The acronym EPIPRE stands for EPIdemic PREdiction and PREvention and is an integrated pest management system for spring and winter wheat based on calculation of costs and benefits of pesticides treatments. Originally developed in the Netherlands but with later further refinements completed in Belgium, EPIPRE advises on the necessity for spraying against the pathogens eyespot, *Pseudocercosporella herpotrichoides, Erysyphe graminis, Puccinia striiformis* and *P. recondita* and the leaf and glume blotches *Mycosphaerella graminicola* and *Lephosphaeria nodorum* respectively and the cereal aphids, *Sitobion avenae, Metopolophium dirhodum* and *Rhopalosiphum podi* (Smeets *et al.*, 1994).

Detailed, verified crop management models now exist for a whole range of cropping systems including apple (van den Ende, 1994), alfalfa, cotton and soybeans (Huffaker, 1980). Most of them provide management advice after inputs of weather data, pest intensity estimates and measures of plant growth and development stage. They represent the direction that IPM will inevitably move for most intensive systems (Croft and Welch, 1984). However, unlike the diffusion of other technologies the adoption of computer-based technologies is dependent on the access conditions (Audirac and Beaulieu, 1986) that are largely controlled by the institutions that develop the software. This could limit the nature and pattern of uptake by farmers.

8.3.4 Communication networks

The extension service does not provide the only means by which farmers learn of new technologies and approaches. Depending on the degree of isolation of a farmer or village, information may be obtained from a range of sources which may include neighbours, friends and relatives. Often these sources are more important than the extension service (Kenmore *et al.*, 1985). A knowledge and use of these informal communication networks is essential in the implementation of IPM systems to prevent dissemination via artificial and inappropriate channels.

The importance of the informal communication networks of farmers has been demonstrated in a number of pest control situations (e.g. Van Huis and Kulto, 1982; Stone, 1983) but few projects have been concerned with the effectiveness of farmer-to-farmer communication. Samonte *et al.* (1987) considered the importance of the extension service, commercial personnel, co-farmers, mass media and 'other' sources as 'first information sources' of rice farmers in the Philippines. Table 8.1 indicates the relative value placed on the different sources for different types of pest control. Although the extension service (government technicians) rates highly in most of the pest control categories, co-farmers are particularly important when it comes to selecting which rice cultivar to use and in methods used for rat control.

Table 8.1 First information sources for crop protection on rice in the Philippines (% frequency of mention by farmers). (From Samonte *et al.*, 1987)

Source	*Village*			
	1	*2*	*3*	*4*
Insect control				
Government technician	56	51	45	34
Commercial personnel	21	16	32	35
Co-farmers	20	28	19	16
Mass media	3	5	2	14
Others	–	–	2	1
Plant disease control				
Government technician	82	41	68	46
Commercial personnel	5	18	23	28
Co-farmers	9	39	6	21
Mass media	–	2	3	5
Others	4	–	–	–
Weed control				
Government technician	50	38	37	29
Commercial personnel	7	11	29	33
Co-farmers	18	31	29	24
Mass media	11	15	5	8
Others	14	5	–	6
Rat control				
Government technician	50	37	50	27
Commercial personnel	–	5	5	38
Co-farmers	29	47	32	14
Mass media	–	–	4	7
Others	21	11	9	14
Rice varieties used				
Government technician	13	35	62	12
Commercial personnel	3	12	20	11
Co-farmers	74	47	10	66
Mass media	–	–	8	–
Others	10	6	–	11

Mass media were rated low in every category. Such information clearly indicates the selective nature of communication networks with some channels preferentially used for some control measures but not others. The implementation process must take these different communication channels into account and devise appropriate means of disseminating information through them. Their role and value should not be underestimated since this would be subsequently reflected in low adoption rates of the pest management system.

8.4 ADOPTION OF IPM

The relative success of the extension programme will be ultimately judged on the adoption rate of the IPM system (or components thereof) and the improvement in production associated with this. Adoption rates of IPM are a function of a number of things: (i) the relative advantage of the IPM system as perceived by the farmer; (ii) the compatibility of IPM with the farming system; (iii) its complexity; (iv) the degree to which it can be subjected to simple, on-farm trials; and (v) the observability of the effect of IPM (Rogers and Shoemaker, 1971).

Given this list it is easy to see why IPM may not be readily adopted by farmers. In comparison with use of a single measure such as a pesticide, IPM appears, and often is, complex. Its effects are rarely immediately observable and may not be readily subjected to small-scale on-farm trials. Also, if the research and development of the IPM programme have not been effectively targeted and has failed to involve farmers throughout its development, then there is also the prospect that the IPM system will be incompatible with the farming system. All of these factors constitute major obstacles to the adoption of IPM by farmers.

The adoption process is rarely a sudden event but a long involved process. One of the schemes used for explaining the adoption process is that of Rogers and Shoemaker (1971) which involves four stages: knowledge, persuasion, decision and confirmation. The 'knowledge' stage of the process occurs when the individual learns of the existence of the innovation and gains some understanding of its function. This is followed by the stage of 'persuasion' when the individual forms either a favourable or unfavourable opinion of the innovation which influences their 'decision' to engage in the activities that will lead to a choice between adoption or rejection. The 'final' 'confirmation' stage occurs when the individual makes a final decision to accept or abandon the use of the technology.

Not every individual is as receptive to new ideas and innovations as others. This is as true of farmers and the adoption of new pest management technologies as it is of individuals in other walks of life. The individuals that are ready to accept new ways of life may have a certain influence and if new ways are seen to benefit those who have adopted them, then the rest

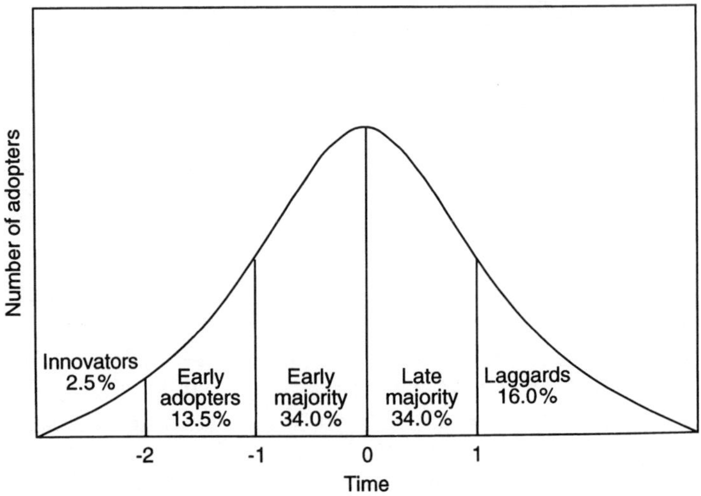

Figure 8.1 An adoption curve divided into segments based on the deviations from the mean time (+ or −). (After Adams, 1982).

of the community may eventually come to accept them (Oakley and Garforth, 1985). These innovator types of farmers represent an obvious first contact point for innovations (they should already have been involved in the development of the IPM system); however, other categories of farmer also have an influence on the adoption process. These are the early adopters, early majority, late majority and laggards (Adams, 1982). Research has shown that adoption of innovations follows a 'normal curve' over time. Sociologists have been able to divide this curve into relative periods during the adoption process when the different categories of adopters have an influence (Figure 8.1). If the cumulative number of adopters is plotted over time then the result is an S-shaped curve. The greater the slope of the adoption curve, the earlier the benefits of the innovations become available (Ruthenberg and Jahnke, 1985).

Adoption is not just a temporal process but also a spatial one. Farmers will often only try out a new technology on part of their crop and only when they are satisfied with the results will they extend its use to greater areas of their crop and farming system. Adoption is also, rarely, once-and-for-all, with farmers often alternating between practising and non-practising of the new technology, often as part of an adoption process where farmers experiment with different aspects of the technology. This process is particularly evident where the farmer is presented with a package of innovations (Figure 8.2) (Ruthenberg, 1975), the farmer switching between different combinations of measures, looking for combinations that suit the particular circumstances.

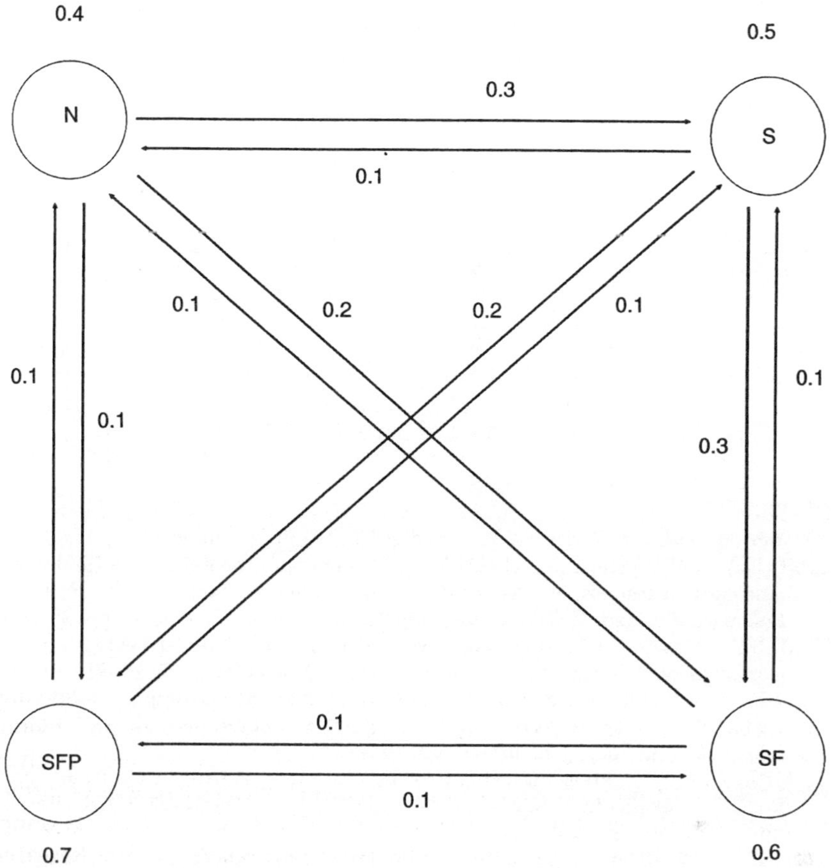

Figure 8.2 The adoption process of a package of innovations. N, non-adoption; S, adoption of modern varieties; SF, adoption of modern varieties and fertilizer; SFP as SF plus adoption of pesticide use. The numbers represent proportions. (After Ruthenberg, 1975).

8.5 IMPLEMENTATION OF IPM SYSTEMS

If the extension service and innovative farmers have been involved from the beginning in the design and development of the IPM system then the implementation process should not represent a major step but rather a gradual transition, a change in emphasis and in utilization of resources. In the development of an IPM R&D programme the emphasis and use of resources should primarily have been with financing scientific research. As the research progresses greater emphasis should be placed on farmer participatory research and on-farm research. Eventually this emphasis should be supported by greater resources for extension and a reduction in scientific support. This perhaps represents an idealistic scenario, but with proper funding and an integrated approach to the development of IPM, there is no reason why IPM systems should not be developed in this way.

REFERENCES

Adams, M.E. (1982) *Agricultural Extension in Developing Countries*, Longman Scientific & Technical, London.

Allen, W.A. and Rajotte, E.G. (1990) The changing role of extension entomology in the IPM era. *Annual Review of Entomology*, **35**, 379–97.

Audirac, I. and Beaulieu, L.J. (1986) Microcomputers in agriculture: a proposed model to study their diffusion/adoption. *Rural Sociology*, **51**(1), 66–77.

Baliddawa, C.W. (1991) Agricultural extension and pest control: a case study of Iganga District. *Insect Sci. Application*, **12**(5/6), 579–83.

Chambers, R. (1983) *Rural Development: Putting the Last First*, Longman Scientific & Technical, Harlow, England.

Chambers, R., Pacey, A. and Thrupp, L.A. (1989) *Farmer First. Farmer Innovation and Agricultural Research*, Intermediate Technology Publications, London.

Croft, B.A. and Welch, S.M. (1984) Use of on-line control systems to implement pest control models, in *Pest and Pathogen Control: Strategic, Tactical and Policy Models*, (ed. G.R. Conway), John Wiley & Sons, Chichester, pp. 352–80.

Croft, B.A., Howes, J.L. and Welch, S.M. (1976) A computer-based extension pest management delivery system. *Environmental Entomology*, **5**, 20–34.

Escalada, M.M. and Heong, K.L. (1993) Communication and implementation of change in crop protection, in *Crop Protection and Sustainable Agriculture*, John Wiley & Sons, Chichester, pp. 191–207.

Garforth, C. (1993) Extension techniques for pest management, in *Decision Tools for Pest Management*, (eds G.A. Norton and J.D. Mumford), CAB International, Oxford, pp. 247–64.

Goodell, G. (1984) Challenges to international pest management research and extension in the Third World: do we really want IPM to work? *Bulletin of the Entomological Society of America*, (Fall), 18–26.

Horn, D.J. (1988) Integrated insect pest management, in *Ecological Approach to Pest Management*, (ed. D.J. Horn), Elsevier Applied Science Publishers, London, pp. 133–49.

Huffaker, C.B. (1980) *New Technology of Pest Control*, John Wiley & Sons, Chichester.

Huus-Bruun, T. (1992) Field vegetables in Denmark: an example of the role of the Danish Advisory Service in reducing insecticide use, in *Biological Control and Integrated Crop Protection: Towards Environmentally Safer Agriculture*, (eds J.C. van Lenteren, A.K. Minks and O.M.B. de Ponti), Pudoc Scientific Publishers, Wageningen, pp. 201–7.

Kenmore, P.E., Heong, K.L. and Putter, C.A. (1985) Political, social and perceptual aspects of integrated pest management programmes, in *Integrated Pest Management in Malaysia*, (eds B.S. Lee, W.H. Loke and K.L. Heong), MAPPS, Malaysia, pp. 47–67.

Norton, G.A. (1982a) Crop protection decision making – an overview. Proceedings 1982 British Crop Protection Symposium, Decision Making in the Practice of Crop Protection, BCPC Monograph No. 25. British Crop Protection Council Publications, Croydon, England, pp. 3–11.

Norton, G.A. (1982b) A decision-analysis approach to integrated pest control. *Crop Protection*, **1** (2), 147–64.

Oakley, P. and Garforth, C. (1985) *Guide to Extension Training*, FAO, Rome.

Percy-Smith, A. and Philipsen, H. (1992) Implementation of carrot fly monitoring in Denmark, in *Proceedings Working Group Meeting – 'Integrated Control in Field Vegetable Crops'*, (eds S. Finch and J. Freuler), IOBC/WPRS pp. 36–42.

Reed, W., Lateef, S.S., Sithanantham, S. and Pawar, C.S. (1989) *Pigeonpea and Chickpea Insect Identification Handbook*, ICRISAT, Andhra Pradesh, India, p. 120.

Rogers, E.M. and Shoemaker, F.F. (1971) *Communication of Innovations*, Free Press, New York.

Ruthenberg, H. (1975) Agricultural extension as an economic investment. *Agricultural Administration*, **2**, 176.

Ruthenberg, H. and Jahnke, H.E. (1985) *Innovation Policy for Small Farmers in the Tropics*, Oxford University Press, Oxford.

Samonte, V.P.B., Obordo, A.S. and Kenmore, P. (1987) The communication and adoption of crop protection technology in rice-growing villages in the Philippines, in *Management of Pests and Pesticides, Farmers' Perceptions and Practices*, (eds T. Tait and B. Napompeth), Westview Press; Boulder, London, pp. 210–18.

Smeets, E., Hencrickx, G. and Geypens, M. (1994) EPIPRE, an up to date link between research and today's farming practice, in *International Symposium on Crop Protection*, Gent University, 3 May 1994, p. 178.

Stone, F.D. (1983) Agricultural change as an adaptive process: adoption of modern methods and responses to pest outbreaks by rice farmers in Chachoengsao Province, Central Thailand. PhD Dissertation, Geography Dept, University of Hawaii, Honolulu, Hawaii.

Swaminathan, M.S. (1993) Perspectives for crop protection in sustainable agriculture, in *Crop Protection and Sustainable Agriculture*, John Wiley & Sons, Chichester, pp. 257–72.

van den Ende, J.E. (1994) Practical use of a computer based advisory system for IPM in apple, in *International Symposium on Crop Protection*, Gent University, 3 May 1994, p. 178.

Van Huis, A. and Kulto, M.E. (1982) *Traditional Pest Management in Maize in Nicaragua: A Survey*, Dept of Entomology, Wageningen Agricultural University, The Netherlands.

Vo, M., Thu Cuc, N.T. and Hung, N.Q. (1993) Farmers' perceptions of rice pest problems and management tactics used in Vietnam. *International Rice Research News*, **18**, 31.

Wardlow, L. R. (1992) The role of extension services in integrated pest management in glasshouse crops in England and Wales, in *Biological Control and Integrated Crop Protection: Towards Environmentally Safer Agriculture*, (eds J.C. van Lenteren, A.K. Minks and O.M.B. de Ponti), Pudoc Scientific Publishers, Wageningen, pp. 193–9.

Wearing, C.H. (1988) Evaluating the IPM implementation process. *Annual Review of Entomology*, **33**, 17–38.

Wightman, J.A. and Rango Rao, G.V. (1993) *A Groundnut Insect Identification Handbook for India*, ICRISAT Information Bulletin no. 39.

Wildbolz, T. (1992) The role of extension in the implementation of integrated fruit production in Switzerland, in *Biological Control and Integrated Crop Protection: Towards Environmentally Safer Agriculture*, (eds J.C. van Lenteren, A.K. Minks and O.M.B. de Ponti), Pudoc Scientific Publishers, Wageningen, pp. 187–91.

CHAPTER 9

Integrated pest management in olives

M.P. Walton

9.1 INTRODUCTION

The olive (*Olea europaea* L.), though occurring world-wide, is widely cultivated in the Mediterranean Basin where it is one of the most distinctive trees to be found. The cultivation of this tree, both for its fruit and the oil derived from the fruit, dates back some 6000 years. Throughout this period, it has been an important crop plant of great nutritional, socioeconomic, cultural and political significance to the people of the area (Hawkes and Wooley, 1963). The olive tree itself belongs to the family Oleaceae and is a small long-lived evergreen, usually 8–12 m high, with grey–green leaves. It is wind pollinated and displays characteristic year-to-year fluctuations in fruit yield (Kochhar, 1986). In general, the trees are robust and may grow and produce a good crop in hilly, rocky and arid areas where other permanent crops cannot survive. Although olive fruits have a high oil content, and the majority are grown for oil production, they are also eaten both green and when ripe (black). The fruits are traditionally harvested by hand picking (green) or, when ripe, by beating the tree with poles. Harvesting the latter is increasingly being done using mechanical shakers. Oil is extracted by crushing and pressing, with so-called 'virgin' oil being the result of the first pressing.

The cultivated form of *O. europaea* known as variety *europaea* is generally thought to be derived from hybridization, probably between *O. laperrinii* and *O. africana* (=*chrysophylla*). It is thought that *O. europaea* originated in the eastern Mediterranean (Kochhar, 1986) perhaps with Lebanon, Syria and/or Israel as the primary region of diversity. Simmonds (1976) suggests it then spread westwards with a second centre of diversity in the Aegean and a third one in Tunisia and southern Italy. Clay tablets

Integrated Pest Management Edited by David Dent. Published in 1995 by Chapman & Hall, London. ISBN 0 412 57370 9

Table 9.1 Distribution and estimated numbers of cultivated olive trees world-wide*

Region	*No. of trees* ($\times 10^6$)
Mediterranean basin	754.2
Americas	8.5
Asia	21.0
Africa (non-Mediterranean)	0.3
Australia	0.2
Total	784.2

*Data from International Olive Council.

show the importance of olive oil to the economy of Crete some 2500 years ago while frequent references have since been made to this commodity from early civilizations to the present day. In, or about, the 16th century Spanish settlers took the olive to Latin America from where it spread to those areas of the New World possessing a Mediterranean climate of long, hot and dry summers.

Well over 90% of the total number of cultivated olive trees in the world occur in the Mediterranean basin (Table 9.1). The remaining trees are to be found in North and South America, Australia, parts of Africa, Iraq and Afghanistan.

9.2 IMPORTANCE OF THE CROP

The olive tree is important for the economic, social and ecological well-being of the Mediterranean region, with more than half of European olive trees occurring in the northern Mediterranean (Table 9.2). In some of these countries, net consumption of olive oil is equal to, or greater than, the total oil production of the country (Table 9.3). Most of the remaining trees are grown in the developing countries of the southern Mediterranean where

Table 9.2 Geographic distribution of olive trees in the European Community *

Country	*No. of olive trees*	*Area (ha)*
Spain	167 000 000	2 087 000
France	5 000 000	44 600
Greece	120 000 000	758 100
Italy	165 000 000	1 176 556
Portugal	49 496 000	1 114 000
EC total	506 496 000	5 180 256

*Data from International Olive Council.

Table 9.3 European olive oil production

Country	*Production (tonnes)*	*Consumption (tonnes)*
Italy	530 000	654 000
Greece	262 000	200 000
France	2000	27 000
Spain	494 000	374 000
Portugal	31 000	35 000
Others		11 000
EC total	1 319 000	1 301 000

* Data from International Olive Council.

they may represent a significant proportion of total agricultural production (FAO, 1991).

In the period 1988–1990, total world production of olives, and olive oil, averaged 9.51 million tonnes and 1.79 million tonnes per year respectively (FAO, 1991) (Figure 9.1). About 9% of production was used for table olives and the remaining 91% for producing olive oil and olive-residue oil. Average world production of olives has increased steadily since the early 1950s, not only due to increased size, and number, of olive groves but also to improvements in cultural practices and crop protection.

Olive production in Europe occupies over 5.2 million hectares, producing approximately 1.8 million tonnes of oil and fruit per annum (Jervis and Kidd, 1993). Insect crop losses can result in losses of some 15% though in some regions and some seasons this figure may be considerably higher. Of the £56 million currently being spent on pest control, nearly half of the total spent relates to pesticides, mainly insecticides. Financial incentives for improved control techniques include not only the lowering of input costs but also the premiums which pesticide-free, biological (i.e. organic) olive products attract for the producer.

In this chapter, olive pest control problems, approaches to their solution and the concept of IPM are all examined in the context of a European Community-funded research programme entitled 'The Development of Environmentally Safe Pest Control for European Olives'. Funding for this research was provided under the European Community programme 'ECLAIR'. This acronym stands for '*E*uropean *C*ollaborative *L*inkage of *A*griculture and *I*ndustry Through *R*esearch', a programme which aims to bring together the expertise and experience of agriculturalists, industrialists and research workers to produce, in the case of the olive IPM programme, pest control systems which have a minimal environmental impact. This programme seeks to integrate collaborative research on a Europe-wide basis to produce environmentally friendly and efficient control measures for the major olive pests found in the northern Mediterranean.

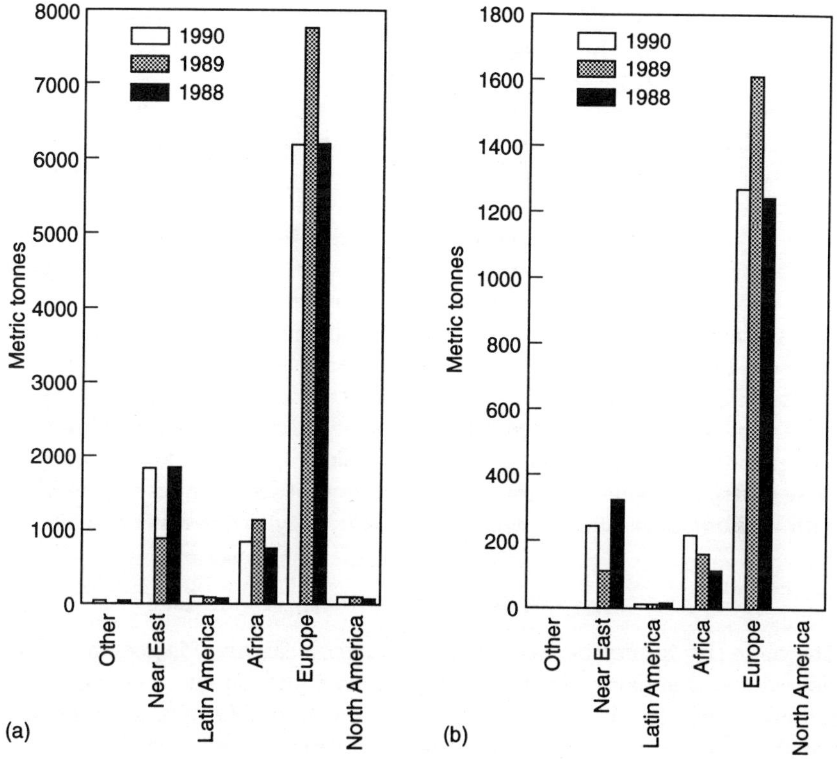

Figure 9.1 World production of (a) olives and (b) olive oil 1988–1990; values are metric tonnes. (After Claridge and Walton, 1992).

9.3 OLIVE PESTS

Although insect pests cause the major losses to the olive crop, weeds and diseases may also result in significant yield reductions (Katsoyannos, 1992). Despite being adapted to survive in semi-arid conditions, many of the weed species to be found in, and around, the olive grove are similarly adapted and therefore provide strong competition for water and nutrients (Ruiz, 1951; Matthews, 1985). Given the highly competitive nature of many of these perennial weeds, weed control is often carried out 4–6 weeks before visible spring growth of the tree and in some regions, a 'bare soil' approach is employed. However, this practice may be detrimental to some natural enemies of insect pests (Ruiz, 1951; Jervis *et al.*, 1992).

Attack by the fungi, *Verticillium dahliae* Kleb., *Cycloconium oleaginum* (Cast) and *Gloeosporium olivarum* (Alm), causes premature fruit and/or leaf fall, dehydration of leaves and fruit, as well as acidification of the extracted oil. The most frequent bacterial infection is by *Pseudomonas savastanoi*, which produces tumours on tree branches.

Eighteen insect pests are well known to attack and damage olive trees in the Mediterranean region. Of these, the key species are generally considered to be the olive fruit fly, *Bactrocera* (=*Dacus*) *oleae* (Gmelin) (White and Elson-Harris, 1992), the olive moth, *Prays oleae* (Bernard), and the olive scale, *Saissetia oleae* (Olivier). All three are widely distributed in the area and regularly cause economic damage to the crop. The losses resulting from attack by these insects are the result of accelerated fruit fall, consumption of fruit pulp and reduction in olive oil quality due to acidification (Delrio, 1992) Of the other insects attacking olives, the olive beetle, *Phloeotribus scarabaeoides* Bern., the olive thrips, *Liothrips oleae* Costa and the pyralid moths, *Margaronia unionalis* Hübn. and *Euzophera pinguis* (Haw.), may cause serious damage on occasion. Damage caused by any of these pests can result in a reduction in the number and/or size of the fruits with a subsequent reduction in yield and quality of the fruit and oil.

9.3.1 *Bactrocera (=Dacus) oleae* (Tephritidae)

The olive fly, found throughout the Mediterranean area, is generally considered the most damaging of the insect pests, especially later in the growing season or in areas of higher temperature and humidity (e.g. near the sea) which are more favourable to its development.

The fly normally has three generations each year, the first occurring from June to August, the second from August to September and the third from October to June. The winter is spent as a puparium either beneath the soil surface or in crevices in the bark of trees. Adult emergence starts in March/April, depending on ambient temperature, but the insects do not reach reproductive maturity and mate until later in the summer. Females lay eggs in developing fruits and 2–6 days later the eggs hatch and the larvae bore a gallery within the fruit. The larval stage consists of three instars, lasts for 10–25 days and, with the exception of the final generation which overwinters in the ground or under bark, pupates beneath the fruit epidermis and emerges later as an adult fly.

Attack by *B. oleae* may potentially account for 50–60% of total insect pest damage and falls into three main categories: (i) premature fruit fall; (ii) decreased yield and quality of oil; and (iii) spoiling of fruit for consumption as table olives.

Control of this pest in recent years has relied mainly on chemical pesticides often applied from the air (Jervis and Kidd, 1993). When applied correctly these methods give good levels of control. However, misuse may lead to pesticide resistance, environmental contamination and/or destruction of beneficial natural enemies.

Two types of insecticide treatment have generally been used, prophylactic and curative. Prophylactic control is used without evaluating whether or not it will produce an economic gain and is often applied in advance of any visible crop damage. Curative treatment is employed after considering the potential economic gain of that action, usually in response to presence of the pest and/or damage caused (see Vandermeer and Andow, 1986). Prophylactic treatment is often used for adult flies and curative action used against larvae already living in the fruit. For adult control, baited sprays may be used which reduce the quantity of pesticide required and the impact on beneficial natural enemies. However, because of the large areas to be covered, and in some cases their inaccessibility at ground level, some treatments are applied by air as 'low volume' or 'ultra-low volume' sprays. Control of larvae is usually by cover spraying of insecticides, such as dimethoate, applied at ground level or from the air. The decision over whether to apply pesticide, and the timing of any such applications, are usually based on counts of living larvae in fruit samples.

Alternatives to wholly insecticidal control methods attempted in the past, or currently being investigated have included: (i) male sterilization and release (Delrio, 1985); (ii) biological control using natural enemies such as the parasitoid *Opius concolor* (Szepl.) (Monastero, 1965); and (iii) use of pheromones for trapping adult flies and for mating disruption (Jones *et al.*, 1985).

9.3.2 *Prays oleae* (Yponomeutidae)

Damage due to larval feeding is often not so obvious as that caused by *B. oleae* but nevertheless significant reduction in oil yield and quality may result. *P. oleae* typically has three generations each year; the first, March/April to May/June, is the 'flower generation', the second, May/June to September/October, is the 'fruit generation' and the third is the 'leaf generation' which occurs from September/October to March/April.

In the flower generation, females from the overwintering leaf generation lay eggs on the flowers. After 7–12 days larvae emerge and feed on the buds and flowers. Although such feeding does not generally destroy either the bud or the flower, it is sufficient to prevent fruit formation and thus, reduce fruit yield. The larvae mature on the flowers, pupate in a loose silken cocoon and 10–12 days later emerge as adults. After mating they lay eggs on the developing fruits to form the fruit generation. Hatching occurs 3–7 days later and the larvae bore into the fruit, feeding on the pulp as they form galleries that penetrate to the still soft developing seed. As with attack by *B. oleae*, this often leads to premature fruit fall and spoiling of the crop. The larval stage of this generation lasts for 3–4 months. The pupal stage lasts for 10–15 days and usually occurs inside the fruit, although in some cases it may take place in the ground or under the bark of the tree (cf. *B. oleae*). In the leaf generation, females from the fruit generation lay their eggs on the leaves, and, depending on prevailing temperatures, hatching takes place 1 week to 2 months later. The larvae then penetrate the leaf, forming galleries in the parenchyma. Later they emerge through, and feed on, the lower leaf surface and terminal buds. The pupal stage lasts from 2 to 4 weeks and adults live for 20–40 days.

Damage by this pest generally accounts for between 30 and 40% of total losses caused by insects. Damage by the leaf generation is of lesser importance whereas that by the flower and fruit generations can often lead to a significant yield loss.

As with *B. oleae*, control of *P. oleae* has relied principally on chemical pesticides, mainly organophosphates and carbamates directed against the larval stage. These are applied on a basis of hostplant developmental stage. In addition, sex-pheromone traps have been introduced for monitoring adult moth populations before insecticide treatment. Widespread use of synthetic pesticides has led to deleterious environmental effects and concerns over possible human health problems, thus prompting research into alternative methods for *Prays* control. Those currently under investigation include:

1. Biological control by natural enemies such as *Chrysoperla carnea* (Steph.) (Jervis *et al.*, 1992; McEwen and Ruiz, 1994), *Chelonus eleaphilus* Silv. and *Ageniaspis fuscicollis praysincola* Silv. (Arambourg, 1985).
2. Control of the flower generation using the microbial pesticide, *Bacillus thuringiensis* (Karamanlidou *et al.*, 1991; Dancer and Varlez, 1992).
3. The use of the sex pheromone of *P. oleae* for monitoring, mass trapping and/or mating disruption (Campion *et al.*, 1979; Jardak *et al.*, 1985; Ramos *et al.*, 1985; Kelly and Mazomenos, 1992).

9.3.3 *Saissetia oleae* (Coccidae)

S. oleae has gained prominence as an olive pest following the indiscriminate use of pesticides to control *B. oleae* and *P. oleae* (Delrio, 1985). These pesticides destroy many of the natural enemies of the scale while being ineffective against the scale itself. This has led to the rise of the olive scale to major pest status.

Reproduction in *S. oleae* is mainly parthenogenetic with females ovipositing beneath their scale and against the leaf surface. The incubation period is dependent upon temperature but in Mediterranean climates is generally about 3 weeks. Two generations can occur during a year, the first from May to June and the second from August to November, although in inland areas only the latter generation is found. Following emergence, the larvae or 'crawlers' remain beneath the female's scale for a few days. They then disperse to young branches and leaves where they pierce, and feed from, the phloem. The larval period consists of three instars and lasts for 35–50 days. Third instars generally metamorphose to females which usually settle on stems, particularly at nodes, and form distinctive black scales. The total length of life-cycle varies greatly between individuals, thus generations are not discrete and adults may be found on the tree throughout the year. Overwintering normally takes place as third instar larvae.

Direct damage by *S. oleae* results from phloem feeding which debilitates the tree, while indirect damage is caused by fungal attack encouraged by the production of honeydew. The latter attack, in severe cases, may lead to death of the tree itself. Also, the honeydew may act as an attractive food source for adult olive flies.

As with both *B. oleae* and *P. oleae*, chemical control of *S. oleae* still predominates in most olive-growing countries, but some biological control methods have been tested (e.g. Daane and Caltagirone, 1989). Organophosphorus insecticides and oil emulsions are used as cover sprays at those times of the year when the most susceptible first, second and early third instars are abundant in the field. However, these sprays are potentially toxic to both animals and plants.

9.4 PEST CONTROL AND INTEGRATED PEST MANAGEMENT

For many years, pest control in olive groves has relied mainly on chemical pesticides, often applied by aerial spraying. Besides being expensive, such pesticides may have serious ecological consequences, in terms of environmental pollution and the destruction of, or great reduction in, numbers of naturally occurring beneficial insects (see Katsoyannos, 1992). The emergence of the olive scale as an important pest in recent years where it had previously been kept naturally below an economic threshold is a good example of this (Delrio, 1985). Furthermore, these pesticides are often highly toxic and as such can pose a significant health risk for personnel involved in olive production. Moreover, since many are lipophilic, there is the risk of accumulation of pesticide residues in oil and fruit (Lentza-Rizos and Avramides, 1991).

A further problem with the relatively indiscriminate use of pesticides is the potential evolution of resistance in the target pests although Delrio (1992) points out that, to date, resistance has not posed a problem in olive pests because of the relatively low levels of pesticide use. Widespread chemical pesticide control is not easily compatible with biological control measures since released natural enemies may be adversely affected by the pesticides (Katsoyannos, 1992) although Montiel Bueno (1992) has shown that the use of certain insecticides may in fact be compatible with successful integrated pest management (IPM) (see below).

Attempts to reduce the input of pesticides and to target their use, where necessary, are an important element of pest control strategies for many crops. This approach to the reduction or elimination of synthetic pesticides for pest control forms part of IPM, defined by Smith and Reynolds (1966) as 'a pest management system that in the context of the associated environment and the population dynamics of the pest species, utilises all suitable techniques and methods in as compatible a manner as possible and maintains the pest population levels below those causing economic injury'. This definition was later modified by Dent (1991) and applied to insect pest management. His definition states that insect pest management is 'a pest management strategy that, in the socio-economic context of the farming systems, the associated environment and the population dynamics of the pest species, utilises all suitable techniques and methods in as compatible manner as possible and maintains the pest population levels below those causing economic injury'. This second definition, though essentially based upon the first, incorporates the need for an understanding of the social and economic context in which the management strategy is to be applied. Thus, in common with many other definitions, it contains statements concerning the need to combine control measures to optimize control, the need for environmentally friendly measures and for maintenance of pest populations below economic threshold levels.

Other definitions may emphasize sustainability, a systems approach, or economics, but they all have in common the recognition that a variety of control options are available, including hostplant resistance, exploitation of beneficial organisms, cultural control and interference methods, as well as pesticides. IPM is the management system whereby the most desirable combination of available methods is selected. However, what constitutes 'desirable' is of course difficult, if not actually impossible, to define in many instances. Furthermore, particular interest groups attach their own specific goals for IPM which will in turn influence the understanding of what constitutes IPM. In the main, three principal groups influence the goals of IPM, namely, political/bureaucratic and community groups, technical/scientific groups and the end-user. Each of these groups has its own set of goals and priorities (some aspects of which overlap with other groups) which influence their perception of precisely what IPM entails. The result of this is that a definition of IPM is considered by many to be a 'moving target'. Nevertheless, although differing in precise definition and wording, it is accepted by many people that IPM represents a 'philosophy' (Dent, 1992) rather than a control technique as such. Therefore, each IPM programme will potentially be different and specific to a particular cropping/pest system, but have in common the potential to utilize one, or more, of the above control measures.

Systems of IPM are now widely accepted as being desirable strategies for sustainable crop protection (Dent, 1991). Properly implemented, they should provide sustainable crop protection systems which centre on environmental concerns and the need for conservation of biological diversity. Unlike the use of broad-spectrum pesticides, the development of IPM strategies requires a more detailed understanding of the biology and life-history of each pest and of its natural enemies within the agroecosystem. Each crop, together with the pests which attack it, requires detailed research in different regions in order to formulate a workable, pragmatic and realistic IPM programme. Continuous monitoring of pests and natural enemies is central to any IPM programme and so it is perhaps not surprising then that few such complete programmes have yet been implemented.

Although ideally involving the harmonious integration of control measures to produce the most effective IPM programme, Pimentel (1985) believed most remain *ad hoc* efforts by individual pest control specialists, each developing so-called integrated pest management programmes independently of one another'. It is precisely the kind of collaborative research effort required for IPM in which the cooperative nature of the ECLAIR programme produces positive benefits.

Coordinated research efforts to date into IPM in olives have been funded mainly by international organizations, including the United Nations Food and Agriculture Organization (FAO), the International Olive Oil Council (IOOC), the International Organization for Biological Control (IOBC) and the Commission for the European Community (CEC). In 1984, over 50 papers were presented at an international joint meeting held in Pisa, supported by the CEC, FAO and IOBC. This meeting resulted in the publication of an extensive proceedings the following year which covered sessions on *B. oleae*, scale insects, *P. oleae*, other pests and future prospects for integrated pest control in olive groves. In each case, the subjects of greatest importance that were dealt with concerned mainly biological and ecological aspects of the pest, population dynamics, pest–plant relationships, crop loss assessment and strategies for pest management systems.

Considerable progress has already been made both in Europe and the United States towards the development and adoption of pest management strategies for olives (e.g. Cirio and Menna, 1985; Cirio and di Cicco, 1990; Cirio, 1992). One of the most well-documented of these schemes is the pilot project which has been running in the Canino area of Italy since 1980. This project is discussed in detail by Cirio and Menna (1985) and Cirio (1992). Cirio (1992) considered that the olive crop itself was particularly suited to IPM for a number of reasons, including:

- a relatively small number of pest species to control;
- a low level of disease;
- a prevalence of monophagous key species;
- stability in the biological community due to prolonged plant/pest co-evolution;
- a high hostplant tolerance to pest damage;
- a good plant capacity to recover from pest damage.

However, despite these advantages, adoption of IPM had been relatively poor due to the four key factors which include:

- the diversity of the Italian olive agroecosystem;

- the difficulty in defining a practical and workable economic injury threshold;
- the lack of technical assistance to farmers;
- the lack of promotion and organization of this new approach.

The methods used to overcome these barriers to adoption were based on the belief that the effective uptake of IPM is dependent upon the farmers' level of experience and/or the degree of applicability of the technique (Table 9.4). Thus, in the first stage, demonstration projects, agricultural assistance/extension and training courses were used to bring the growers up to the second stage of experience. During this second stage, adoption and rationalization of IPM were encouraged through the standardization of methodologies and the carrying out of cost/benefit analyses. Finally, the third stage of the programme was aimed at large-scale adoption of IPM when technical understanding and methodologies have been fully developed.

After 8 years of operation, the results of this work show that the involvement of farmers and the use of information technologies at the local level are very important, not only for making the transition from individual to collective pest control but also the effective integration of all information.

Table 9.4 Stages in application and adoption of IPM. (After Cirio, 1992)

Stage	*Experience/complexity level*	*Characteristics*
1. Understanding and assimilation of the technology	Low	Farmers recognize the potential of the technology Pilot projects Training courses to develop technical abilities
2. Adoption and rationalization	Medium	The technology is reasonably well understood Standard procedures are developed (monitoring, management) Cost/benefit analysis Subsequent projects are studied
3. Maturity	High	Farmers gain technical know-how, achieve awareness of the technology's full potential Pests are brought under control The technology is correctly assimilated into society Widespread application

Application of IPM in the Canino area has resulted in a significant reduction in the quantity of chemicals used for pest control and an increase in the value of the crop. Almost all local olive growers participated in the programme with the result that natural enemies are once again becoming important for reducing levels of *S. oleae*.

Despite the success of this programme, Cirio does point out that IPM must be undertaken with realism and unless state-supported is only likely to be embraced by the more advanced olive-growing areas which have the infrastructure to deal with the problems encountered during the development and adoption of an IPM strategy.

On a Mediterranean scale, Delrio (1992) discussed the current status of IPM in olive growing, concentrating on what are generally considered the key components. These include:

1. Biological control by the conservation and augmentation of naturally occurring and exotic predators, parasitoids and disease organisms. For example, several non-Mediterranean exotic parasitoids from North Africa have been imported in an attempt to control *B. oleae* in the northern Mediterranean (Jervis *et al.*, 1992).
2. Sampling and monitoring of pest and natural enemy populations to allow prediction of the onset, and period, of infestation as well as the optimal time for treatment should it be required (Kelly and Mazomenos, 1992).
3. Cultural practices. These include the use of cultivars resistant to particular pests and farming practices that may encourage beneficial organisms and discourage pests (Daane and Caltagirone, 1989). Changes in cropping practices directly affect the dynamics of olive tree/arthropod associations. Irrigation, pruning, fertilization and a shift in harvesting dates can all be used to help reduce pest damage.
4. Behaviour-modifying chemicals. Pheromones in particular may be used selectively to sample pest populations in monitoring programmes. They may also be used in mass trapping programmes to lure and kill pests selectively. There are also possibilities in some species to disrupt mating behaviour and thus reduce pest populations (Jardak *et al.*, 1985; Jones *et al.*, 1985; Ramos *et al.*, 1985).
5. Selective pesticides. There are few agroecosystems where pesticide application can easily be totally eliminated in the near future. Therefore, selective pesticides should be sought which can be applied in minimal concentrations. Montiel Bueno (1992) has shown that the judicious use of certain insecticides may be compatible with successful olive IPM.
6. Damage thresholds. In order to determine if pest control intervention is necessary, it is essential to know how pest density and duration of attack relate to crop loss. These relationships are not as yet fully understood for most olive pests but practical action thresholds are established in many areas, especially for *B. oleae* (Delrio, 1992).

The current CEC-funded ECLAIR 209 programme (1991–1994), involving some 60 scientists in 10 different organizations (Table 9.5) from four European countries, aims to develop an overall IPM programme for olives. This has required strategic research and development effort in four main areas:

1. Microbial technology, for the production, development, testing, formulation and application of microbial pesticides which will be specific to lepidopterous and dipterous pests.
2. The development, production and field evaluation of behaviour-modifying chemicals for use in monitoring, mating disruption and mass-trapping systems for pest forecasting and control.
3. The development, testing and field application of techniques for biological control of pests by the conservation, augmentation, and manipulation of natural enemies.
4. Research on fruit and oil biochemistry in relation to the effects of pest attack on oil quality.

The technical developments from this programme of research are currently being integrated into an overall management system (Figure 9.2). This is being achieved through the use of theoretical computer models which will be adapted and further developed to provide practical guidelines for use by growers (see Kidd and Gazziano, 1992). Accurate simulation models allow rapid, cost-effective analysis of pest population dynamics under differing control regimes. These models rely on sound experimentation to provide the basic data for understanding the pest population/control interaction but can then add value to the existing experimental data base.

Table 9.5 Participants in the ECLAIR 209 programme (1991–1994)

Participant	*Type of organization*	*Country*
University of Wales, Cardiff	University	UK
Consejo Regulador Sierra de Segura, Andalucia	Oil-producing cooperative	Spain
Consejeria de Agricultura y Pesca, Andalucia	Government ministry	Spain
CSIC, Estacion Exp. Zaidin, Granada	Government research institute	Spain
Energia e Industrias Aragonesas SA	Commercial firm	Spain
CSIC, Instituto de la Grasa, Seville	Government research institute	Spain
AgriSense-BCS Ltd	Commercial firm	UK
Institute of Biology, National Research Centre, Athens	Government research institute	Greece
Division of Agrobiotechnology, ENEA, Rome	Government research institute	Italy
Co-operative Energia e Territorio, Viterbo	Commercial firm	Italy

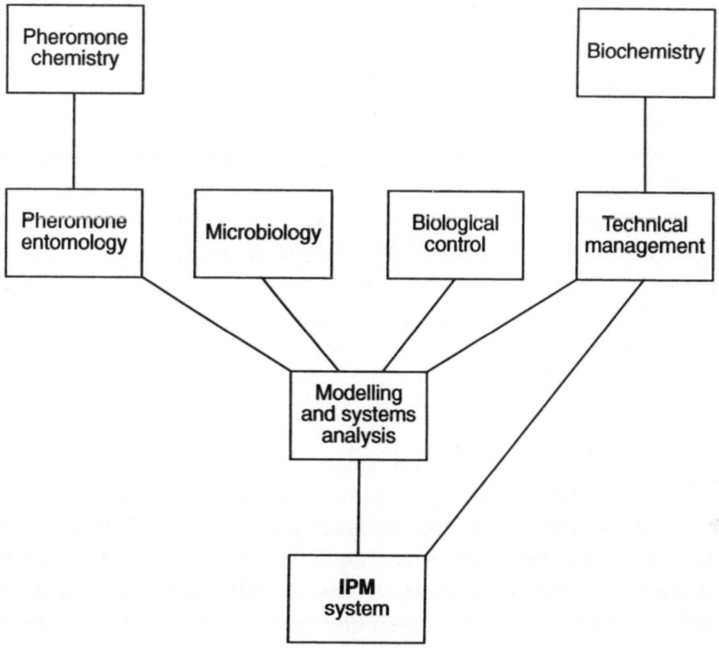

Figure 9.2 R&D matrix for ECLAIR 209 programme.

The involvement and cooperation of growers and processors is a vital element in the programme and is being encouraged to ensure that the research is effectively targeted. Two industrial partners are also involved in this project to facilitate development and commercialisation of any resulting new products or technologies (see Jones and Gonzalez Estebaranz, 1992). Each of the four main areas of research are summarised below.

(a) Microbial insecticides

Microbial insecticides such as *B. thuringiensis* (*Bt*) are effective against the larvae of the olive moth and have the major advantage that they are selective, killing only the pest without harming beneficial insects (Yamvrias, 1985). However, their main drawback, like contact synthetic pesticides, is the comparatively short time that larvae are vulnerable, i.e. during the flower-feeding stage. At other times, the larvae are protected within the plant tissue (fruit or leaf). *Bt*s generally show weak toxicity to the olive fly which is similarly difficult to treat since the larvae are concealed within the fruit while the adults are highly mobile and widely dispersed.

(b) Pest population monitoring

An efficient and reliable method is necessary to monitor the phenology and abundance of pests. In the olive grove, pheromone traps have been widely used to monitor both fly and moth populations (Jardak *et al.*, 1985; Quaglia *et al.*, 1985) and developments of these methods for some of the other insect pests will play an important role in future olive IPM schemes. Semiochemicals have also been used in mating disruption, 'lure and kill' and mass trapping systems (Jones *et al.*, 1985; Ramos *et al.*, 1985).

(c) Predators and parasitoids

Since early this century, attempts have been made to control olive pests through the introduction of exotic predators and parasitoids. Jervis *et al.* (1992) reported that over 63 introductions have been performed involving 35 species of parasitoid and three species of predator. Unfortunately, as they point out, although establishment rates have been quite good, rates of control have been poor and no cases of complete control have been reported. One possible reason for these failures may be the vegetation management practised in some olive groves. Although traditionally intercropped with cereals, many modern olive groves are managed more intensively with herbicides or cultural control methods (ploughing, raking or harrowing) used to suppress weeds. The advisability of this approach has recently been called into question since many of the plants formerly regarded as weeds may in fact prove to be beneficial by promoting water retention in the soil rather than competing with the olives for water and nutrients (Jervis and Kidd, 1993). Also, they may provide food and shelter for beneficial insects and help prevent soil erosion.

Natural enemy populations are also affected by cultural practices such as the type of pruning carried out, the spacing of trees and the type of irrigation practised. Pruning is used both to give the tree its shape and secondly to maintain productivity since only shoots of the previous year's growth bear fruit. Daane and Caltagirone (1989) suggested that differences in such practices may affect black scale populations in Californian olive groves with a subsequent effect on parasitoids released as biological control agents.

(d) Oil quality

The quality of an edible oil such as olive oil can be affected by various environmental factors (Harwood, 1989) including pesticide treatment (Harwood, 1991). Therefore, in a study of the role of pesticides in an IPM system it is important to understand any possible effects of treatments on oil quantity and/or quality. Consequently, studies have been performed on the biochemical characteristics of triacylglycerol formation in the olive fruit, with particular reference to those factors influencing the quality of the final storage oil (Harwood *et al.*, 1992).

All of the above areas of research are aimed at providing the necessary data to build robust and realistic computer-based simulation models, and models are, at the time of writing, under development for *P. oleae*, *B. oleae* and *S. oleae*. These can then be used, together with experimental results, to answer two questions posed by Jervis and Kidd (1993). Firstly, what is the best combination of available control methods and secondly, how can these methods be integrated to best effect?

Deciding upon the best combination of methods depends on a multitude of factors including obviously any effects of control measures on the pests and any beneficial organisms. Attempts to consider the interactions of these factors are often referred to as a systems approach (Norton and Mumford, 1993). This is generally associated with mathematical models used to describe the interactions among major components of the cropping system. Such a system has been used (Shoemaker *et al.*, 1979; Shoemaker, 1980) to produce a simple static model based on the best combination of cultural (i.e. pruning), biological (two entomophagous parasitoids) and chemical (three types of pesticide) methods for controlling three insect pests and one fungal pathogen attacking olives. They concluded that the most cost-effective method of controlling olive pests was to rely on biological control of *Parlatoria oleae* Col. and frequent pruning to suppress *S. oleae* and the pathogen causing the disease, olive knot.

Dent (1992, 1994) reports not only on the scientific achievements of the ECLAIR 209 programme but also draws attention to the significant role played by strategic research management and the importance of providing an effective, pro-active management infrastructure which can make best use of the technical achievements of multidisciplinary research efforts. The work of the ECLAIR 209 programme shows what can be achieved if such a systems approach is taken to R&D and when sufficient resources are available to develop a 'complete' IPM programme utilizing expertise from different disciplines and different countries. One of the lasting lessons of the ECLAIR 209 programme will be that successful R&D programmes depend upon both good science and good research management.

ACKNOWLEDGEMENTS

I wish to thank my many colleagues working on the ECLAIR 209 project for their help and support during the preparation of this chapter. In particular, I would like to thank the programme coordinator, Professor Mike Claridge, for his encouragement during the course of this work. Also, David Dent for his helpful comments and discussion. This work was supported by ECLAIR 209 and funded by the European Community.

REFERENCES

Arambourg, Y. (1985) Control of *Prays oleae* (Bern.), in *Proceedings of the CEC/FAO/IOBC International Joint Meeting, Integrated Pest Control in Olive-Groves,* Pisa, 3–6 April 1984 (eds R. Cavalloro and A. Crovetti), A.A. Balkema, Rotterdam, pp. 195–8.

Campion, D.G., McVeigh, L.J. and Polyrakis, J. (1979) Laboratory and field studies of the female sex pheromone of the olive moth, *Prays oleae. Experientia,* **35**, 1146–7.

Cirio, U. (1992) Integrated pest management for olive groves in Italy, in *Research Collaboration in European IPM systems,* (ed. P.T. Haskell), BCPC Monograph No. 52, BCPC, Farnham, pp. 47–55.

Cirio, U. and di Cicco, G. (1990) Integrated pest control in olive cultivations. *Acta Horticulturae,* **286**, 323–37.

Cirio, U. and Menna, P. (1985) Progress on the integrated pest management for olive groves in the Canino area, in *Proceedings of the CEC/FAO/IOBC International Joint Meeting, Integrated Pest Control in Olive-Groves,* Pisa, 3–6 April 1984 (eds R. Cavalloro and A. Crovetti), A.A. Balkema, Rotterdam, pp. 348–356.

Claridge, M.F. and Walton, M.P. (1992) The European Olive and its pests – management strategies, in *Research Collaboration in European IPM systems,* (ed. P.T. Haskell), BCPC Monograph No. 52. BCPC, Farnham. pp. 3–12.

Daane, K.M. and Caltagirone, L.E. (1989) Biological control of black scale in olives. *California Agriculture,* **43**, 9–11.

Dancer, B. and Varlez, S. (1992) Application of microbial pesticides in integrated pest management of pests of olives, in *Research Collaboration in European IPM systems,* (ed. P.T. Haskell), BCPC Monograph No. 52, BCPC, Farnham, pp. 13–21.

Delrio, G. (1985) Biotechnical methods for olive pest control, in *Proceedings of the CEC/FAO/IOBC International Joint Meeting, Integrated Pest Control in Olive-Groves,* Pisa, 3–6 April 1984 (eds R. Cavalloro and A. Crovetti), A.A. Balkema, Rotterdam, pp. 394–410.

Delrio, G. (1992) Integrated control in olive groves, in *Biological Control and Integrated Crop Protection: Towards Environmentally Safer Agriculture,* (eds J.C. Van Lenteren, A.K. Minks and O.M.B. de Ponti), Pudoc Scientific Publishers, Wageningen, pp. 67–76.

Dent, D.R. (1991) *Insect Pest Management,* CAB International, Wallingford, 604pp.

Dent, D.R. (1992) Scientific programme management in collaborative research, in *Research Collaboration in European IPM systems,* (ed. P.T. Haskell), BCPC Monograph No. 52, BCPC, Farnham, pp. 69–76.

Dent, D.R. (1994) *Scientific Management in an Olive IPM Programme,* International Symposium on Crop Protection, Ghent, Belgium, 3 May, 1994.

FAO (1991) *Production Yearbook 1990,* Vol. 44. Rome.

Harwood, J.L. (1989) Lipid metabolism in plants. *CPC Critical Review Plant Science,* **8**, 1–43.

Harwood, J.L. (1991) Lipid synthesis, in *Target Sites for Herbicide Action,* (ed. R.C. Kirkwood), Plenum, New York, pp. 57–94.

Harwood, J.L., Rutter, A.J., del Cuvillo, M.T., de la Vega, M. and Sanchez, J. (1992) Biochemical research relevant to olive oil quality and IPM, in *Research Collaboration in European IPM Systems,* (ed. P.T. Haskell), BCPC Monograph No. 52, BCPC, Farnham, pp. 57–63.

Hawkes, J. and Wooley, L. (1963) *History of Mankind Cultural and Scientific*

Development, Vol. 1. Prehistory and the Beginnings of Civilisation, Allen and Unwin, London, i–xlviii, 1–873.

Jardak, T., Moalla, M., Khalfallah, H. and Laboudi, M. (1985) Sexual traps for *Prays oleae* (Lepidoptera, Hyponomeutidae) as a prediction and forecasting method, in *Proceedings of the CEC/FAC/IOBC International Joint Meeting,* Integrated Pest Control in Olive-Groves, Pisa, 3–6 April 1984 (eds R. Cavalloro and A. Crovetti), A.A. Balkema, Rotterdam, pp. 204–29.

Jervis, M. and Kidd, N. (1993) Integrated pest management in European olives – new developments. *Antenna,* **17**, 108–14.

Jervis, M., Kidd, N.A.C., McEwen, P., Campos Aranda, M. and Lozano, C. (1992) Biological control strategies in olive pest management, in *Research Collaboration in European IPM Systems,* (ed. P.T. Haskell), BCPC Monograph No. 52, BCPC, Farnham, pp. 31–9.

Jones, O.T. and Gonzalez Estebaranz, J.P. (1992) The role of industry in IPM systems development, in *Research Collaboration in European IPM systems,* (ed. P.T. Haskell), BCPC Monograph No. 52, BCPC, Farnham, pp. 65–8.

Jones, O.T., Lisk, J.C., Baker, R., Mitchell, A.W. and Ramos, P. (1985) A sex pheromone baited trap which catches the olive fly (*Dacus oleae*) with a measurable degree of selectivity, in *Proceedings of the CEC/FAO/IOBC International Joint Meeting, Integrated Pest Control in Olive-Groves,* Pisa, 3–6 April 1984, (eds R. Cavalloro and A. Crovetti) A.A. Balkema, Rotterdam, pp. 104–9.

Karamanlidou, G., Lambropoulos, A.F., Koliais, S.I., Manousis, T., Ellar, D. and Kastritsis, C. (1991) Toxicity of *Bacillus thuringiensis* to laboratory populations of the olive fruit fly (*Dacus oleae*). *Applied and Environmental Microbiology,* **57**, 2277–82.

Katsoyannos, P. (1992) *Olive Pest Problems and their Control in the Near East,* FAO Plant production and protection paper.

Kelly, D.R. and Mazomenos, B. (1992) The identification and synthesis of olive pest semiochemicals, in *Research Collaboration in European IPM Systems,* (ed. P.T. Haskell), BCPC Monograph No. 52, BCPC, Farnham, pp. 23–30.

Kidd, N.A.C. and Gazziano, S. (1992) Development of population models for olive pest management, in *Research Collaboration in European IPM Systems,* (ed. P.T. Haskell), BCPC Monograph No. 52, BCPC, Farnham, pp. 41–6.

Kochhar, S.L. (1986) *Tropical Crops: a Textbook of Economic Botany,* Macmillan, London.

Lentza-Rizos, C. and Avramides, E.J. (1991) Organophosphorus insecticide residues in virgin Greek olive oil, 1988–1990. *Pesticide Science,* **32**, 161–71.

Matthews, L.J. (1985) Weed control in olives, in *Proceedings of the CEC/FAO/IOBC International Joint Meeting, Integrated Pest Control in Olive Groves,* Pisa, 3–6 April 1984 (eds R. Cavalloro and A. Crovetti), A.A. Balkema, Rotterdam, pp. 324–7.

McEwen, P.K. and Ruiz, J. (1994) Relationship between non-olive vegetation and green lacewing eggs in a Spanish olive orchard. *Antenna,* **18**, 148–50.

Monastero, S. (1965) Risultati della lotta biologica contro il *Dacus oleae* Gmel. a mezzo dell' *Opius concolor* Sz. *siculus* Mon. *Entomophaga,* **10**, 335–8.

Montiel Bueno, A. (1992) Field testing semiochemicals in IPM systems, in *Research Collaboration in European IPM Systems,* (ed. P.T. Haskell), BCPC Monograph No. 52, BCPC, Farnham, pp. 77–80.

Norton, G.A. and Mumford, J.D. (1993) *Decision Tools for Pest Management,* CAB International, Oxford.

Pimentel, D. (1985) Insect pest management. *Antenna,* **9**, 168–71.

Quaglia, F., Conti, B. and Rossi, E. (1985) Competitive comparison of the biological activity of two pheromone blends for *Dacus oleae* (Gmel) adult monitoring, in *Proceedings of the CEC/FAO/IOBC International Joint Meeting, Integrated*

Pest Control in Olive-Groves, Pisa, 3–6 April 1984 (eds R. Cavalloro and A. Crovetti), A.A. Balkema, Rotterdam, pp. 113–16.

Ramos, P., Campos, M., Ramos, J.M. and Jones, O.T. (1985) Field experiments with *Prays oleae* sex pheromone traps, in *Proceedings of the CEC/FAO/IOBC International Joint Meeting, Integrated Pest Control in Olive-Groves*, Pisa, 3–6 April 1984 (eds R. Cavalloro and A. Crovetti), A.A. Balkema, Rotterdam, pp. 247–56.

Ruiz, A. (1951) Fauna Entomologica del Olivo en España. II (Hemiptera, Lepidoptera y Thysanoptera). CSIC Trabajos del instituto español de entomologia, Madrid.

Shoemaker, C.A. (1980) The role of systems analysis in integrated pest management, in *New Technology of Pest Control*, (ed. C.B. Huffaker), John Wiley, Chichester.

Shoemaker, C.A., Huffaker, C.B. and Kennet, C.E. (1979) A systems approach to the integrated pest management of a complex of olive pests. *Environmental Entomology*, **8**, 182–9.

Simmonds, N.W. (ed.) (1976). *Evolution of Crop Plants*, Longman, London.

Smith, R.F. and Reynolds, H.T. (1966) Principles, definitions and scope of integrated pest management. *Proceedings of FAO (United Nations Food and Agriculture Organisation) Symposium on Integrated Pest Control*, **1**, 11–17.

Vandermeer, J. and Andow, D.A. (1986) Prophylactic and responsive components of an integrated pest management programme. *Journal of Economic Entomology*, **79**, 299–302.

White, I.M. and Elson-Harris, M.M. (1992) *Fruit Flies of Economic Significance: Their Identification and Bionomics*. CAB International, Wallingford.

Yamvrias, C. (1985) Present status of microbiological control of olive pests, in *Proceedings of the CEC/FAO/IOBC International Joint Meeting, Integrated Pest Control in Olive-Groves*, Pisa, 3–6 April 1984 (eds R. Cavalloro and A. Crovetti), A.A. Balkema, Rotterdam, pp. 380–5.

CHAPTER 10

Integrated pest management in wheat

S.D. Wratten, N.C. Elliott and J.A. Farrell

10.1 INTRODUCTION

If world-wide insecticide use can be used as a guide, wheat is attacked by a wide range of insect pests. Wheat is ranked highly in terms of world pesticide markets and in the mid 1980s, only cotton, maize and rice received significantly higher insecticide inputs worldwide. In 1984, for instance, global wheat production was 510 million tonnes and attracted an insecticide market of 108 million US dollars. Data for cotton, for comparison, were 17 million tonnes and 969 million US dollars. The insecticide market for wheat comprised (in US $ millions): USA, 16; Western Europe, 34; East Asia, 23; Rest of World, 35 (Anon, 1985; FAO, 1986). In practice, however, pesticide use is unlikely closely to reflect pesticide need because farmers are 'risk-averse' and do not commonly use pesticides with a clear appreciation of pests' economic threshold levels (Wratten *et al.*, 1990). The economic and environmental costs of prophylactic and ill-timed pesticide use have been well documented (e.g. Greig-Smith *et al.*, 1992; Higley and Wintersteen, 1992). Integrated pest management (IPM) aims to use pesticides only when needed and alongside other ways of reducing pest numbers. This review is based on the presumption that pesticides are an important aspect of world-wide crop protection but that a greater adherence to economic thresholds and other components of IPM would limit their use to situations where they are really needed. The review takes its case studies from four cereal ecosystems. These have been selected because they differ markedly in their agricultural histories, research base and grower attitudes; these countries are the UK, the Netherlands, the USA and New Zealand. In the first two, agriculture is intensive and yields/unit area are high; research activity in cereals IPM is also fairly well advanced, with some good examples of its

Integrated Pest Management Edited by David Dent. Published in 1995 by Chapman & Hall, London. ISBN 0 412 57370 9

application at grower level, but there are big gaps, hostplant resistance being one of them. In the USA, low profit margins are associated with wheat production, and non-insecticidal methods have dominated control tactics as a result. For this reason, host-plant resistance has been an important factor there, while in Britain and the Netherlands there is no current wheat cultivar which has been bred for insect resistance. In New Zealand, geographical isolation means that native pests are few but non-native aphids have attracted research attention and have become good models for programmes in which their natural enemies have been introduced.

It is difficult to find a country or a region in which all the potential components of IPM have been applied to wheat; local economics, the pests' ecology, the extent of crop losses caused have all led to particular aspects of IPM being given higher profiles in different places. This review does not cover all aspects of the IPM research in wheat carried out in the four countries; this would be too big a task for the space available and would duplicate some recent thorough reviews of certain aspects. For instance, a wide-ranging research programme on the ecology and enhancement of native natural enemies of cereal aphids in the UK has been reviewed recently by Wratten (1987), Wratten *et al.* (1995) and Wratten and van Emden (1994). Biological control work in the USA and New Zealand is covered below, however. Similarly, because the potential for hostplant resistance to wheat pests in Europe has not been realized, the papers by van Emden (1987), Niemeyer (1988, 1991), and Nicol *et al.* (1993) are suggested reading to provide an insight into recent work; this topic is not covered further here for European work. In terms of pest damage thresholds and models, USA, UK and the Netherlands provide good examples of working systems while the potential for New Zealand is indicated below, but not yet realized.

As pointed out by Wratten (1987), the declining value of wheat as a commodity for European farmers, because of subsidy reductions, has forced farmers to examine more closely the variable costs associated with its production. This change, and the political and social pressures on pesticide use (Wratten and van Emden, 1994) have made growers and research bodies more open to IPM approaches. In contrast, wheat prices have always been relatively low in the USA and New Zealand, so it is not surprising that non-insecticidal methods have had a longer history of use in those countries. The examples below, and how the approaches differ between the countries selected, illustrate this underlying cost factor which drives research programmes along particular IPM pathways.

10.2 WHEAT IN THE UK AND THE NETHERLANDS

10.2.1 Reducing unnecessary pesticide use

One of the best-known examples of the need to reduce or change pesticide use comes from attempts to control the cereal pest, the wheat bulb fly (*Delia coarctata*). The use of cyclodiene seed-dressings against this insect caused widespread deaths of wild birds in the 1960s and early 1970s; concern over their effect higher up the food chain led to their restriction, and finally in 1974 to their complete withdrawal for use on cereals.

Despite the relatively minor importance of pests in the economics of cereal growing in Europe, the extent of insecticide use on this crop has a disproportionately large effect. Following a bad world harvest in 1972, cereal growing has expanded, and now covers about 3.5×10^6 ha in the UK and about 25×10^6 ha in the rest of Europe. Figure 10.1 shows the increase in the use of insecticides in cereals in Britain since 1974. Similar increases have occurred in France, where in 1983 1.5×10^6 ha were treated against aphids in spring, and 4×10^5 ha in autumn (Lescar, 1984).

There is little evidence that such intensive pesticide use has led to more frequent pest outbreaks, although this possibility has been demonstrated experimentally (Burn, 1992; Duffield and Aebischer, 1994). Other recent changes in farming practice have, however, led to changes in pest status. The trend towards earlier sowing of winter crops has increased the risk of barley yellow dwarf virus (BYDV), and of attack by frit fly (*Oscinella frit*) and yellow cereal fly (*Opomyza florum*). To achieve early sowing, minimum tillage is practised, and this, especially where straw burning is banned, or following a crop such as oilseed rape, has led to greater problems from pests such as slugs. Nearly half of the 'insecticidal' applications in England and Wales in 1982 were against these pests, although by 1990, insecticides dominated this group of agrochemicals, almost certainly due to the frequency of pyrethroid use to control BYDV in autumn cereals (Figure 10.1).

Intensification of cereal growing has been accompanied by fewer crop rotations and increased use of nitrogenous fertilizer, both of which have

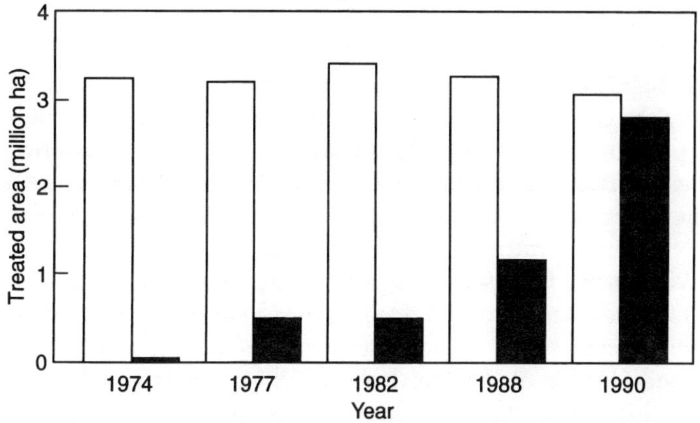

Figure 10.1 The area of cereals grown in England and Wales (□), and the cereals area sprayed with insecticides (■) from 1974 to 1990. (From Sly, 1977, 1986; Davis *et al*., 1991).

increased weed problems, and susceptibility to diseases. Nitrogen may be one factor contributing to increased severity of aphid attacks, although evidence that aphids have increased in importance is equivocal (Vickerman and Wratten, 1979). The relative importance of aphids has, however, grown with crop value, leading to more aphicide applications.

The pressure for the introduction of insect pest management into cereal growing seems to derive more from environmental considerations than from either the economic benefits of reduced pesticide use, or the prospect of pesticide resistance. But it is the scale of pesticide use in this crop that makes the 'environmental' argument a particularly powerful one. In practice a complete IPM 'package' is difficult to achieve in cereals in the UK partly because of the low importance of invertebrate pests compared with that of weeds and diseases, but progress is being made towards this end.

Most recent insecticide use has been against aphids in the UK, so the account below will concentrate on these. IPM approaches to managing BYDV and dipteran problems in the UK and Europe were reviewed by Burn (1987). The above outline in this section is derived from the useful introduction in Burn (1987).

Sitobion avenae (F.) is the aphid species most frequently causing direct damage, with *Metopolophium dirhodum* Wlk. reaching damaging levels only sporadically. *Rhopalosiphum padi* (L.) is important as a vector of BYDV and is the target of heavy pesticide use in the autumn. This species is also the main BYDV vector in the USA, the Netherlands and in New Zealand (see below).

The account below will concentrate on *S. avenae*; this species has attracted a large volume of IPM-related work (compared with very little on *R. padi*, for example) but as explained earlier, only damage relationships and control strategies will be dealt with here.

10.2.2 Yield loss caused by *S. avenae*

Outbreaks have been recorded and quantified since the late 1960s, although historically the species reached high populations in some years before that time (Vickerman and Wratten, 1979; Carter *et al.*, 1980). Quantification of the damage caused and appropriate insecticide control tactics was carried out by the Agricultural Development and Advisory Service (MAFF) and at some universities (Wratten and George, 1985).

Early work in the UK used field cages to create 'outbreaks' of *S. avenae*. Figure 10.2 illustrates how aphid populations can be manipulated in this way, using insecticides to remove the pest population at particular crop growth stages. Figure 10.2b is the best model for this type of manipulation because in this case aphid population size is fixed and only crop growth stage is varied; Figures 10.2a and 10.2c involve changing pest population size and growth stage. This type of manipulative work led to an understan-

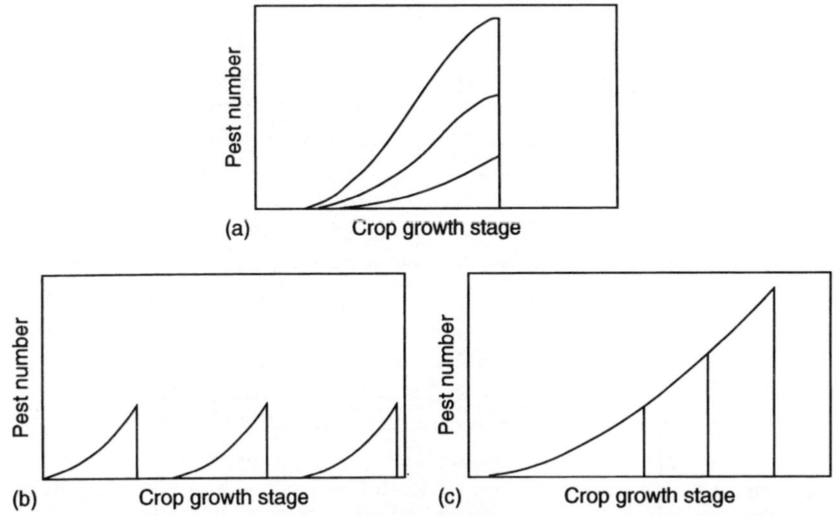

Figure 10.2 Three ways of experimentally manipulating insect populations using field cages. Vertical parts of the figure represent the removal of the population with an insecticide. (a) Inoculations at three different population levels at the same crop growth stage; (b) one population level inoculated at three growth stages; (c) one inoculation but three spray dates, each in a different cage or group of cages. (From Wratten and George, 1985.)

ding of the interactions between crop growth stage, aphid number and feeding site. *S. avenae* colonizes the flag leaf and ear as soon as they have emerged; *M. dirhodum*, on the other hand, moves up the plant as the lower leaves senesce and causes significant yield losses only when feeding on the flag leaf (Wratten, 1975; Lee *et al.*, 1981) (Figure 10.3). Yield loss was brought about by assimilate removal, acceleration of flag-leaf senescence (in the absence of virus (Figure 10.4) (Wratten, 1975), while Dutch data showed how honeydew on leaf surfaces, and the fungi which grow on it, reduce leaf photosynthesis rates (Drenth *et al.*, 1989; see below).

Grain quality was also affected, and aspects of quality such as the quantity of high-molecular-weight glutenin proteins, flour extractability during milling and flour colour were all changed by aphid feeding (Lee *et al.*, 1981).

Field-cage and open-field data were synthesized in a model developed by Mann *et al.* (1986). The model was based on Bayesian algebra and used data on aphid damage, control costs, grain prices and expected yield in the absence of damage. Aphid populations were fed into the model, rather than simulated from reproductive and developmental data. The model calculated the aphid population for each day (by interpolation), daily and cumulative

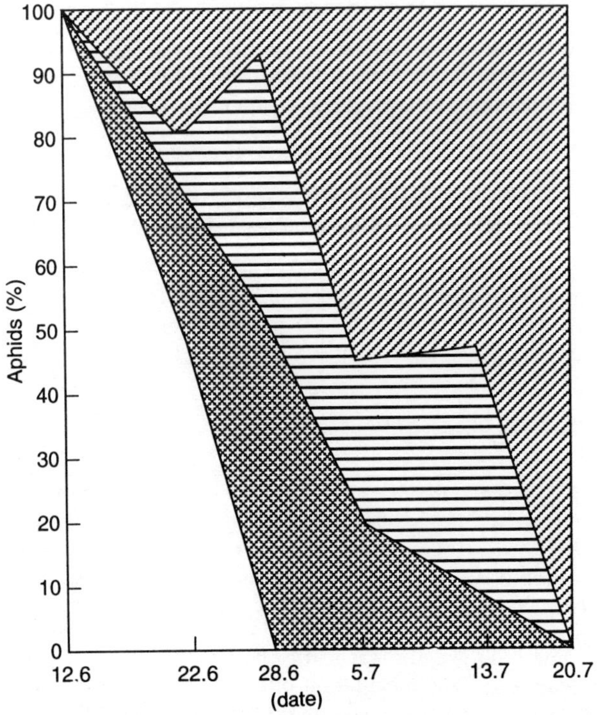

Figure 10.3 Changes in feeding site of the rose-grain aphid as the crop develops. □, Lowest of four leaves; ▩, leaf three; ▤, leaf two; ▨, flag-leaf. (From Wratten, 1975).

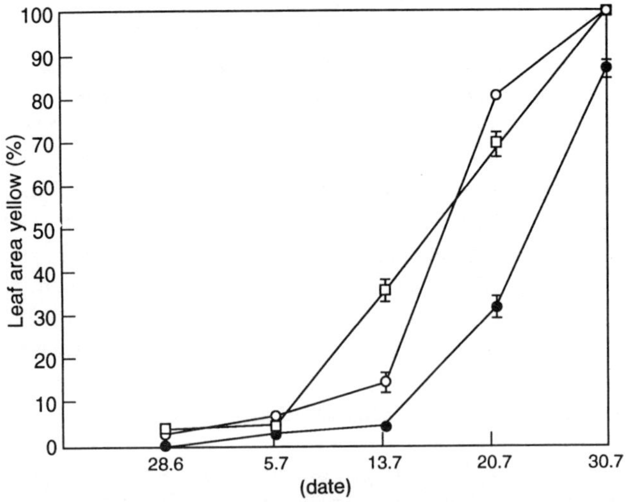

Figure 10.4 Flag-leaf senescence induced by cereal aphids. ●, Control plants; ○, grain-aphid-infested, □, rose-grain aphid-infested. Where two standard errors are greater than the symbol size, they are shown. (From Wratten, 1975).

yield losses, together with the amount of damage saved by an insecticide spray. It also calculated the cumulative and avoidable damage in monetary terms. An example of one 'run' of the model is given in Figure 10.5. This shows that as the amount of damage increased, the amount of expected (preventable) damage declined as, therefore, did the net value of an insecticide spray. In this example the return from spraying became a net loss 1 day before the aphid population reached its peak and 1 week before it declined to zero.

The total gross damage caused by aphids in this case was 13.3%. If an insecticide had been applied at the beginning of flowering (when there were seven aphids/stem) a 12.8% gross level of damage would have been avoided. This figure is extremely close to a published average of 12.5% derived from field trials.

The model was used subsequently in two ways. One was to analyse, using questionnaire data, the economics of farmers' spraying decisions. This revealed that farmers often mis-timed their spray, applying it too late; Figure 10.5 shows the consequences of this in terms of cumulative yield loss. Depending on spray date, farmers either made a net loss from spraying or made a profit which was much smaller than that actually achievable (Figure 10.6) (Watt *et al.*, 1984). A later survey used a simulation model to calculate what aphid populations would have been in the absence of a spray, and

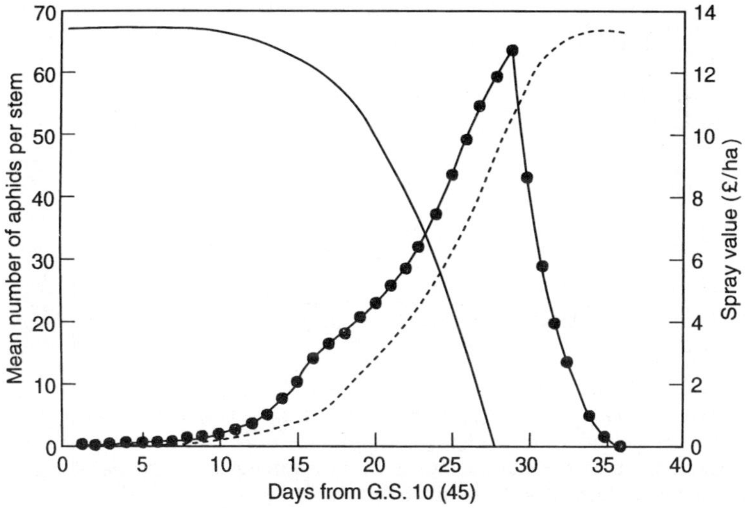

Figure 10.5 Cumulative percentage crop damage (———) and the value (£/ha) of a crop spray (· · · · · ·) against an outbreak of aphids (●—●). Analysis based on four populations in two years in Norfolk, UK. Treatment by tractor-applied insecticide costing £5.9/ha; grain price £117.50/t; expected yield 6t/ha. (From Watt *et al.*, 1984). G.S. = Growth stage.

demonstrated similar sub-optimum spray applications (Wratten *et al.*, 1990).

These analyses suggested that farmers were either receiving poor or no advice or were ignoring threshold recommendations. Therefore the second way the model was used was to convert the original model of Mann *et al.* (1986) into interactive software for use by farmers and advisers. This software became available directly to farmers via a telephone-line 'videotext' system as part of a commercial company called 'Farmlink' (Wratten *et al.*, 1986). With the demise of that company, the software again became available to farmers, this time on computer disk, distributed, with documentation, by the agrochemical company Hoechst UK Ltd. The software

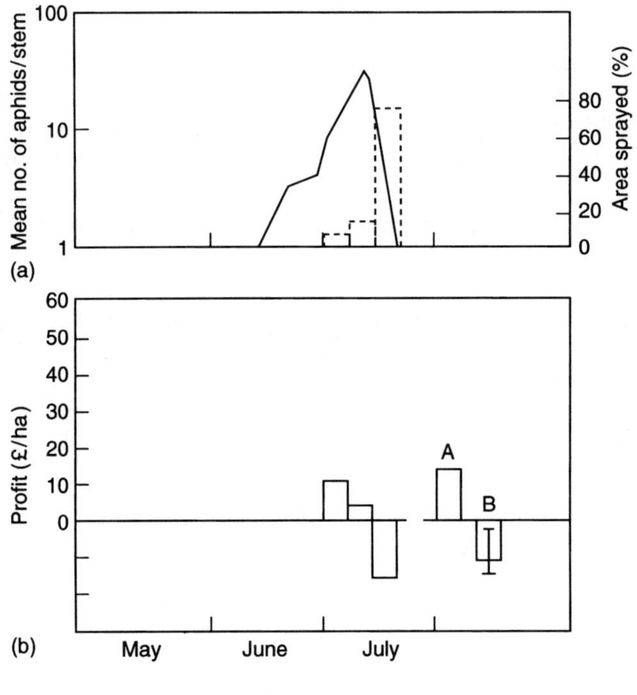

Figure 10.6 (a) Mean aphid numbers (———) and percentage area of crop sprayed each week (--------). (b) Profit gained for each week; the average profit for the whole period of treatment (adjusted according to the area sprayed during each week of the infestation) with an estimate of the range of the average profit obtained, and the maximum profit possible. (Grain price and expected yield as Figure 10.5; control costs based on the range of insecticides and methods of application used by the farmers surveyed.) A, maximum profit possible; B, average profit achieved. Data in (a) were obtained from a farmer questionnaire and from aphid population sampling; data for (b) were obtained from model. (From Watt *et al.*, 1984.)

did not recommend particular pesticides, and in fact the commonest compound used, as revealed by the questionnaire surveys, was dimethoate (Wratten *et al.*, 1990), a cheap, broad-spectrum organophosphorus insecticide. This kills many beneficial arthropods (Vickerman and Sunderland, 1977), so is a far from ideal compound for use in IPM schemes, even with the above software-guided decision making. The aphicide pirimicarb is a much 'softer' product (Wratten *et al.*, 1988) but is much more expensive than dimethoate. The software was evaluated via a series of field-cage and open-field trials from 1985 to 1988 (Mann and Wratten, 1991). The treatments used in these trials were:

1. Model's advice, in which the spray advice given by the model was followed.
2. No spray, in which aphid populations were allowed to increase at their natural rate.
3. Prophylaxis, in which the plots were sprayed at intervals of approximately 4 to 14 days in order to maintain a very low aphid population, which was intended to cause no significant yield loss.
4. The older type of UK MAFF Agricultural and Development Advisory Service (ADAS) advice for *S. avenae* based on the threshold of George and Gair (1979), i.e. that of five or more aphids at the start of flowering with numbers increasing. This advice was still being published by ADAS at least as late as 1986 despite changes made in 1984 (e.g. Anon., 1986). The advice for *M. dirhodum* was 30 aphids or more per flag-leaf between flowering and milky-ripe stages (Anon., 1984).
5. The ADAS advice first issued in 1984 (Anon., 1984) which differed from that above in that for *S. avenae* the advice was to spray if there were five or more aphids and increasing at any stage between the start of flowering and milky-ripe stages. The advice for *M. dirhodum* was as above.
6. The ADAS advice first given in 1988 for *S. avenae* (Anon., 1988) which differed from all previous advice in that for *S. avenae* spraying was advised if 66% of stems were infested at any stage between the start of flowering and milky-ripe stages.

The sampling data were entered into the model together with the relevant economic data, to determine if a spray application should be made, or if and when the next aphid assessment should be made. For all years, undetermined variables fed in to the model were kept constant at £100/t for the expected selling price, 7.5 t/ha for expected yield and £5/ha for spraying costs, with no allowance being made for application costs, or wheeling losses. Keeping these variables constant enabled comparisons between years to be made easier since the number of variables was reduced. Harvests were taken and economics calculated. In 1986, the model's advice and prophylac-

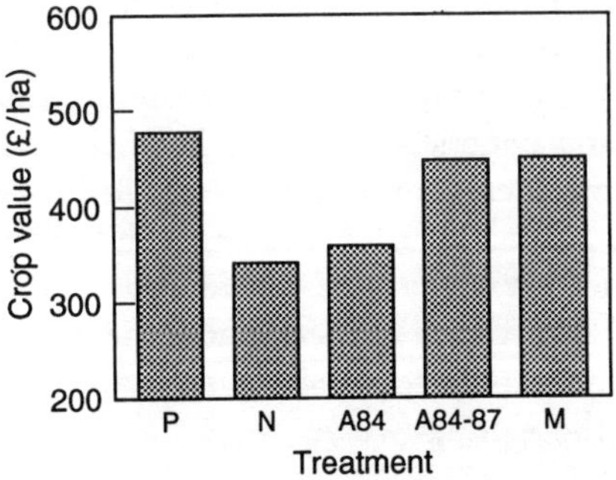

Figure 10.7 Crop value (£/ha) obtained from field trials from 1985 and 1987 in which the model (M) of Watt *et al.* (1984) was compared with other advice and with prophylaxis P and no control (N). A84, ADAS pre-1984 advice; A84–87, ADAS 1984–1987 advice. (For details, see text). (From Mann and Wratten, 1991).

tic pesticide use gave significantly higher crop values than did 'no control' or either of the two ADAS recommendations tested. In the 1987 cage trials, 'pre-1984 ADAS' and 'no control' were significantly worse, economically, than the three other treatments, which did not differ (Figure 10.7). In 1988, only in one field did aphid populations reach double figures and in this field, all treatments were better than 'no control', with little difference between them. In 1985, 1987 and in two of the three 1988 fields, no treatments recommended spraying. Overall, the model performed as well as or better than ADAS advice. Its main advantage is that its output explains to the user the economic justification for the advice given. It was also continuously available to farmers on disk. A related model for pests of oilseed rape shares these advantages (Mann and Wratten, 1991). A further advantage is that as economics change (e.g. grain price, pesticide cost, etc.) the model can accommodate these instantly, whereas an 'ADAS' threshold would need to be recalculated and distributed whenever new information became available. The model's availability to the grower differs from that of the Dutch system (see below) in which orthodox mail and telephone calls are used to disseminate information.

Another UK model is that of Carter *et al.* (1989), which uses similar algebra to that developed by Mann *et al.* (1986) but requires detailed aphid counts, as its predictions are based on aphid population growth rates in the field. This level of detailed recording makes it unlikely to be used by farmers, although it could be added to existing 'packages' for use by advisers.

The Dutch system, well-known by its acronym EPIPRE (Epidemics Prediction and Prevention) is a system of supervised control of five diseases and three cereal aphid species on spring and winter wheat. Its development was stimulated not by pest problems, but by the very heavy epidemics of yellow rust (*Puccinia striiformis* Westend.) in 1975 and 1977 in the Netherlands. It is a computer-based system which generates field-specific recommendations. Like the UK model, it is based on calculations of costs and benefits of chemical control, using farmers' inputs on crop growth stage and disease and aphid incidence. It is made available to farmers via mail and telephone media, so is not so rapid and accessible as the UK system. Field experiments showed EPIPRE to result in financial returns which were sometimes higher than those from normal practice (Table 10.1), although on average, they were equal to but more variable than those from conventional practice (Reinink, 1986). Less tangible benefits were that farmers became more aware of pest damage thresholds, and exposure of farmers to pesticides was likely to have been reduced (Zadoks, 1984; Rossing *et al.*, 1985). Forty-five person-years were required up to 1989 for development and support of the system; these costs were thought to be in proportion to the savings accruing. At the farm level, cost savings were certainly achieved and at national level, environmental damage was considered to have been reduced (Drenth *et al.*, 1989). At its peak EPIPRE was used by 650 farmers and in 1380 fields (Drenth *et al.*, 1989). Numbers declined later and this was considered to be because the farming community had absorbed the philosophy of supervised control.

The above models, when converted to advisory 'packages', still permitted the farmer to make his/her own selection of pesticide. As mentioned above, cheap, broad-spectrum compounds were often used. One way of minimizing the toxicity of those broad-spectrum products is to apply less active ingredient per hectare (in an unchanged water volume). The economics and environmental advantages of this have been explored by Poehling (1989) and Mann *et al.* (1991) (Figure 10.8).

Table 10.1 Comparison between net yield of fields treated according to EPIPRE and of fields treated more frequently than EPIPRE recommendations. (Modified from Rabbinge and Rijsdijk, 1983)

	Yield class		
	<6 tonnes/ha	*6–8 tonnes/ha*	*>8 tonnes/ha*
Net yield EPIPRE	5.34	6.90	8.16
Net yield 'more than EPIPRE'	4.93	6.77	8.13
Costs of treatment (EPIPRE) in tonnes/ha	0.15	0.23	0.31
Cost of treatment 'more than EPIPRE' in tonnes/ha	0.45	0.41	0.45
% of fields treated in accordance with EPIPRE	75	57	41
No. of fields	69	524	263

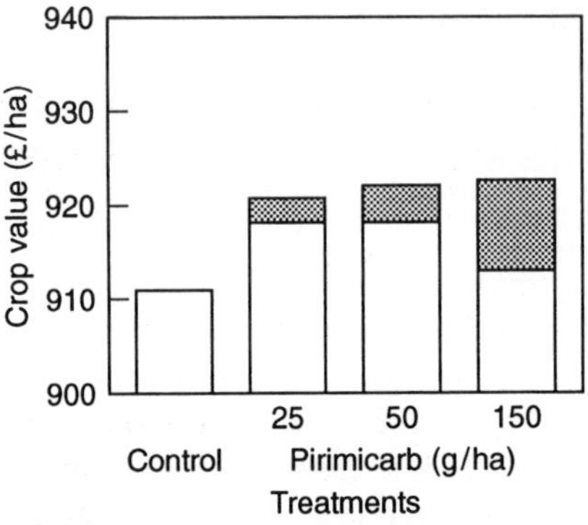

Figure 10.8 Effects of different rates of insecticide application on crop value. □, Net crop value; ■, chemical cost. (From Mann *et al.*, 1991).

These and other ways of minimizing pesticide use in UK cereals are available to farmers, including predator-enhancement schemes (Wratten and van Emden, 1994), and 'conservation headlands' (Sotherton, 1991), in which pesticide use is reduced in the outer 6 m of cereal fields. UK cereal farmers are using these approaches increasingly and the main factor inducing them to use pesticides rationally in these ways seems to be the desire to minimize variable costs because of declining grain prices, coupled with a genuine environmental concern.

10.3 WHEAT IN THE USA

Wheat is second only to maize in the United States in terms of total area planted and production at approximately 30 million ha and 65 Tg annually (Briggle and Curtis, 1987). Wheat in the USA is grown in a variety of climates and soils using a wide range of agronomic practices (Cook and Veseth, 1991). Wheat is grown in widely different environments which vary from northern temperate to sub-tropical, and from arid (under irrigation) to humid. Cropping systems vary from double cropping with late-season crops such as soybeans to an alternate wheat–fallow rotation where a crop is harvested every other year, while tillage practices range from no-till to mouldboard ploughing (Cook and Veseth, 1991). In regions with low yield potential, which includes the majority of the wheat production region of the USA, wheat is usually grown using low levels of external inputs, whereas in regions with high potential yields wheat is more intensively managed (Cook and Veseth, 1991).

The diversity of management practices, environments, and economic situations under which wheat is grown create some unique problems for those developing IPM programmes in the USA. First, the generally low profit margin associated with wheat production requires that IPM programmes be relatively inexpensive if they are to be of practical value to producers. Second, because of the wide geographic distribution of wheat production, IPM practices must vary regionally in concert with variations in the major pest complex associated with a crop. Finally, economic injury levels vary among regions due to effects on plant growth of different environments, cultivars, and management practices, and also on economics.

10.3.1 Insect pests of wheat in the USA

Over 30 insect and mite species attack wheat in the USA. Most species are of minor economic significance because they rarely cause damage to wheat or because they inhabit very localized areas. Table 10.2 lists the important insect and mite pests of wheat in the USA which can be classified as major or potentially important pests. In this classification, major pests severely reduce yield over a large portion of the wheat-growing region on a fairly regular basis. Potentially important pests only occasionally cause widespread damage or are restricted to a small portion of the geographic region in which wheat is grown. Major pests include the greenbug, *Schizaphis graminum*, Russian wheat aphid, *Diuraphis noxia*, and the Hessian fly, *Mayetiola destructor*. The greenbug and Russian wheat aphid are the most important economic pests of wheat in the Great Plains region and Pacific north-west, whereas the Hessian fly is the most important pest in the south-eastern United States. The wheat stem sawfly, *Cephus cinctus*, has historically been an important wheat pest in the Northern Great Plains, but its importance has diminished with an increase in the use of resistant cultivars and changes in cultural practices.

Some pests are important primarily because of the plant diseases they transmit. For example, wheat yield losses due to feeding by the bird cherry-oat aphid, *Rhopalosiphum padi*, can be of economic importance in some regions of the USA, especially if large populations occur during early growth stages (Kieckhefer and Kantack, 1980, 1988; Pike and Schaffner, 1985). However, the economic importance of this species is due primarily to barley yellow dwarf disease, the causative virus of which it readily transmits (Kieckhefer and Stoner, 1967; Wyatt *et al.*, 1988). Similarly, the economic importance of the wheat curl mite, *Eriophyes tulipae*, is due primarily to transmission of the virus causing wheat streak mosaic disease (Hatchett *et al.*, 1987).

10.4 INTEGRATED PEST MANAGEMENT IN WHEAT IN THE USA

Estimated annual losses caused by insects and mites average about 6% of the total value of the USA wheat crop (Hatchett *et al.*, 1987). Non-

Table 10.2 Major and potentially important insect and mite pests of wheat in the United States and the region in which each species is a significant pest

Pest species	*Region of economic importance*
Major pests	
Mayetiola destructor (Say) Hessian fly (Diptera)	Most of the USA
Diuraphis noxia (Mordvilko) Russian wheat aphid (Homoptera)	Great Plains and north-west
Schizaphis graminum (Rondani) Greenbug (Homoptera)	Great Plains and north-west
Oulema melanopus (L.) Cereal leaf beetle (Coleoptera)	Mid-western states
Rhopalosiphum padi (L.) Bird cherry-oat aphid (Homoptera)	Most of the USA
Eriophyes tulipae (Keifer) Wheat curl mite (Acari)	Great Plains region
Potential pests	
Cephus cinctus Norton Wheat stem sawfly (Hymenoptera)	Northern Great Plains
Cephus pygmaeus (L.) European wheat stem sawfly (Hymenoptera)	North-eastern USA
Pseudaletia unipuncta (Haworth) Armyworm (Lepidoptera)	Most of the USA
Spodoptera frugiperda (Smith) Fall armyworm (Lepidoptera)	Southern states
Euxoa auxiliaris (Grote) Army cutworm (Lepidoptera)	Central Great Plains and north-west
Agrotis orthogonia Morrison Pale western cutworm (Lepidoptera)	Central Great Plains
Ctenicera and Agriotes spp. Wireworms (Coleoptera)	Most of the USA
Blissus leucopterus (Say) Chinch bug (Heteroptera)	Central Great Plains
Grasshoppers Several species (Orthoptera)	Great Plains region
Petrobia latens (Müller) Brown wheat mite (Acari)	Central Great Plains
Meromyza americana Fitch Wheat stem maggot (Diptera)	Great Plains and western USA

insecticidal methods of pest control have been heavily relied upon to control insect and mite pests of wheat in the USA. Emphasis of non-insecticidal methods is perhaps due to the low profit margins associated with wheat production brought about by a combination of climatic conditions in most wheat-growing regions that preclude the use of intensive, high yield technology, and the generally low prices obtained for wheat. Hostplant resistance has been a major focus of pest management research on major wheat pests, and cultural and physical controls and biological control have also received attention. Still, chemical control is important in USA wheat production, and an average of 7% of wheat hectarage is treated with insecticide at least once annually to control insect or mite pests (Hatchett *et al.*, 1987).

10.4.1 IPM of aphid pests of wheat

(a) Economic importance of aphids as wheat pests

The greenbug and Russian wheat aphid are the most economically important aphid pests of wheat in the Great Plains region, but the English grain aphid, *Sitobion avenae*, and bird cherry-oat aphid typically infest wheat and sometimes cause economic damage in the Northern Great Plains and north-west (Kieckhefer and Kantack, 1980, 1988; Pike and Schaffner, 1985; Johnston and Bishop, 1987). In the Northern Great Plains bird cherry-oat aphid populations develop sporadically in winter wheat shortly after emergence and persist throughout autumn (Kieckhefer and Gustin, 1967). Greenbugs, bird cherry-oat aphids, and English grain aphids typically invade wheat in the Northern Great Plains in the spring, and occur throughout the growing season (Kieckhefer *et al.*, 1974; Kieckhefer, 1975). Each species can cause economic damage if large populations develop before heading (Kieckhefer and Kantack, 1980, 1988; Pike and Schaffner, 1985; Kieckhefer and Gellner, 1992; Kieckhefer *et al.*, 1994), and the English grain aphid and bird cherry-oat aphid can cause economic damage as late as the soft dough stage in years when cool summers permit aphids to persist in the developing grain heads (Johnston and Bishop, 1987; Noetzel, 1994).

Only the greenbug and Russian wheat aphid are considered important economic pests in the Southern Great Plains. Other cereal aphids infest wheat there but are not considered economically important. In most years damage to wheat by the greenbug is limited to localized areas. However, widespread outbreaks occur at approximately 5 to 10-year intervals. In outbreak years, losses in the Southern Great Plains can exceed $250 million, while in non-outbreak years losses average $67 million.

The Russian wheat aphid is a new pest, first detected in the USA in 1986. The Russian wheat aphid has caused an estimated $670 million in direct and indirect losses in wheat production from 1986 to 1991, primarily in the Great Plains.

(b) Managing cereal aphids in the Northern Great Plains

The extent of injury to wheat caused by cereal aphids varies depending on the aphid species and its population density, the growth stage of the crop, feeding duration, plant variety, and the general health of the crop (Kieckhefer and Kantack, 1980, 1988; Pike and Schaffner, 1985; Johnston and Bishop, 1987; Kieckhefer and Gellner, 1992; Kieckhefer *et al.*, 1994). As a result, accurate economic thresholds are difficult to establish. Cereal aphids usually occur in mixed species populations in the Northern USA, and this complicates the aphid number–yield loss relationship (Kieckhefer 1975; Pike and Schaffner 1985; Johnston and Bishop, 1987).

Elliott *et al.* (1990) adopted 12.5 aphids per tiller as an economic threshold for cereal aphids in wheat in the Northern Great Plains because it represented a compromise among published estimates. This threshold is probably acceptable in both autumn and spring (up to boot stage). Populations that develop to high levels after boot stage often do not cause economic damage to wheat in the Northern Great Plains (Kieckhefer and Kantack, 1980, 1988). Therefore, the typical recommendation has been to forego treatment in fields after head emergence regardless of aphid populations. However, under some environmental conditions economic injury results from densities as low as 10–15 aphids per tiller in wheat during the flowering to milky ripe stages (Johnston and Bishop, 1987), and (Noetzel, 1994) found that peak populations of 50 bird cherry-oat or English grain aphids per head during the milky ripe stage caused losses ranging from 650–1200 kg/ha. Because of uncertainty in economic injury levels, Minnesota recently lowered their economic threshold for flowering to milky ripe wheat to 12.5 aphids per tiller (Noetzel, 1994).

Sampling cereal aphids in the Northern Great Plains for the purpose of making control decisions is accomplished using a truncated incidence-count sequential sampling scheme (Elliott *et al.*, 1990; Noetzel, 1994). This scheme is adequate for sampling mixed species aphid populations (Elliott *et al.*, 1990; Hein *et al.*, 1995). Using the scheme, 25 tillers are inspected for the presence or absence of cereal aphids before consulting a decision table (Table 10.3). After 25 tillers have been inspected the number of aphid-infested tillers is compared with the upper and lower stop-limits. Sampling continues until a decision is reached or a total of 100 tillers is inspected. If after 100 tillers are inspected the number of infested tillers still falls between the stop limits the standard recommendation is to treat the field.

Organophosphate insecticides are most frequently used for cereal aphid control (Noetzel, 1994). While insecticides are sometimes necessary to control cereal aphids, their widespread use causes some concerns. There are concerns for wildlife safety, particularly for waterfowl (Grue *et al.*, 1988; Flickinger *et al.*, 1991), and the small profit margins often associated with wheat production reduce the economic benefits of insecticides. Further research is needed to more reliably define economic injury levels for cereal aphids.

Coccinellids play an important role in the biological control of aphids in wheat in the Northern Plains (Daniels, 1975; Elliott and Kieckhefer, 1990; Rice and Wilde, 1988). Because of the unpredictability of coccinellid populations in wheat fields in the region (Elliott and Kieckhefer, 1990), sampling methods for coccinellids are required in order to account for their

Table 10.3 Decision table for incidence count (presence/absence sampling for cereal aphids in small rains in the northern Great Plains

Number of tillers inspected (n)	*Lower stop-limit*	*Upper stop-limit*
25	19	24
30	23	29
35	28	34
40	32	39
45	36	43
50	41	48
55	45	53
60	49	58
65	54	62
70	58	67
75	62	72
80	67	77
85	71	81
90	76	86
95	80	91
100	if <84 tillers infested do not treat	Otherwise treat with insecticide

impact on cereal aphids. Simple, yet relatively precise sampling methods are available for estimating coccinellid populations in spring wheat (Elliott *et al.*, 1991); these methods have not been used in aphid management programmes because knowledge of the relationship of coccinellid abundance to cereal aphid population growth is inadequate to permit reliable prediction.

(c) *Managing barley yellow dwarf disease in the Pacific North-west*

A network of Allison–Pike type suction traps (Allison and Pike, 1988) was established in 1986–1987 in the western USA and Great Plains to monitor aerial populations of economically important aphids (Figure 10.9). In eastern Washington the traps are stationed on farms or agricultural research stations and monitored weekly from July to November. Trap catch results are summarized weekly and distributed to growers, consultants, and extension agents. Interpretive comments are provided when appropriate. Data on the seasonal abundance of migrating bird cherry-oat aphids are used to determine appropriate dates for planting winter grains to avoid infestations. In the Pacific North-west, the seasonal distribution and abundance of migrating bird cherry-oat aphids is sufficient information for IPM purposes because manipulating planting dates is an agronomically feasible approach for controlling BYDV (Pike *et al.*, 1990).

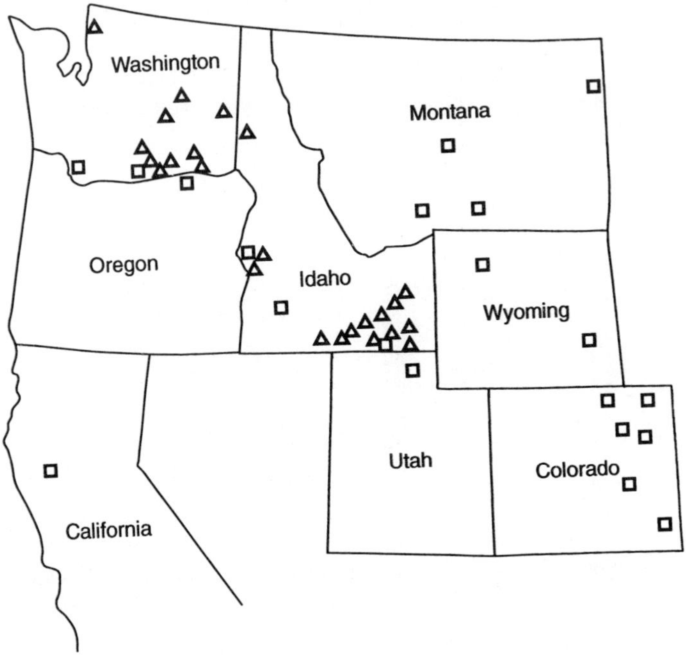

Figure 10.9 Suction trap network in the western United States for monitoring migrant aphids. △, Location of traps stationed before 1987; □, traps stationed in 1987. (Redrawn from Pike *et al.*, 1990).

(d) Managing greenbugs in the Southern Great Plains

Insecticides have been the primary means of controlling greenbugs in wheat in the Southern Great Plains. However, the potential for development of insecticide resistance by the greenbug has become a concern. During the mid-1970s resistance to organophosphate insecticides (primarily to disulfoton and dimethoate) was detected at several locations in the Great Plains (Peters *et al.*, 1975; Teetes *et al.*, 1975). Major problems in controlling the aphid were avoided because resistant greenbugs showed only a low level of resistance to other commonly used insecticides such as parathion and because development of biotype-E greenbug-resistant sorghum cultivars in the late 1970s significantly reduced the use of insecticides for greenbug control and resistance levels decreased (Sloderbeck, 1992). A resurgence of insecticide resistance occurred recently, and resistant greenbugs are now widespread in the Southern Great Plains (Shotkoski *et al.*, 1990; Sloderbeck *et al.*, 1991; Sloderbeck, 1992). Widespread use of insecticides for Russian wheat aphid control in wheat and for control of a new greenbug biotype (biotype I) virulent to biotype-E-resistant sorghum has probably contributed to the resurgence (Sloderbeck, 1992).

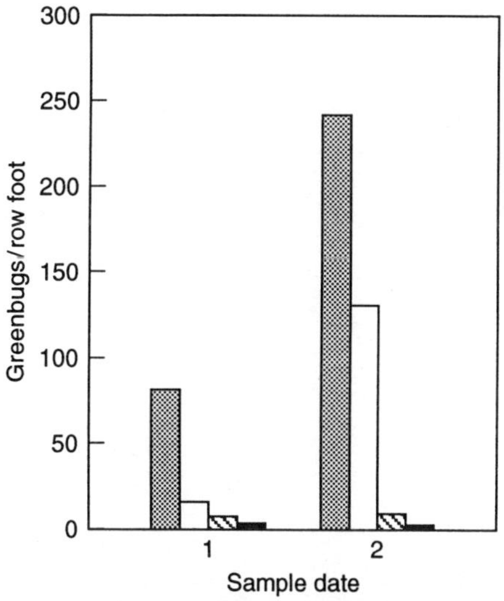

Figure 10.10 Effect of two dates of tillage on greenbug populations in winter wheat. ■, mouldboard; □, disc; □, V-blade; □, no tillage. (Redrawn from Burton and Krenzer, 1985).

Although resistance genes to greenbug damage have been discovered and incorporated into wheat cultivars during the last 20 years, hostplant resistance has had minimal impact on greenbug management in wheat because greenbug biotypes virulent to new resistance sources became widespread in the field before resistant cultivars could be released commercially. There are no wheat cultivars in production with resistance to biotype-E greenbugs, the predominant biotype in the field since the mid-1980s. However, recent development and release of germplasm resistant to biotypes-B, -C, -E, -G, and- I as well as other important wheat pests and diseases provides potential for developing greenbug-resistant cultivars (Porter *et al.*, 1991 ; Porter, 1993).

Several cultural practices can be used to reduce greenbug damage to winter wheat. Conservation tillage provides increased crop residue on the soil surface and reduces immigration into wheat fields by migrating greenbugs (Burton and Krenzer, 1985), presumably because surface residue reduces the attractiveness of the field to the aphids compared with bare soil. Reductions in greenbug populations resulting from conservation tillage are proportional to the amount of residue left on the soil surface, with no-till plots providing the greatest protection from damaging greenbug infestations (Figure 10.10). Nitrogen fertilization at recommended rates reduces injury caused by greenbugs. Under proper fertilization the rate of greenbug population growth is reduced relative to the growth rate of wheat plants, and the more rapid growth allows plants to escape some injury (Daniels, 1957). Grazing cattle on wheat during winter, a common practice in much of the Southern Great Plains, also reduces greenbug populations (Figure

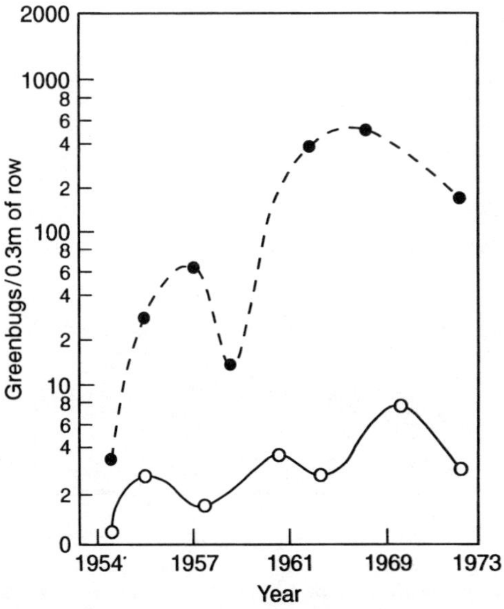

Figure 10.11 Effect of winter cattle grazing on winter wheat and greenbug populations in the Texas Panhandle. --------, Grazed; ———, not grazed. (Adapted from Daniels, 1975).

10.11) (Daniels, 1975). Grazing after initiation of jointing reduces wheat yields, so cattle are typically removed from fields in late winter. Greenbug populations tend to recover after cattle are removed and can still require insecticidal treatment (Daniels, 1975).

Aphid-specific natural enemies play an important role in greenbug population dynamics in wheat and other small grain crops in the Southern Great Plains (Daniels, 1975; Kring *et al.*, 1985). Extension recommendations for insecticidal control of greenbugs sometimes incorporate natural enemies in the decision-making process. For example, Texas guidelines advise farmers to delay control measures in wheat fields with greenbug populations near the economic threshold until it can be determined whether greenbug populations will be controlled by natural enemies (Patrick and Boring, 1990). This approach is advised when there are one or more coccinellids per 0.3-m of row or when 15% or more of greenbugs have been mummified by the parasitoid *Lysiphlebus testaceipes* (Cresson).

(e) Managing Russian wheat aphids in the Great Plains

The relationship of plant injury caused by Russian wheat aphids and yield loss is complex, and varies depending on plant growth stage and environmental interactions. The Russian wheat aphid introduces damaging phytotoxins into plants when it feeds (Fouche *et al.*, 1984) which interfere with

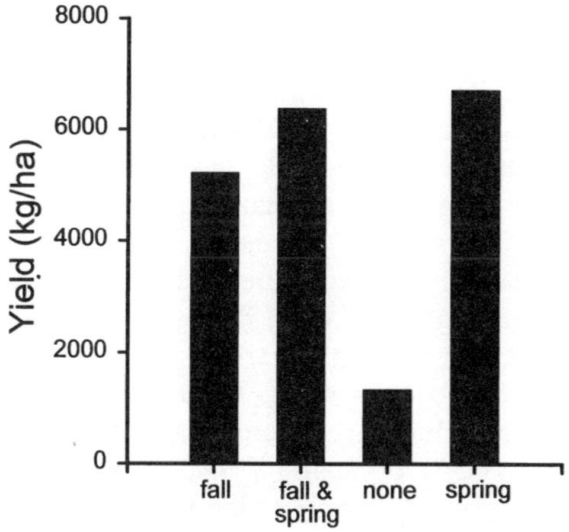

Figure 10.12 Yield of winter wheat in plots treated with insecticides to control Russian wheat aphids during autumn, spring autumn and spring, or untreated. (Adapted from Hammon and Peairs, 1992.)

photosynthesis (Burd and Todd, 1992), delay leaf initiation and tillering, and reduce shoot, root, and leaf biomass, turgor, and grain yield (Gray *et al.*, 1990; Archer and Bynum, 1992; Burd *et al.*, 1993; Girma *et al.*, 1993). Typically, injury caused by Russian wheat aphids in autumn results in small yield reductions in the Southern Plains, whereas spring infestations cause greater losses (Archer and Bynum, 1992; Hammon and Peairs, 1992). However, large Russian wheat aphid infestations before tillering in autumn can result in stand reductions due to reduced tillering (Johnson and Kammerzell, 1991a). In the Northern Great Plains autumn infestations increase susceptibility to winterkill (Butts, 1992a). Except in fields with large infestations before tillering or with infestations large enough to kill plants, insecticide applications during autumn are not usually economically justified in the Southern Plains because plants recover from the injury sustained (Figure 10.12). In the Northern Great Plains, insecticidal control in autumn is more often justified because of greater risk of winterkill (Johnson and Kammerzell, 1991a; Butts, 1992b). Spring Russian wheat aphid infestations often cause minimal yield loss if controlled before flag-leaf emergence (Hammon and Peairs, 1992). This relationship has also been observed in South Africa (Kriel *et al.*, 1986). However, if Russian wheat aphids are present at flag-leaf emergence the injury they cause reduces yield due primarily to flag-leaf injury and head trapping (Archer and Bynum, 1992; Hammon and Peairs, 1992) and removal of plant assimilates required for grain filling. Russian wheat aphids often overwinter in wheat fields in the Southern Great Plains. Although survival is often low and influenced by micro- and macroclimatic factors (Hammon and Peairs, 1991; Armstrong *et al.*, 1992; Archer and Bynum, 1993), overwintering populations can contribute markedly to the potential for economically damaging populations the following spring (Archer and Bynum, 1992, 1993).

Perhaps because it is a new pest of wheat in the USA, there are relatively few effective methods for controlling the Russian wheat aphid. Insecticides, primarily organophosphates and synthetic pyrethroids, are relied upon as the primary means for controlling Russian wheat aphids (Peairs, 1988). Even though Russian wheat aphid injury incurred before jointing normally does not result in economic yield loss, insecticide applications are often recommended before jointing (Patrick and Boring, 1990; Peairs *et al.*, 1992; Legg and Bennett, 1993) because populations typically increase rapidly from boot through the soft dough stage. Natural enemies have minimal impact on Russian wheat aphid populations, and even though abiotic factors such as rainfall can cause populations to decrease (Legg and Brewer, in press), they are unlikely to reduce populations enough to prevent yield loss later in the growing season. Effective population suppression is necessary if insecticides are applied before jointing because poor or ephemeral control can result in population resurgence before wheat plants mature enough to escape economic injury. A second insecticide application on dryland wheat often cannot be justified economically. A single treatment with chlorpyrifos often provides effective control that persists until plants have matured enough to escape serious injury (Peairs, 1988; Richardson, 1989; Johnson and Kammerzell, 1991b; Peairs *et al.*, 1992). Recent data suggest that recommended rates of chlorpyrifos for control of Russian wheat aphids, typically about 480 g (AI)/ha, can be reduced by about one-half and still achieve effective control (Hill *et al.*, 1993).

Tools have been developed to aid crop managers and consultants in making decisions concerning the need for insecticide application. A sequential sampling scheme based on incidence counts (percent infested tillers) was developed (Legg *et al.*, 1991, 1994). A novel procedure termed the 'sequential interval procedure' was used to construct non-linear sequential sampling stop-curves (Brewer *et al.*, 1992; Legg *et al.*, 1992). Using this procedure, stop-curves were developed with variable error rates predetermined so that a specified maximum allowable error rate occurs for population densities equal to the economic threshold. Stop-curves were constructed through an iterative process of simulated sampling and curve-fitting.

A decision-support system to aid managers in making Russian wheat aphid control decisions was developed and implemented on a hand-held computer for use in the field (Figure 10.13a) (Legg and Bennett, 1992, 1993). The system has components for selecting an appropriate economic injury level or threshold, for automating decision tables for a sequential sampling scheme (see above), and for providing management decision support in cases where sampling continues to a user-defined maximum number of samples without having reached a control decision. The decision-support

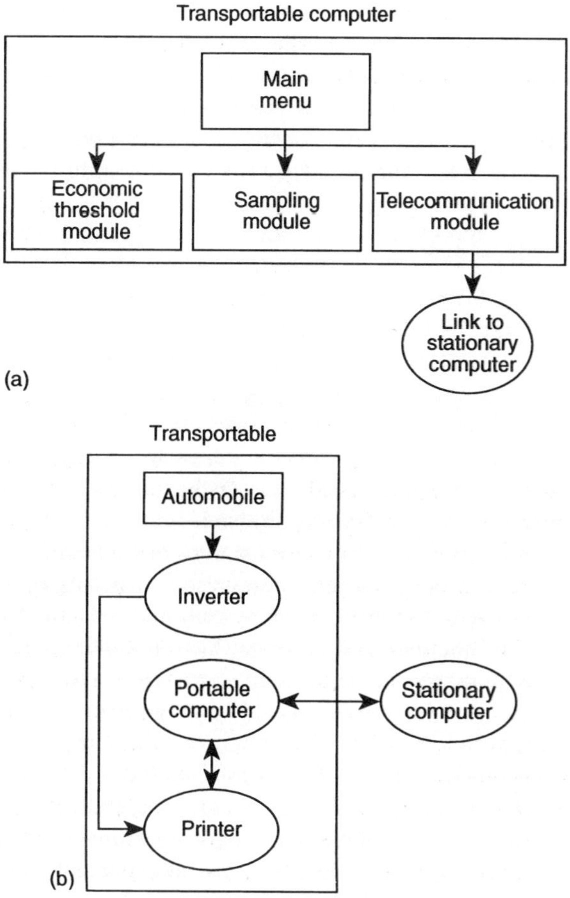

Figure 10.13 Schematic drawing of (a) software and (b) hardware components of a Russian wheat aphid sampling and IPM decision support system. (Adapted from Legg and Bennett, 1993).

module allows managers to select among three management strategies: (i) risk averse (always spray when uncertain); (ii) passive (never spray when uncertain); and (iii) spray if the calculated proportion of infested tillers exceeds the economic threshold. Economic injury levels are calculated by the following equations in the decision-support system:

$$EIL_h = CC/(0.2 \times YI \times VA) \quad (10.1)$$

$$EIL_{ph} = CC/(0.5 \times YI \times VA) \quad (10.2)$$

In these equations, EIL_h and EIL_{ph} are economic injury levels, expressed as percent infested tillers at head emergence and post head emergence, respectively; *CC* is control cost, *VA* is the market value of wheat, and *YI* is expected yield in the absence of Russian wheat aphids (Legg and Bennett, 1993). The constants 0.2 and 0.5 represent percent yield reduction for each 1% infested tillers at heading and post-heading, respectively (Archer and Bynum, 1992). The economic threshold for wheat during the jointing stage is calculated using the model of du Toit (1986):

$$ET = EIL_{\mathrm{h}}(C^{-X}) \tag{10.3}$$

where *ET* is the economic threshold (percent infested tillers), *C* is the ratio of percent infested tillers at head emergence to that at first joint, and *X* is time in weeks between jointing and full head emergence. Values of $C = 1.4$ and $X = 4.0$ are acceptable for dryland winter wheat in Wyoming (Bennett, 1990). The portable computer can be connected to a printer powered by a car battery via a 12 V to 110 V power converter (Figure 10.13b) so that sampling results and recommendations can be summarized and delivered to the farm manager before leaving the site.

A comprehensive IPM decision support system for Russian wheat aphid management is under development. The system consists of two modular expert systems: an aphid identification system and a tactical and strategic advice system. The tactical and strategic advice system was constructed using a case-based reasoning approach rather than the usual rule-based approach because Russian wheat aphid management is a complex and poorly understood endeavour, and a case-based system can be more easily updated with new knowledge as it becomes available (Berry *et al.*, 1993).

Two major objectives of a multi-agency programme for IPM of the Russian wheat aphid are breeding for hostplant resistance and classical biological control. Foreign exploration for natural enemies has been a major focus of the programme since its initiation in 1987. Natural enemies imported from Europe, Asia, Africa and South America have been released in 16 Russian wheat aphid-infested states (McKinnon *et al.*, 1992). *Aphelinus asychis* (Walker) was recently reported to have established in Texas (Michels and Whitaker-Deerberg, 1993) and there are other unpublished reports of establishment.

Using uniform screening methods described by Webster *et al.* (1987) over 8000 wheat accessions have been evaluated for resistance to Russian wheat aphid; 13 wheat accessions originating from various countries have demonstrated significant resistance levels (Quick *et al.*, 1992). Resistance in these wheats is believed to be conferred by several different genes. Variety development in Colorado using T-57 (Turcikum-57) has proceeded to the F_6 generation (Quick *et al.*, 1992); other states are making similar progress.

Delayed planting of winter wheat is effective in reducing autumn populations of Russian wheat aphids in some regions of the Great Plains (Figure

10.14). Planting late often allows wheat to escape economic injury during autumn. In years with mild winters when overwintering survival is high, the reduced autumn populations in late-planted wheat result in smaller spring populations (Hammon *et al.*, 1991; Walker *et al.*, 1991; Butts, 1992b). Late planting is also recommended for control of Hessian fly, wheat curl mite and barley yellow dwarf disease (Hatchett *et al.*, 1987), and though it is unlikely to completely alleviate Russian wheat aphid damage (because in most regions the aphids invade fields again in spring) it may significantly reduce injury in some years.

10.4.2 IPM of the Hessian fly

The Hessian fly damages wheat from the Atlantic coast to the Great Plains, and also along the Pacific coast (Hatchett *et al.*, 1987). Widespread outbreaks occur at irregular intervals, while local outbreaks occur almost every year. The Hessian fly typically has two generations in the northern USA. Larvae of the autumn generation infest seedling and tillering plants, where they stunt or kill tillers and predispose plants to winterkill. Larvae of the spring generation typically feed above the nodes in jointed stems which

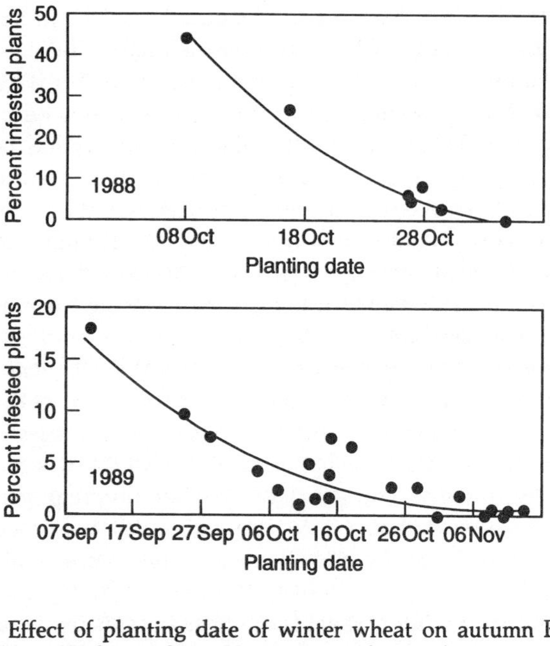

Figure 10.14 Effect of planting date of winter wheat on autumn Russian wheat aphid infestations. (Redrawn from Hammon *et al.*, 1991).

reduces grain filling and causes lodging. In the south the insect typically has four to six generations (Hoelscher and Turney, 1985; Chapin *et al.*, 1989; Buntin and Chapin, 1990). For example, in the coastal region of Georgia two generations typically occur in the autumn, one in the winter, and one or two in the spring (Buntin and Chapin, 1990).

The primary tactics available for Hessian fly management are plant resistance, delayed planting date, ploughing under wheat stubble, destruction of volunteer wheat, and in-furrow application of systemic insecticides at planting time (Hatchett *et al.*, 1987). Some of these tactics are inapplicable in certain regions either because ecological conditions are inappropriate or technology has not developed to an acceptable point.

Breeding wheat cultivars with Hessian fly resistance has been very successful in reducing economic losses in some regions in the USA. This is particularly true in the midwestern USA where the insect has a long history as a pest (Foster *et al.*, 1991). During the last 10–20 years the insect has become a major pest in some southern regions (Hatchett *et al.*, 1987). Locally-adapted varieties in which resistance is compatible with other crop requirements such as yield potential and disease resistance are not currently available in some southern areas (Buntin and Raymer, 1989; Buntin *et al.*, 1992). Superior agronomic performance is important for farmer acceptance of Hessian fly-resistant cultivars (Patterson *et al.*, 1990; Foster *et al.*, 1991).

The genetic basis of resistance to Hessian fly in wheat is well understood. A gene-for-gene relationship exists between resistance in wheat and virulence in the Hessian fly (Gallun and Khush, 1980). Nineteen genes conferring resistance in wheat have been identified along with the 'Kawvale' type of resistance (Obanni *et al.*, 1989). Eleven Hessian fly biotypes that possess virulence to various resistance genes are known from field populations (Hatchett *et al.*, 1987; Obanni *et al.*, 1989).

The strategy for managing resistance genes in breeding programmes in the midwest has been sequential release of cultivars with different individual resistance genes (Hatchett *et al.*, 1987). In order to maintain high levels of resistance in the field three genes, H3, H6 and H5 were incorporated into cultivars and released in sequence at approximately 8-year intervals (Foster *et al.*, 1991). Although Hessian fly biotypes exhibiting virulence to cultivars with single gene resistance developed (or increased in frequency) over time, infestations were reduced to low levels in fields of resistant cultivars for periods of 6–10 years before resurgence (Foster *et al.*, 1991). Foster *et al.* (1991) suggest that development and release of cultivars with single resistance genes may be adequate for achieving long-term resistance because cultivars with a particular gene maintain high levels of resistance for 6–10 years, and because it may be possible to re-deploy a particular gene after several years' absence from the field and achieve resistance of similar duration to the original release. Other approaches such as pyramiding genes in individual cultivars may also be effective in maintaining durable resistance (Cox and Hatchett, 1986; Gould, 1986; Foster *et al.*, 1991).

Cultural practices play an important role in Hessian fly management. Delaying planting until adult activity declines in autumn is an effective strategy for reducing fly populations in winter wheat in the northern USA. In most northern states, where Hessian fly has only two generations, long-term records of adult fly seasonal activity have provided useful guidelines for 'fly-free' planting dates (Metcalf *et al.*, 1962). In the south, delayed planting is of limited use as a management tool because dates of onset of weather cold enough to curtail adult activity are unpredictable or do not occur (Hoelscher and Turney, 1985; Buntin *et al.*, 1990). Ploughing-in wheat stubble soon after harvest to destroy pupating insects, and destruction of alternative hosts such as volunteer wheat are proven practices for reducing Hessian fly populations (Metcalf *et al.*, 1962).

In-furrow application of systemic insecticides at planting time is effective in controlling the autumn Hessian fly generation (Brown, 1960; Nelson and Morrill, 1978; Chapin *et al.*, 1991; Buntin *et al.*, 1992). The autumn population in a field is usually the main source of Hessian fly infestation for the same field in the spring. Therefore, effectively suppressing the autumn population often results in a non-economic population the following spring (Brown, 1960; Nelson and Morrill, 1978; Chapin *et al.*, 1991; Buntin *et al.*, 1992). Although systemic insecticides have not been widely used in the south-eastern USA to manage Hessian fly they appear to provide net economic returns across a broad range of Hessian fly densities. Prophylactic use of insecticides may provide a stop-gap control measure in the south-east, where infestations occur in most years and fly-free dates are often well after optimal planting dates for winter wheat (or do not occur at all), until agronomically competitive resistant varieties become available (Buntin *et al.*, 1992). Insecticide applications at planting time provide effective control of Hessian fly in the mid-west, but are generally not economically justifiable because of availability of suitable resistant varieties and the unpredictable nature of Hessian fly infestations (Foster *et al.*, 1991). Multiple applications of foliar-applied insecticides can be used to control Hessian fly populations in spring, but are not usually economically justified (Buntin and Hudson, 1990).

10.5 WHEAT IN NEW ZEALAND: BIOLOGICAL CONTROL SUCCESSES

New Zealand produces a small wheat crop – 181 thousand tonnes in 1991 – of which 70% is produced in Canterbury, on the east coast of the South Island. The agricultural area of the Canterbury plain comprises approximately 1 million ha of arable and intensively managed pasture land, and is the site of most cereal insect studies carried out in New Zealand over the past 40 years. The climate is cool temperate, with monthly mean temperatures of ca. 5°C in July and ca. 17°C in January, and 500–1000 mm of rain annually. In spring and summer, approximately 85 000 ha of wheat, barley and oat crops are scattered through this area, followed by 20 000 ha of forage barley and oats in autumn and winter. Specialist cereal insects are thus provided with a 'green bridge' of hostplants throughout the year, while ryegrass pastures and hay or seed crops provide permanent habitats for more generalist pests of Gramineae. The following account refers to wheat insect pests in Canterbury.

10.5.1 Insect pests

An endemic lygaeid bug, *Nysius huttoni* White (wheat bug) fills the role of the pentatomid genera *Eurygaster* and *Aelia* in eastern Europe, and like them injects proteolytic enzymes into the developing wheat grain, breaking down high molecular weight glutenins (Every *et al.*, 1989). The affected grain makes poor dough, and is useful only for stock feed. Diapausing adult wheat bugs overwinter in aggregations under dead pine bark, in rolled dead leaves of weeds, or in gorse hedges. In September, they migrate to annual weeds on fallow land, where their offspring feed on developing seed and become adult in December or January. Some adults migrate after emergence, and may colonize wheat ears if the grain is suitable - at the early grain-filling stage - for their feeding (Farrell and Stufkens, 1993). Outbreaks of wheat bug damage to grain occurred at intervals of 3–17 years in the last 60 years, usually after exceptionally warm and dry spring weather (Swallow and Cressey, 1987). The most recent outbreak occurred in 1988 after a warm, dry spring, with early wheat bug flights from mid-November, and populations of >500 bugs/m^2 in some wheat crops with 50% damaged grain. Outbreaks may be predicted on the basis of meteorological data from September onwards and on the basis of developmental stage of immature wheat bugs in November (Farrell and Stufkens, 1993).

Two pasture pests may damage wheat sown after grass. *Costelytra zealandica* White is an endemic New Zealand scarabaeid whose larvae feed on the roots of grasses and clovers. Granular insecticide drilled with the seed is used to control known infestations. The introduced Argentine stem weevil (*Listronotus bonariensis* Kuschel), whose larvae are borers in ryegrass tillers, will also attack stems of wheat, causing them to break off as the crop matures. The parasitoid *Microctonus hyperodae* (Loan) was recently imported from South America and released in Canterbury as a biological control agent against the stem weevil (Goldson *et al.*, 1994), and may reduce damage to wheat. This damage is often mistaken for that caused by the Hessian fly *Mayetiola destructor* (Say), whose larvae also mine wheat tillers. A suite of parasitoids were either self-introduced or imported as biological control agents against the fly in the late 19th century. Recently, 40–50% parasitism of Hessian fly larvae and pupae has been reported (Macfarlane, 1989).

Three aphid species have achieved pest status on wheat in the last 40 years. The bird cherry-oat aphid (*Rhopalosiphum padi* L.) infests grasses and cereals, and is a vector of BYDV. Outbreaks of this virus occurred in 1960–1966 and 1985–1990, when yield losses of 20–35% were recorded in severely infected cereals (Farrell and Stufkens, 1992). *Sitobion* sp. caused damage to wheat in 1967–1969, when 100–150 aphids per ear reduced yields by 20% (Sanderson and Mulholland, 1969). Unfortunately, no aphid specimens were collected from Canterbury wheat during the outbreak, so it is not known whether the species involved was *S. miscanthi* Takahashi (oriental grain aphid), which was recorded in New Zealand before the outbreak, or *S. fragariae* (Walker) (blackberry–cereal aphid), which is the only member of the genus *Sitobion* to be recorded in New Zealand since the outbreak. *S. fragariae* is now found in low numbers in grasses and cereals, but on three occasions in the last decade it has been found in dense infestations on wheat ears in single crops. *Metopolophium dirhodum* (Walker) (rose–grain aphid) appeared in New Zealand in 1982, and briefly achieved pest status in cereals, particularly spring barley, where yield losses of about 15% were recorded in 1984 (Stufkens and Farrell, 1985). Insecticides were used for control of these aphid pests. Recently, two aspects of cereal IPM – aphid biological control and BYDV forecasting – have been developed at Lincoln, Canterbury.

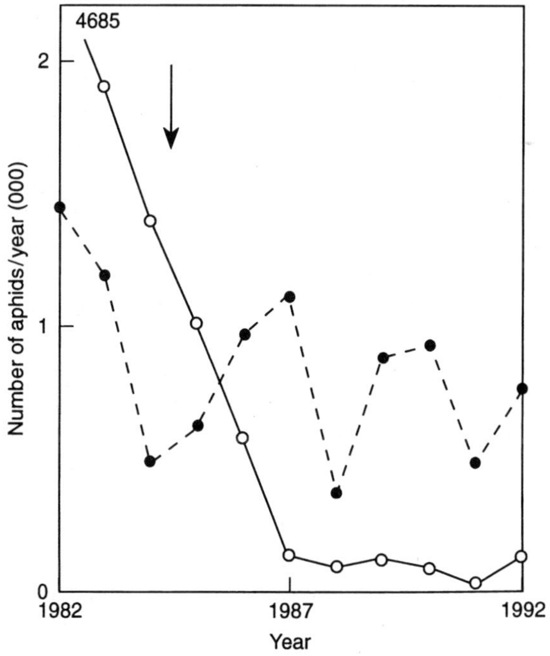

Figure 10.15 Annual suction trap catches of winged aphids between 1982 and 1992 at Lincoln, Canterbury, New Zealand. ○, numbers of *Metapolophium dirhodum*; ●, numbers of *Rhopalosiphum padi*. Vertical arrow, release of *Aphidius rhopalosiphi*.

10.5.2 Biological control

Aphidius rhopalosiphi was imported from France (Dr J.-M. Rabasse) and England (Dr W. Powell) in May 1985. The parasitoids were reared at Lincoln and released at three sites in Canterbury in winter and spring of the same year. By December 1986, *A. rhopalosiphi* was found throughout the Canterbury plain, and further releases elsewhere were followed by establishment of the parasitoid throughout the cereal growing areas of New Zealand by 1988. Dense populations of parasitized mummies of both *M. dirhodum* and *R. padi* were seen in Canterbury and *A. rhopalosiphi* comprised 98–99% of primary parasitoids reared from both aphid species. Figure 10.15 shows that annual suction trap catches of *R. padi* varied around the mean value of 834 throughout the study whereas catches of *M. dirhodum* fell steeply from 1982 to 1987, when catches reached a stable equilibrium varying around a mean value of 106. In spring barley, population maxima of 100–200 *M. dirhodum* per tiller between 1982 and 1984 declined to maxima of 1–5 per tiller between 1986 and 1989 (Farrell and Stufkens, 1990). Few or no *M. dirhodum* were found in ryegrass, the main habitat of *R. padi*, whereas numbers of the latter species rarely exceeded one aphid per tiller in barley.

The rate of parasitism of *M. dirhodum* by *A. rhopalosiphi* was determined in barley field trials in 1988 and 1989, using the method of rearing 4th instar nymphs (Farrell and Stufkens, 1990). Figure 10.16 shows that rates of parasitism quickly rose above 50% in winter barley and reached 100% in spring barley, at the times of aphid maxima. Similar ratios of the aphids and parasitoids were in flight, and recorded in a suction trap, during the build-up of aphids on spring barley. Recent work indicates that *A. rhopalosiphi* migrates to spring barley from winter barley and ryegrass

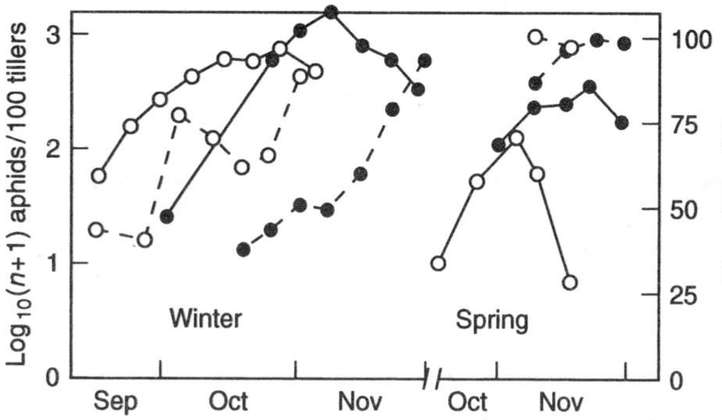

Figure 10.16 Aphid population density and parasitism in winter barley (left) and spring barley (right) at Lincoln, Canterbury, New Zealand. ○, 1988 data; ●, 1989 data. Solid line: $\log_{10}(n+1)$ *Metopolophium dirhodum* per 100 tillers; broken line: % 4th instar nymphs parasitized by *Aphidius rhopalosiphi*.

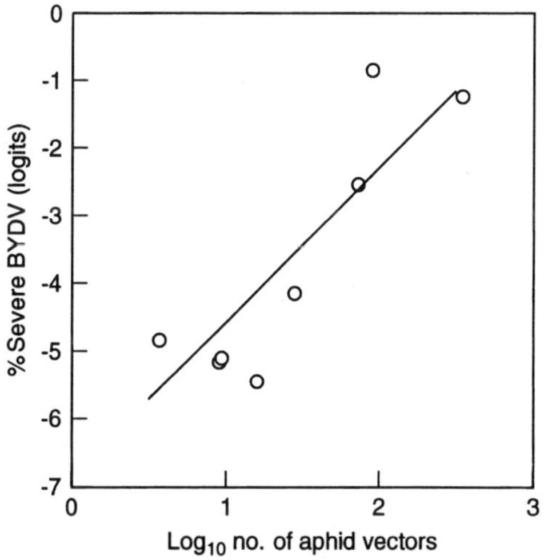

Figure 10.17 Relationship between $\log_{10}$ number of aphid vectors of BYVD trapped in June–July at Lincoln (x) and logit % Canterbury cereal crops with >5% BYDV in the following early December (y). $y = 2.311\,x(\pm 0.523) - 6.91$ (± 1.15).

(J.A. Farrell, unpublished data) and it is likely that the concentration of large numbers of parasites on the young offspring of immigrant aphids led to the truncation of *M. dirhodum* population development on spring barley. In contrast, the grassland habitat of *R. padi* may provide a refuge from attack by *A. rhopalosiphi*, in that the ratio of parasitoids to hosts is substantially lower in ryegrass than in spring barley (J.A. Farrell, unpublished data). Thus the ecological factors of migration and habitat selection may play a part in determining the effectiveness of biological control of *M. dirhodum* and the seeming lack of effect on *R. padi*.

REFERENCES

Allison, D. and Pike, K.S. (1988) An inexpensive suction trap and its use in an aphid monitoring network. *Agricultural Entomology*, 5, 103–7.

Anon. (1984) Use of fungicides and insecticides in cereals. Ministry of Agriculture, Fisheries and Food (ADAS), Booklet B2257 (84).

Anon. (1985) A look at world pesticide markets. *Farm Chemicals (September)*, 26–34.

Anon. (1986) Use of fungicides and insecticides in cereals. Ministry of Agriculture, Fisheries and Food (ADAS), Booklet B2257(86).

Anon. (1988) Grain aphid control. ADAS Divisional Bulletin 144, June 1988.

Archer, T.L. and Bynum, Jr. E.D. (1992) Economic injury level for the Russian wheat aphid (Homoptera: Aphididae) on dryland winter wheat. *Journal of Economic Entomology*, **85**, 987–92.

Archer, T.L. and Bynum, Jr. E.D. (1993) Ecology of the Russian wheat aphid

(Homoptera: Aphididae) on dryland winter wheat in the Southern United States. *Journal of the Kansas Entomological Society*, **66**, 60–8.

Armstrong, S., Peairs, F., Nielsen, D., Roberts, E., Holtzer, T. and Stushnoff, C. (1992) The overwintering biology of *Diuraphis noxia* on the northeastern plains of Colorado, in *Proceedings of the 5th Russian Wheat Aphid Conference, 26–28 January 1992*, (ed. W.P. Morrison), Fort Worth, Texas, USA, pp. 211–21.

Bennett, L.E. (1990) Preliminary economic thresholds for Russian wheat aphid (RWA), *Diuraphis noxia* (Mordvilko), on dryland winter wheat in southeastern Wyoming. Final Report to the Wyoming Wheat Growers Association, 20 August.

Berry, J., Lanier, W. and Belote, D. (1993) A decision support system for management of Russian wheat aphid in the Western United States. *AI Applications in Natural Resource Management*, **7**, 49–52.

Brewer, M.J., Legg, D.E. and Kaltenbach, J.E. (1992) Presence/absence sampling for Russian wheat aphid: a comparison of the sequential probability ratio test, sequential interval procedure, and fixed-sample inspection plan, in *Proceedings of the 5th Russian Wheat Aphid Conference, 26–28 January 1992*, (ed. W.P. Morrison), Fort Worth, Texas, USA, pp. 260–72.

Briggle, L.W. and Curtis, B.C. (1987) Wheat worldwide, in *Wheat and Wheat Improvement. Agronomy Monograph 13*, 2nd Edn, (ed. E.G. Heyne), American Society of Agronomy, Madison, Wisconsin, USA.

Brown, H.E. (1960) Insecticidal control of the Hessian fly. *Journal of Economic Entomology*, **53**, 501–3.

Buntin, G.D. and Chapin, J.W. (1990) Biology of the Hessian fly (Diptera: Cecidomyiidae) in the southeastern United States: geographic variation and temperature-dependent phenology. *Journal of Economic Entomology*, **83**, 1015–24.

Buntin, G.D. and Hudson, R.D. (1990) Spring control of the Hessian fly (Diptera: Cecidomyiidae) in winter wheat using insecticides. *Journal of Economic Entomology*, **84**, 1913–19.

Buntin, G.D. and Raymer, P.L. (1989) Susceptibility of winter wheat and triticale to the Hessian fly. *Georgia Agricultural Experiment Station Research Bulletin*, **389**, Athens, Georgia, USA.

Buntin, G.D., Bruckner, P.L. and Johnson, J.W. (1990) Management of Hessian fly (Diptera: Cecidomyiidae) in Georgia by delayed planting of winter wheat. *Journal of Economic Entomology*, **83**, 1025–33.

Buntin, G.D., Ott, S.L. and Johnson, J.W. (1992) Integration of plant resistance, insecticides, and planting date for management of the Hessian fly (Diptera: Cecidomyiidae) in winter wheat. *Journal of Economic Entomology*, **85**, 530–8.

Burd, J.D. and Todd, G.W. (1992) Total chlorophyll and chlorophyll fluorescence profiles of Russian wheat aphid resistant and susceptible wheat, in: *Proceedings of the 5th Russian Wheat Aphid Conference, 26–28 January 1992*, (ed. W.P. Morrison), Fort Worth, Texas, USA, pp. 101–6.

Burd, J.D., Burton, R.L. and Webster, J.A. (1993) Evaluation of Russian wheat aphid (Homoptera: Aphididae) damage on resistant and susceptible hosts with comparisons of damage ratings to plant measurements. *Journal of Economic Entomology*, **86**, 974–80.

Burn, A.J. (1987) Cereal crops, in *Integrated Pest Management*, (eds A.J. Burn, T.H. Coaker and P.C. Jepson), Academic Press, London, 474 pp.

Burn, A.J. (1992) Interactions between cereal pests and their predators and parasites, in Greig-Smith, P.W., Frampton, G.K. and Hardy, A.R. (eds). loc. cit., pp. 110–31.

Burton, R.L. and Krenzer, E.G. Jr (1985) Reduction of greenbug (Homoptera: Aphididae) population by surface residues in wheat tillage studies. *Journal of Economic Entomology*, **78**, 390–4.

Butts, R.A. (1992a) Factors influencing the overwintering ability of the RWA in western Canada, in *Proceedings of the 3rd Russian Wheat Aphid Conference, 25–27 October*, (ed. D. Baker), Albuquerque, New Mexico, USA, pp. 148–50.

Butts, R.A. (1992b) The influence of seeding dates on the impact of fall infestations of Russian wheat aphid in winter wheat, in *Proceedings of the 5th Russian Wheat Aphid Conference, 26–28 January 1992*, (ed. W.P. Morrison), Fort Worth, Texas, USA, pp. 120–122.

Carter, N., McLean, I.F.G., Watt, A.D. and Dixon, A.F.G. (1980) Cereal aphids: a case study and review. *Applied Biology*, **5**, 271–348.

Carter, N., Entwistle, J.C., Dixon, A.F.G. and Payne, J.M. (1989) Validation of models that predict the peak density of grain aphid (*Sitobion avenae*) and yield loss in winter wheat. *Annals of Applied Biology*, **115**, 31–7.

Chapin, J.W., Grant, J.F. and Sullivan, M.J. (1989) Hessian fly (Diptera: Cecidomyiidae) infestation of wheat in South Carolina. *Journal of Agricultural Entomology*, **6**, 137–46.

Chapin, J.W., Sullivan, M.J. and Thomas, J.S. (1991) Disulfoton application methods for control of Hessian fly (Diptera: Cecidomyiidae) on southeastern winter wheat. *Journal of Agricultural Entomology*, **8**, 17–28.

Cook, J.R. and Veseth, R.J. (1991) *Wheat Health Management*, APS Press, St. Paul, Minnesota, USA.

Cox, T.S. and Hatchett, J.H. (1986) Genetic model for wheat/Hessian fly (Diptera: Cecidomyiidae) interaction: strategies for deployment of resistance genes in wheat cultivars. *Environmental Entomology*, **15**, 24–31.

Daniels, N.E. (1957) Greenbug populations and their damage to winter wheat as affected by fertilizer applications. *Journal of Economic Entomology*, **50**, 793–4.

Daniels, N.E. (1975) Factors influencing greenbug infestations in irrigated winter wheat. *Texas Agricultural Experiment Station Publication MP-1187*, College Station, Texas, USA.

Davis, R.P., Garthwaite, D.G. and Thomas, M.R. (1991) Pesticide usage survey report 85. *Arable farm crops in England and Wales, 1990*. Ministry of Agriculture, Fisheries and Food, London.

Drenth, H., Hoek, J., Daamen, R.A., Rossing, W.A.H., Stol, W. and Wignands, F.G. (1989) An evaluation of the crop physiological and epidemiological information in EPIPRE, a computer-based advisory system for pests and diseases in winter wheat in the Netherlands. *Bulletin OEPP/EPPO*, **19**, 417–24.

Duffield, S.J. and Aebischer, N. (1994) The effect of spatial scale of treatment with dimethoate on invertebrate population recovery in winter wheat. *Journal of Applied Ecology*, **31**, 263–81.

du Toit, F. (1986) Economic thresholds for *Diuraphis noxia* (Hemiptera, Aphididae) on winter wheat in the Orange Free State, South Africa. *Phytophylactica*, **18**, 107–10.

Elliott, N.C. and Kieckhefer, R.W. (1990) The dynamics of aphidophagous coccinellid assemblages in small grains. *Environmental Entomology*, **19**, 1320–9.

Elliott, N.C., Kieckhefer, R.W. and Walgenbach, D.D. (1990) Binomial sequential sampling methods for cereal aphids in small grains. *Journal of Economic Entomology*, **83**, 1381–7.

Elliott, N.C., Kieckhefer, R.W. and Kauffman, W.C. (1991) Estimating adult coccinellid populations in wheat fields by removal, sweepnet, and visual count sampling. *Canadian Entomologist*, **123**, 13–22.

Every, D., Farrell, J.A. and Stufkens, M.W. (1989) Effect of *Nysius huttoni* on the protein and baking qualities of two New Zealand wheat cultivars. *New Zealand Journal of Crop and Horticultural Science*, **17**, 55–60.

FAO (1986). *FAO Production Yearbook Vol. 39*. Food and Agriculture Organisation of the United Nations.

Farrell, J.A. and Stufkens, M.W. (1990) The impact of *Aphidius rhopalosiphi* (Hymenoptera: Aphidiidae) on populations of the rose-grain aphid (*Metopolophium dirhodum*) (Hemiptera: Aphididae) on cereals in Canterbury, New Zealand. *Bulletin of Entomological Research*, **80**, 377–83.

Farrell, J.A. and Stufkens, M.W. (1992) Cereal aphid flights and barley yellow dwarf virus infection of cereals in Canterbury, New Zealand. *New Zealand Journal of Crop and Horticultural Science*, **20**, 407–12.

Farrell, J.A. and Stufkens, M.W. (1993) Phenology, diapause and overwintering of the wheat bug, *Nysius huttoni* (Hemiptera, Lygaeidae) in Canterbury, New Zealand. *New Zealand Journal of Crop and Horticultural Science*, **21**, 123–31.

Flickinger, E.L., Juenger, G., Roffe, T.J., Smith, M.R. and Irwin, R.J. (1991) Poisoning of Canada geese in Texas by parathion sprayed for control of Russian wheat aphid. *Journal of Wildlife Diseases*, **27**, 265–8.

Foster, J.E., Ohm, H.W., Patterson, F.L. and Taylor, P.L. (1991) Effectiveness of deploying single gene resistances in wheat for controlling damage by the Hessian fly (Diptera: Cecidomyiidae). *Environmental Entomology*, **20**, 964–9.

Fouche, A.M., Verhoeven, R.L., Hewitt, P.H., Walters, M.C., Kriel, C.F. and De Jager, J. (1984) Russian aphid *Diuraphis noxia* feeding damage on wheat, related cereals and a *Bromus* grass species, in *Progress in Russian Wheat Aphid* (Diuraphis noxia *Mordw.*) *Research in the Republic of South Africa*, (ed. M.C. Walters), South African Department of Agriculture Technical Communication 191, pp. 22–3.

Gallun, R.L. and Khush, G.S. (1980) Genetic factors affecting expression and stability of resistance, in *Breeding Plants Resistant to Insects*, (eds F.G. Maxwell and P.R. Jennings), John Wiley and Sons, New York, pp. 63–85.

George, K.S. and Gair, R. (1979) Crop loss assessment on winter wheat attacked by the grain aphid, *Sitobion avenae* (F.), 1974–1977. *Plant Pathology*, **28**, 143–9.

Girma, M., Wilde, G.E. and Harvey, T.L. (1993) Russian wheat aphid (Homoptera, Aphididae) affects yield and quality of wheat. *Journal of Economic Entomology*, **86**, 594–601.

Goldson, S.L., McNeill, M.R., Proffitt, J.R., Barker, G.M., Addison, P.J., Barratt, B.I. and Ferguson, C.M. (1994) Systematic mass rearing and release of *Microctonus hyperodae* Loan (Hymenoptera: Braconidae: Euphorinae), a parasite of the Argentine stem weevil *Listronotus bonariensis* (Kuschel) (Coleoptera: Curculionidae) and records of its establishment in New Zealand. *Entomophaga*, **38**, 527–36.

Gould, F. (1986) Simulation models for predicting durability of insect-resistant germplasm: Hessian fly (Diptera: Cecidomyiidae)-resistant winter wheat. *Environmental Entomology*, **15**, 11–23.

Gray, M.E., Hein, G.L., Walgenbach, D.D. and Elliott, N.C. (1990) Effects of Russian wheat aphid (Homoptera: Aphididae) on winter and spring wheat infested during different plant growth stages under greenhouse conditions. *Journal of Economic Entomology*, **83**, 2434–42.

Greig-Smith, P.W., Frampton, G.K. and Hardy, A.R. (1992) *Pesticides, Cereal Farming and the Environment. The Boxworth Project*. HMSO, London, 288 pp.

Grue, C.E., Tome, M.W., Swanson, G.A., Borthwick, S.A. and DeWeese, L.R. (1988) Agricultural chemicals and the quality of prairie-pothole wetlands for adult and juvenile waterfowl – what are the concerns?, in *Proceedings of the National Symposium on Protection of Wetlands from Agricultural Impacts*, in

Proceedings of the National Symposium on Protection of Wetlands from Agricultural Impacts, (ed. P.J. Stuber), USDA, Fish and Wildlife Service Biological Report 88, pp. 55–66.

Hammon, R.W. and Peairs, F.B. (1991) Microclimate and RWA winter survival, in *Proceedings of the 4th Russian Wheat Aphid Workshop, 10–12 October 1990* (ed. G.D. Johnson), Bozeman, Montana, USA, pp. 113–8.

Hammon, R.W. and Peairs, F.B. (1992) Identification of critical stages for Russian wheat aphid control in western Colorado, in *Proceedings of the 5th Russian Wheat Aphid Conference, 26–28 January 1992*, (ed. W.P. Morrison), Fort Worth, Texas, USA, pp. 56–61.

Hammon, R.W., Judson, F.M., Sanford, D. and Peairs, F.B. (1991) Date of planting and Russian wheat aphid populations, in *Proceedings of the 4th Russian Wheat Aphid Workshop, 10–12 October 1990*, (ed. G.D. Johnson), Bozeman, Montana, USA, pp. 49–53.

Harris, M.O. (1993) Virulence of a Manawatu population of the Hessian fly to several wheat cultivars. *Proceedings of the 46th New Zealand Plant Protection Conference*, 334–7.

Hatchett, J.H., Starks, K.J. and Webster, J.A. (1987) Insect and mite pests of wheat, in *Wheat and Wheat Improvement, Agronomy Monograph 13, 2nd edn*, American Society of Agronomy, Madison, Wisconsin, USA.

Hein, G.L., Elliott, N.C., Michels, Jr. G.J., and Kieckhefer, R.W. (1995). A general method for estimating cereal aphid populations in small grain fields based on frequency of occurrence. *Canadian Entomologist*, **127**, 59–63.

Higley, L.G. and Wintersteen, W.K. (1992) A new approach to environmental risk assessment of pesticides as a basis for incorporating environmental costs into economic injury levels. *American Entomologist*, **38**, 34–9.

Hill, B.D., Butts, R.A. and Schaalje, G.B. (1993) Reduced rates of foliar insecticides for control of Russian wheat aphid (Homoptera: Aphididae) in western Canada. *Journal of Economic Entomology*, **86**, 1259–65.

Hoelscher, C.E. and Turney, H.A. (1985) *Hessian Fly in Texas Wheat*, Texas Agricultural Extension Service Publication, College Station, Texas, USA.

Johnson, G.D. and Kammerzell, K.J. (1991a) Impact of RWA on winter wheat tiller production and winter hardiness, in *Proceedings of the 4th Russian Wheat Aphid Conference, 10–12 October 1990*, (ed. G.D. Johnson), Bozeman, Montana, USA, pp. 46–8.

Johnson, G.D. and Kammerzell, K.J. (1991b) Seasonal performance of selected insecticides against Russian wheat aphid in Montana, in *Proceedings of the 4th Russian Wheat Aphid Conference, 10–12 October 1990*, (ed. G.D. Johnson), Bozeman, Montana, USA, pp. 143–9.

Johnston, R.L. and Bishop, G.W. (1987) Economic injury levels and economic thresholds for cereal aphids (Homoptera: Aphididae) on spring-planted wheat. *Journal of Economic Entomology*, **80**, 478–82.

Kendall, D.A. and Chinn, N.E. (1990). A comparison of vector population indices for forecasting barley yellow dwarf virus in autumn cereal crops. *Annals of Applied Biology*, **116**, 87–102.

Kieckhefer, R.W. (1975) Field populations of cereal aphids in South Dakota spring grains. *Journal of Economic Entomology*, **68**, 161–3.

Kieckhefer, R.W. and Gellner, J.L. (1992) Yield losses in winter wheat caused by low-density cereal aphid populations. *Agronomy Journal*, **84**, 180–3.

Kieckhefer, R.W. and Gustin, R.D. (1967) Cereal aphids in South Dakota. I. Observations of autumnal bionomics. *Annals of the Entomological Society of America*, **60**, 514–61.

Kieckhefer, R.W. and Kantack, B.H. (1980) Losses in yield in spring wheat in South Dakota caused by cereal aphids. *Journal of Economic Entomology*, **73**, 582–5.

Kieckhefer, R.W. and Kantack, B.H. (1988) Yield losses in winter grains caused by cereal aphids (Homoptera: Aphididae) in South Dakota. *Journal of Economic Entomology*, **81**, 317–21.

Kieckhefer, R.W. and Stoner, W.N. (1967) Field infectivities of some aphid vectors of barley yellow dwarf virus. *Plant Disease Reporter*, **51**, 981–5.

Kieckhefer, R.W., Lytle, W.F. and Spuhler, W. (1974) Spring movement of cereal aphids into South Dakota. *Environmental Entomology*, **3**, 347–50.

Kieckhefer, R.W., Elliott, N.C., Riedell, W.E. and Fuller, B.W. (1994) Yield of spring wheat in relation to level of infestation by greenbugs (Homoptera: Aphididae). *Canadian Entomologist*, **126**, 61–66.

Kriel, C.F., Hewitt, P.H., van der Westhuizen, M.C. and Walters, M.C. (1986) The Russian wheat aphid, *Diuraphis noxia* (Mordvilko) – Population dynamics and effect on grain yield in the western Orange Free State. *Journal of the Entomological Society of South Africa*, **49**, 317–35.

Kring, T.J., Gilstrap, F.E. and Michels, Jr. G.J. (1985) Role of indigenous coccinellids in regulating greenbugs (Homoptera: Aphididae) on Texas grain sorghum. *Journal of Economic Entomology*, **78**, 269–73.

Lee, G., Stevens, D.J., Stokes, S. and Wratten, S.D. (1981) Duration of cereal aphid populations and the effects on wheat yield and bread making quality. *Annals of Applied Biology*, **98**, 169–79.

Legg, D.E. and Bennett, L.E. (1992) A mobile workstation for use in an integrated pest management program for the Russian wheat aphid, in *Proceedings of the 5th Russian Wheat Aphid Conference, 26–28 January 1992*, (ed. W.P. Morrison), Fort Worth, Texas, USA, pp. 66–9.

Legg, D.E. and Bennett, L.E. (1993) Development of a transportable computing system for on-site management of the Russian wheat aphid in wheat. *Trends in Agricultural Science – Entomology*, **1**, 31–9.

Legg, D.E. and Brewer, M.J. (in press) Relating within-season Russian wheat aphid (Homoptera: Aphididae) population growth to heat units and rainfall. *Journal of the Kansas Entomological Society*,

Legg, D.E., Hein, G.L. and Peairs, F.B. (1991) Sampling Russian wheat aphid in the western Great Plains. Russian Wheat Aphid Task Force, Great Plains Agricultural Council Report GPAC-138, Fort Collins, Colorado, USA.

Legg, D.E., Kroening, M.E. and Peairs, F.B. (1992) A new procedure for developing binomial sequential sampling models for the Russian wheat aphid, in *Proceedings of the 5th Russian Wheat Aphid Conference, 26–28 January 1992*, (ed. W.P. Morrison), Fort Worth, Texas, USA, pp. 254–7.

Legg, D.E., Nowierski, R.M., Feng, M.G., Peairs, F.B., Hein, G.L., Elberson, L.R. and Johnson, J.B. (1994) Binomial sequential sampling plans and decision support algorithms for managing the Russian wheat aphid (Homoptera: Aphididae) in small grains. *Journal of Economic Entomology*, **87**, 1513–1533.

Lescar, L. (1984) Development of the protection of cereals against pests and diseases in France. *Proceedings of the 1984 British Crop Protection Conference – Pests and Diseases*, pp. 159–68.

Macfarlane, R.P. (1989) *Mayetiola destructor* (Say), Hessian fly (Diptera: Cecidomyiidae), in *A Review of Biological Control of Invertebate Pests and Weeds in New Zealand 1874 to 1987*, (eds P.J. Cameron, R.L. Hill, J. Bain and W.P. Thomas), Technical Communication No. 10, CAB International, Wallingford, UK.

Mann, B.P. and Wratten, S.D. (1991) A computer-based advisory system for cereal aphids – field-testing the model. *Annals of Applied Biology*, **118**, 503–12.

Mann, B.P., Wratten, S.D. and Watt, A.D. (1986) A computer-based advisory system for cereal aphid control. *Computers and Electronics in Agriculture*, **1**, 263–70.

Mann, B.P., Wratten, S.D., Poehling, H.M. and Borgemeister, C. (1991) The economics of reduced-rate insecticide applications to control aphids in winter wheat. *Annals of Applied Biology*, **119**, 451–64.

McKinnon, L.K., Gilstrap, F.E., Gonzalez, D., Woolley, J.B., Stary, P. and Warton, R.A. (1992) Importations of natural enemies for biological control of Russian wheat aphid, 1988–1991, in *Proceedings of the 5th Russian Wheat Aphid Conference, 26–28 January 1992*, (ed. W.P. Morrison), Fort Worth, Texas, USA, pp. 136–145.

Metcalf, C.L., Flint, W.P. and Metcalf, R.L. (1962) *Destructive and Useful Insects, Their Habits and Control*, McGraw-Hill, New York, USA.

Michels, G.J., and Whitaker-Deerberg, R.L. (1993) Recovery of *Aphelinus asychis*, an imported parasitoid of Russian wheat aphid, in the Texas Panhandle. *Southwestern Entomologist*, **18**, 11–17.

Nelson, L.R. and Morrill, W.L. (1978) Hessian fly control in wheat by suppression of the fall generation with seed and soil treatments using the systemic insecticide carbofuran. *Agronomy Journal*, **70**, 139–41.

Nicol, D., Wratten, S.D., Eaton, N. and Copaja, S.V. (1993) Effects of DIMBOA levels in wheat on the susceptibility of the grain aphid (*Sitobion avenae*) to deltamethrin. *Annals of Applied Biology*, **122**, 427–33.

Niemeyer, H.M. (1988) Hydroxamic acids (4-hydroxy-1, 4-benzoxazin-3-ones), defence chemicals in the Gramineae. *Phytochemistry*, **27**, 3349–58.

Niemeyer, H.M. (1991) Secondary plant chemicals in aphid–host interactions, in *Aphid–plant Interactions: Populations to Molecules*, (eds D.C. Peters, J.A. Webster and C.S. Chouber), Oklahoma State University, pp. 101–11.

Noetzel, D. (1994) *Revised Aphid Thresholds in Small Grains*, Minnesota Agricultural Extension Service Publication, University of Minnesota, St. Paul, Minnesota, USA.

Obanni, M., Ohm, H.W., Foster, J.E. and Patterson, F.L. (1989) Genetics of resistance of PI 422297 durum wheat to the Hessian fly. *Crop Science*, **27**, 49–52.

Patrick, C.D. and Boring III, E.P. (1990) Managing Insect and Mite Pests of Texas Small Grains, Texas Agricultural Extension Service Publication B-1251, College Station, Texas, USA.

Patterson, F.L., Shanner, G.E., Ohm, H.W. and Foster, J.E. (1990) A historical perspective for the establishment of research goals for wheat improvement. *Journal of Production Agriculture*, **3**, 30–8.

Peairs, F.B. (1988) Chemical control of the Russian wheat aphid, in *Proceedings of the 2nd Russian Wheat Aphid Workshop, 11–12 October*, (compilers F.B. Peairs and S.D. Pilcher), Denver, Colorado, USA, pp. 134–9.

Peairs, F.B., Beck, G.K., Brown, Jr, W.M., Schwartz, H.F. and Westra, P. (1992) Russian wheat aphid, in *1992–1993 Colorado pesticide guide–Field crops*, Colorado State University Cooperative Extension Service and Agricultural Experiment Station. Bulletin XCM-45, Fort Collins, Colorado, USA, p. 46.

Peters, D.C., Wood, Jr, E.A. and Starks, K.J. (1975) Insecticide resistance in selections of the greenbug. *Journal of Economic Entomology*, **68**, 339–40.

Pike, K.S. and Schaffner, R.L. (1985) Development of autumn populations of cereal aphids, *Rhopalosiphum padi* (L.) and *Schizaphis graminum* (Rondani) (Homoptera: Aphididae) and their effects on winter wheat in Washington State. *Journal of Economic Entomology*, **78**, 676–80.

Pike, K.S., Allison, D., Low, G., Bishop, G.W., Halbert, S. and Johnston, R. (1990) Cereal aphid vectors: western regional USA monitoring system, in *World Perspectives on Barley Yellow Dwarf*, (ed. P.A. Burnett), CIMMYT, Mexico D.F., Mexico, pp. 282–5.

Poehling, H.M. (1989) Selective application strategies for insecticides in agricultural crops, in *Pesticides and Non-target Invertebrates*, (ed. P.C. Jepson), Intercept, Wimborne, Dorset, pp. 151–75.

Porter, D.R. (1993) Host plant resistance to greenbugs in wheat: status and prospects, in *Proceedings of the Greenbug Workshop and Symposium, 9–10 February*, (compiler P.E. Sloderbeck), Albuquerque, New Mexico, USA.

Porter, D.R., Webster, J.A., Burton, R.L., Puterka, G.J. and Smith, E.L. (1991) New sources of resistance to greenbug in wheat. *Crop Science*, **31**, 1502–4.

Quick, J.S., Nkongolo, K.K. and Peairs, F.B. (1992) Breeding wheat for resistance to the Russian wheat aphid, in *Proceedings of the 5th Russian Wheat Aphid Conference, 26–28 January*, (ed. W.P. Morrison), Fort Worth, Texas, USA, pp. 74–78.

Rabbinge, R. and Rijsdijk, F.H. (1983) EPIPRE: a disease and pest management system for winter wheat, taking account of micrometeorological factors. *Bulletin OEPP/EPPO*, **13**, 297–305.

Reinink, K. (1986) Experimental verification and development of EPIPRE. A supervised disease and pest management system for wheat. *Netherlands Journal of Plant Pathology*, **92**, 3–14.

Rice, M.E. and Wilde, G.E. (1988) Experimental evaluation of predators and parasitoids in suppressing greenbugs (Homoptera: Aphididae) in sorghum and wheat. *Environmental Entomology*, **17**, 836–41.

Richardson, J.M. (1989) Insecticidal control of Russian wheat aphid in California, in *Proceedings of the 3rd Russian Wheat Aphid Conference, 25–27 October*, (compiler D. Baker), Albuquerque, New Mexico, USA, p. 77.

Rossing, W.A.H., Schans, J. and Zadoks, J.C. (1985) Het project Epiprø. Een proeve van projektanalyse in de gewasbescherming. *Landbouwkundig Tijdschrift*, **97**, 29–33.

Sanderson, F.R. and Mulholland, R.I. (1969) Effect of the grain aphid on yield and quality of wheat. *Proceedings of the New Zealand Weed and Pest Control Conference*, **22**, 227–35.

Shotkoski, F.A., Mayo, Z.B. and Peters, L.L. (1990) Induced disulfoton resistance in greenbugs (Homoptera: Aphididae). *Journal of Economic Entomology*, **83**, 2147–52.

Sloderbeck, P.E. (1992) Discovery of pesticide resistant greenbugs in Kansas, in *Proceedings of the Greenbug Workshop, 4 February*, (compiler P.E. Sloderbeck), Garden City, Kansas, USA.

Sloderbeck, P.E., Chowdhury, M.A., Depew, L.J. and Buschman, L.L. (1991) Greenbug (Homoptera: Aphididae) resistance to parathion and chlorpyrifos-methyl. *Journal of the Kansas Entomological Society*, **64**, 1–4.

Sly, J.M.A. (1977) *Review of Usage of Pesticides in Agriculture in England and Wales, 1965–74*, Ministry of Agriculture, Fisheries and Food, London.

Sly, J.M.A. (1986) *Review of Usage of Pesticides in Agriculture, Horticulture and Animal Husbandry in England and Wales, 1980–1993*. Ministry of Agriculture, Fisheries and Food, London.

Sotherton, N.W. (1991) Conservation headlands: a practical combination of intensive cereal farming and conservation, in *The Ecology of Temperate Cereal Fields 32nd Symposium of the British Ecological Society*, (eds L.G. Firbank, N. Carter, J.F. Darbyshire and G.R. Potts), Blackwell Scientific Publications, Oxford, pp. 373–98.

Stufkens, M.W. and Farrell, J.A. (1985) Yield responses in cereals treated with pirimicarb for rose-grain aphid control. *Proceedings of the 38th New Zealand Weed and Pest Control Conference*, pp. 188–90.

Swallow, W.H. and Cressey, P.J. (1987) Wheat bug damage in New Zealand wheats 3. An historical overview. *New Zealand Journal of Agricultural Research*, **30**, 341–4.

Teetes, G.L., Schaefer, C.A., Gipson, J.R., McIntyre, R.C. and Latham, E.E. (1975) Greenbug resistance to organophosphorus insecticides on the Texas high plains. *Journal of Economic Entomology*, **68**, 214–16.

van Emden, H.F. (1987) Cultural methods: the plant, in *Integrated Pest Management*, (eds A.J. Burn, T.H. Coaker and P.C. Jepson), Academic Press, London.

Vickerman, G.P. and Sunderland, K.D. (1977) Some effects of dimethoate on arthropods in winter wheat. *Journal of Applied Ecology*, **14**, 767–77.

Vickerman, G.P. and Wratten, S.D. (1979) The biology and pest status of cereal aphids (Hemiptera: Aphididae) in Europe: a review. *Bulletin of Entomological Research*, **69**, 1–32.

Walker, C.B., Peairs, F.B. and Ham, D. (1991) The effect of planting date in southeast Colorado on RWA infestations in winter wheat, in *Proceedings of the 4th Russian Wheat Aphid Conference, 10–12 October*, (compiler G.D. Johnson), Montana, USA, pp. 54–62.

Watt, A.D., Vickerman, G.P. and Wratten, S.D. (1984) The effect of the grain aphid, *Sitobion avenae* F., on winter wheat in England; an analysis of the economics of control practice and forecasting systems. *Crop Protection*, **3**, 209–22.

Webster, J.A., Starks, K.J. and Burton, R.L. (1987) Plant resistance studies with *Diuraphisnoxia* (Mordvilko) (Homoptera: Aphididae), a new United States wheat pest. *Journal of Economic Entomology*, **80**, 944–9.

Wratten, S.D. (1975) The nature of the effects of the aphids *Sitobion avenae* and *Metopolophium dirhodum* on the growth of wheat. *Annals of Applied Biology*, **79**, 27–34.

Wratten, S.D. (1987) The effectiveness of native natural enemies, in *Integrated Pest Management*, (eds A.J. Burn, T.H. Coaker, and P.C. Jepson), Academic Press, London.

Wratten, S.D. and George, K.S. (1985) Effect of cereal aphids on yield and quality of barley and wheat, in *International Organisation of Biological Control/West Palaearctic Regional Section. Working Group 'Integrated Control of Cereal Pests'*, (ed. C. Dedryver), pp. 10–35.

Wratten, S.D. and van Emden, H.F. (1995) Habitat management for enhanced activity of natural enemies of insect pests, in *Ecology and Integrated Farming Systems*, (eds D.M. Glen, M.P. Greaves and H. Anderson), Wiley, London, pp. 117–145.

Wratten, S.D., Mann, B.P. and Wood, D. (1986) Information exchange in future crop protection–interactions within the independent sector: Prestel Farmlink. *Proceedings of the 1986 British Crop Protection Conference – Pests and Diseases*, pp. 703–11.

Wratten, S.D., Mead-Briggs, M.A., Vickerman, G.P. and Jepson, P.C. (1988) Effects of the fungicide pyrazophos on predatory insects in winter barley, in *Field Methods for the Study of Environmental Side-effects of Pesticides, British Crop Protection Council Monograph 14*, pp. 327–34.

Wratten, S.D., Watt, A.D., Carter, N. and Entwistle, J.C. (1990). Economic consequences of pesticide use for grain aphid control on winter wheat in 1984 in England. *Crop Protection* **9**, 73–77.

Wratten, S.D., van Emden, H.F. and Thomas, M.B. (1995) Within-field and border refugia for the enhancement of natural enemies, in *Enhancing Natural Control of Arthropod Pests through Habitat Management*, (eds C.H. Pickett and R. Bugg), Ag Access Corporation, California pp. 117–45.

Wyatt, S.D., Seybert, L.J. and Mink, G. (1988) Status of the barley yellow dwarf virus problem of winter wheat in eastern Washington. *Plant Disease*, **72**, 110–13.

Zadoks, J.C. (1984) Analysing the cost-effectiveness of EPIPRE. *EPPO Bulletin*, **14**, 61–65.

CHAPTER 11

Integrated pest management in cotton

A.P. Gutierrez

11.1 INTRODUCTION

Cotton (primarily *Gossypium hirsutum* L. and to a lessor extent *G. barbadense* L.) is the most important fibre crop grown world-wide (Figure 11.1). Domesticated *G. hirsutum* cottons were bred from wild progenitors that evolved in the neotropics (Fryxell, 1979). The breeding increased yields and enhanced fibre quality, but little progress has been made in breeding against most arthropod pests, and recent attempts to introduce toxin-producing genes into genomes to protect against pests seems unwise. Cotton has accumulated a wide assortment of herbivores and diseases, greatly increasing the complexity of developing integrated pest management (IPM) strategies (see Frisbie *et al.*, 1989).

In this chapter a brief overview of pest control problems in cotton is given, the characteristics of plant growth types and their resultant ability to compensate for pest damage are examined, and a concept of the economic threshold for pests that attack demand or supply sides of the plant's allocation strategy is outlined. Compendia of more classical approaches to cotton production and pest management are Kohel and Lewis (1984), Matthews (1989) and Frisbie *et al.* (1989).

11.1.1 Historical overview of pest control in cotton

In many areas where cotton is grown, many of the pests (diseases, insects, mites, nematodes and weeds) are new associations (*sensu* Hokkanen and Pimentel, 1984), and all too often massive quantities of pesticides are applied to control them. The reliance on insecticides in cotton pest control has only succeeded in inducing the three scourges of agricultural pest con-

Integrated Pest Management Edited by David Dent. Published in 1995 by Chapman & Hall, London. ISBN 0 412 57370 9

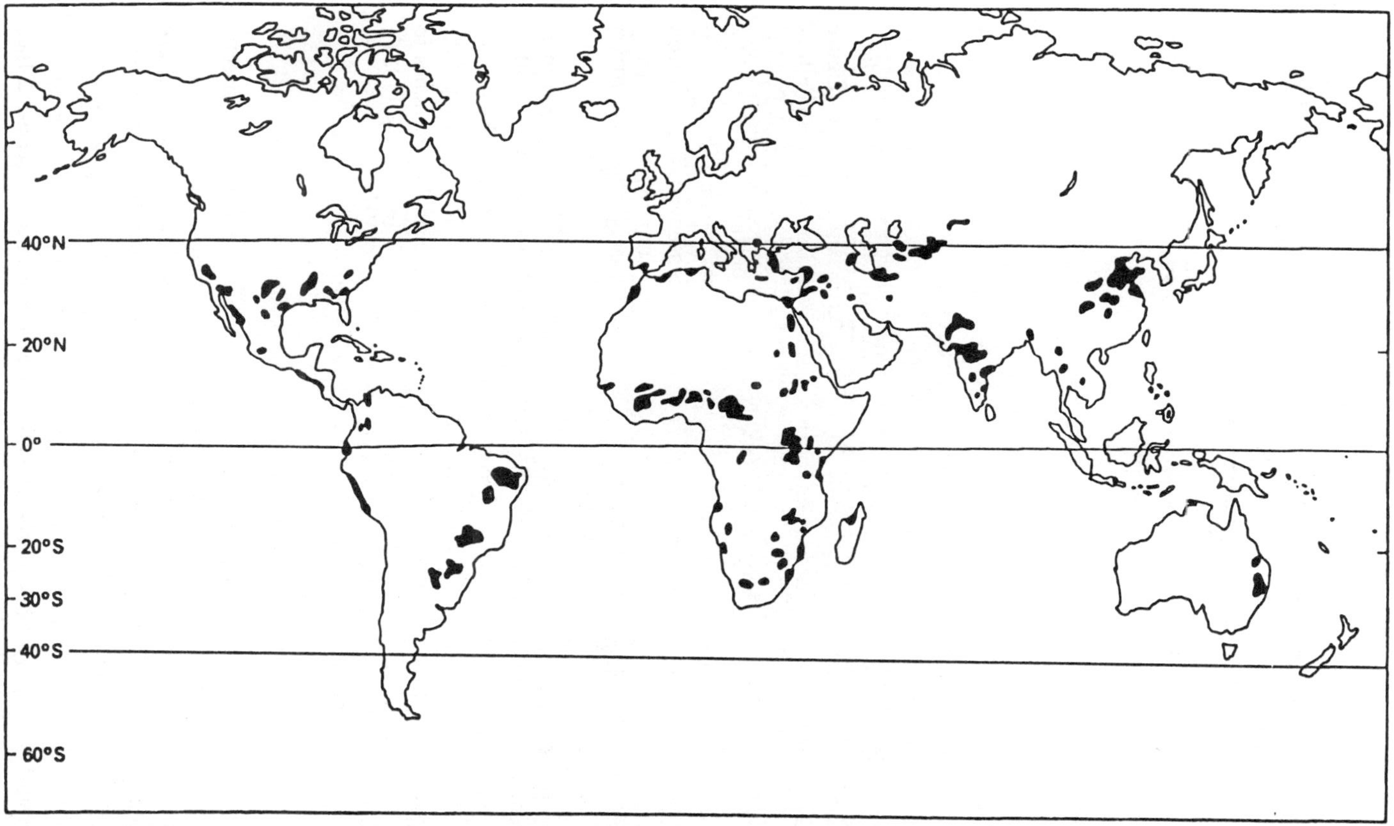

Figure 11.1 Distribution of cotton production (shaded areas) world-wide. (After Matthews, 1989).

trol so colourfully described by van den Bosch (1978): namely *pest resurgence* of the original *target* or *key pest*, outbreaks of secondary pests (mostly high-fecundity noctuid moths, white flies and spider mites) and *resistance to insecticides*, and of course contamination of the air we breathe, the water we drink and the food we eat.

Key pests are those that cause economic damage in the absence of control, and induced pests are those that cause damage when natural enemies controlling their densities are destroyed by pesticides. The most important key pest of cotton in many areas of the Americas is the cotton boll weevil (*Anthonomus grandis* Boh.). This herbivore is thought to have evolved on species of *Hampea* in Central America and later moved to cotton (Burke, 1976). It occurs in Central America, Mexico, Northern South America, the Southern United States (Falcon *et al.*, 1986), and has recently invaded Brazil (Dos Santos, 1989) where it promises to reshape the cotton economy. Other important key insect pests are the ubiquitous pink bollworm (*Pectinophora gossypiella* Saunders) and the important stem borer (*Eutinobothrus brasiliensis* Hambleton) in Brazil. The exotic pink bollworm, originally from Papua New Guinea–North Australia, is the key pest over the range of boll weevil as well as Australia, the desert valleys of Southern California, the Middle East, Africa and Asia. The Brazilian stem borer is found only in South America and its economic impact has only recently been assessed (Dos Santos *et al.*, 1989). The plant pathogen, verticillium wilt (*Verticillium dahliae* Kelb.), is a cosmopolitan species, but it is not a major problem where crop rotation is practised.

Most insect pests of cotton are either pesticide-induced or are minor pests, including the highly damaging bollworms. Pesticide-induced disasters have occurred world-wide in cotton, and among the more spectacular cases are the Ord River of Australia, the Cañet Valley of Peru, the Rio Grande Valley of Mexico, the Nile Delta of Egypt, the Gaziera of the Sudan, South-east Asia, Southern Africa, the South-eastern United States and the sophisticated cotton production systems of California (van den Bosch, 1978). The failure of modern, pesticide-based pest control is best illustrated by reviewing two examples in technologically advanced California. Weed and nematode control are not emphasized in this chapter.

11.2 CALIFORNIA AS AN EXAMPLE

11.2.1 The San Joaquin Valley

Nearly a million acres of cotton are grown in the fertile San Joaquin Valley of California. During the early 1960s, *Lygus hesperus* Knight was considered the key pest, and the disruption caused by pesticide used to suppress it quickly induced other more serious pests such as the cotton bollworm (*Heliothis zea* Hübner) and an array of defoliating noctuids (beet armyworms (*Spodoptera exigua* Hübner), cabbage looper (*Trichoplusia ni*

Hübner) and salt marsh caterpillars (*Estigmene acraea* (Drury)). This period was called the time of the *Lygus* bug wars because of the vehement stances taken by the agrochemical industry, farmers, university researchers and extension agents and anyone else that wanted to get involved (van den Bosch, 1978). In the final analysis, the late Professor Robert van den Bosch and colleagues (Falcon *et al.*, 1968, 1971; Gutierrez *et al.*, 1975) showed that contrary to the perceptions of farmers and the majority of the research establishment, *Lygus* was not a pest under most circumstances; that yields in untreated cotton fields were higher, but not significantly so than those in the best insecticide treatments; that farmers were spending money on insecticides to lose money via increased pest damage; that the per capita damage rate of *Lygus* to cotton fruits was relatively small (Gutierrez *et al.*, 1977b); that only young fruits were attacked and hence the plant lost little time and energy in replacing them; that *Lygus* densities in the field were generally low and the methods used to assess them were grossly inadequate (Byerly *et al.*, 1978); that the pest did not prefer cotton (Cave and Gutierrez, 1983) and could be easily diverted from cotton to crops such as alfalfa and safflower where its damage was trivial (Stern *et al.*, 1964); that its survival and reproduction rates were low on cotton (Cave and Gutierrez, 1983); that cotton plants have a considerable capacity to compensate for lost fruit buds and hence the basis of the economic threshold (i.e. 10 *Lygus* per 50 sweeps) used to assess the need for control was ill advised (Gutierrez *et al.*, 1979).

So why did farmers ignore the evidence? In hindsight, the likely explanation was that most considered the maximum yield a realistic estimate of yearly potential, and hence assumed that departures from that must have been due to pest damage. The feeding damage caused by *Lygus* bugs is difficult to assess because the bug's tiny feeding styles leave imperceptible marks that can be distinguished only with great difficulty even by experts. Through this imbroglio, the facts that cotton yields are strongly influenced by planting densities that may vary from 20 000 to 65 000 plants per acre (49 000–160 000 per ha) and that weather patterns, edaphic factors and cultural practices have large important influences on yields were largely ignored. The fact that secondary pests were induced was largely unrecognized, this despite the fact that it had been demonstrated in other crops (DeBach, 1975). All of these factors added complications that clouded the assessment of the economic status of this pest. Most important of all was the misperception of yield potential.

11.2.2 The desert valleys of Southern California

The pink bollworm is not a pest in the San Joaquin Valley because it is frost limited and cannot to date successfully overwinter there (Gutierrez *et al.*, 1977a). Despite this biology, millions of dollars have been spent on misguided eradication programmes there and in the Desert Valleys of

Southern California, where it is a serious key pest. Problems with this pest in the Desert Valleys have been exacerbated by farmers who ignored non-chemical control recommendations such as the early termination of the crop and stalk destruction to avoid the build-up of overwintering diapause larvae, the elimination of the practice of growing ratoon cotton, the use of short-season cultivars to reduce the within season build-up of the pest, the abandonment of disruptive season-long pest control tactics, and the use of sex pheromones for mating disruption (Gaston *et al.*, 1977; Gutierrez *et al.*, 1977a; Burrows *et al.*, 1982, 1984; Stone and Gutierrez, 1986a,b).

Farmers again assumed that the maximum yield before the introduction of pink bollworm in the 1960s was a realistic annual goal, and they were unwilling to accept the one-quarter to one-half-bale lower yields associated with early termination of the crop. Instead they sought special use permits for a suspected carcinogenic insecticide with known poor insecticidal qualities because they thought it stimulated the plant to produce higher yields. Private profit was the prime motivator and the public good was the sacrificial lamb. As one farmer put it during the late 1970s . . . 'we like to meet at the Elks Club to drink beer and lie about our yields'. It may have been the Elks Club in Southern California, but it could just as easily be the RSL Clubs in Australia or any social club frequented by the local cotton barons in other parts of the world.

The myopic pest control practices induced outbreaks of budworms, whiteflies and mites followed by a rapid increase in insecticide resistance in tobacco budworm (*Heliothis virescens* (Fabricius)) (see van den Bosch *et al.*, 1978), economic ruin and a 90% decrease in the cotton-growing area by the late 1980s. As pest problems increased, more and other insecticides were used, further exacerbating the pest control problem as multiple resistance to several insecticides occurred. They had stepped onto a treadmill of more insecticides, in different combinations, at higher dosages and with increased frequency of application (c.f., van den Bosch, 1978). The industry collapsed, falling in the Imperial Valley from ca. 44 000 ha to ca. 6000 ha.

The industry has since recovered, and has been greatly sobered by this fiasco as farmers are more willing to accept the old recommendations, and in addition the methodology of dispensing mating disruption pheromones for pink bollworm control has greatly improved providing a realistic alternative to pesticides. However, I have no doubt that short memories and poor judgment will cause some to return to failed pest control practices to begin anew the cycle of pesticide misuse. The few will exploit the common property of natural enemies for their advantage (*sensu* Regev, 1984). What will limit the duration of the next episode of pesticide misuse are the genetic relics of past folly – widespread resistance to insecticides.

Irrational pest control practices have occurred because of greed and ignorance, but also because the biological constraints on yield and the consequences of pesticide misuse have not been well understood. A complicating factor is that until recently, the scientific methods of analysis have tended to be piecemeal and not holistic, static rather than dynamic, the effects of plant growth characteristics and their ability to compensate for pest damage and the important effects of weather and edaphic factors on yields and plant–pest interactions have been largely ignored. Modelling provides the mechanism for linking these separate areas so that their interactions may be examined and analysed. Here we review some of that progress.

11.3 A PHYSIOLOGICAL BASIS FOR PEST CONTROL

Several groups have developed mechanistic models for cotton and some pests (Baker *et al.*, 1972; Ephrath and Marani, 1993), and other groups have developed demographic models (Gutierrez *et al.*, 1975, 1981; Wang *et al.*, 1977; Curry and Cate, 1984; Gutierrez and Curry 1989). Gutierrez and

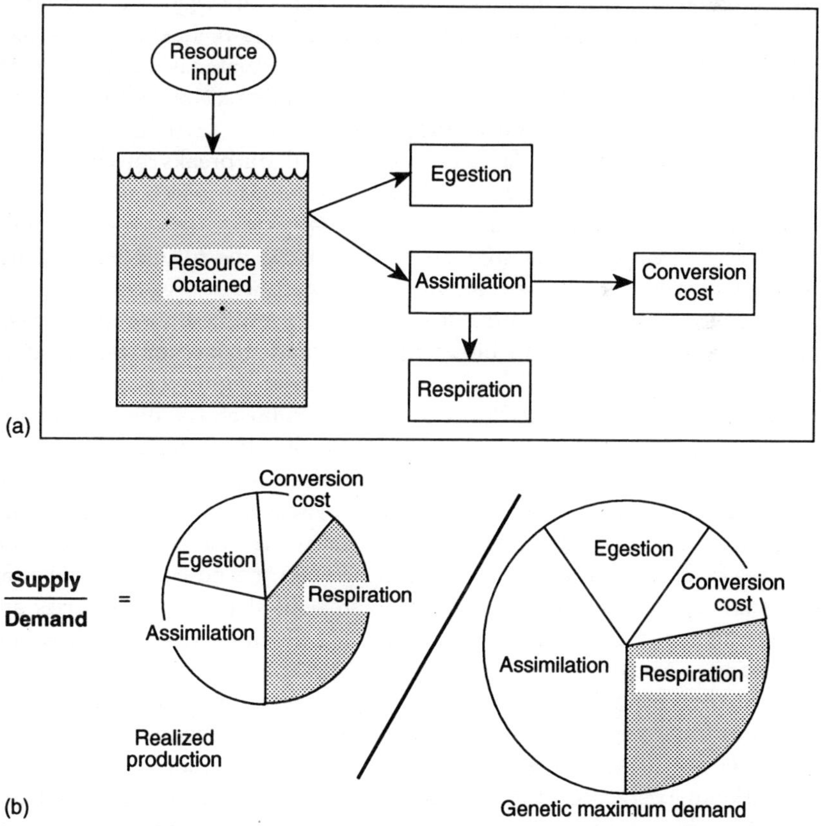

Figure 11.2 (a) The metabolic pool model and (b) the ratio of actual resources acquired and the genetic maximum demand. (After Gutierrez and Wang, 1976).

Wang (1976), Gutierrez *et al.* (1984, 1987) and Gutierrez and Wilson (1989) generalized the demographic model across trophic levels. Another important innovation was the use of resource allocation schemes in plant modelling (DeWit and Goudriaan, 1978) (e.g. Figure 11.2), and Baker *et al.* (1972, 1983) pioneered the use of this paradigm in cotton.

Gutierrez and Wang (1976) unified this paradigm across all trophic levels by linking resource allocation schemes and demographic models (see also Gutierrez and Baumgaertner, 1984; Gutierrez *et al.*, 1988b). This model assumes that under optimal conditions, an organism (i.e. a cotton plant) has a maximum *genetic potential* (demand, D) for respiration, vegetative growth and fruit growth as well as the costs of tissue conversion that can be estimated for any state of the organism (Figure 11.2a), and that this demand drives the resource acquisition process (e.g. photosynthesis in plants or food acquisition in animals). In practice, the maximum photosynthates that may be produced are roughly those required for maximum vegetative growth, and like prior schemes, allocation is first to respiration, then reproduction and lastly vegetative growth. Failure to meet the demand may be due to abiotic and biotic factors that affect the plant's photosynthetic rate (i.e. the supply, S). A shortfall in photosynthates is summarized by the supply/demand ratio (i.e. $0 < S/D < 1$; Figure 11.2b) which may be viewed as a survivorship term that regulates all birth, death and net growth rates in all trophic levels. In plants, the growth rates of vegetative parts, the production rate of new fruit buds, the shedding rates of susceptible fruits, the growth rates of survivors, the viability of the seed, and other factors are affected by S/D (Gutierrez and Wang, 1976). Gutierrez (1992) provides a concise overview of this paradigm.

In plants, the capacity to produce photosynthates is maximized when sufficient leaves are present to intercept most of the solar radiation incident in the area where they grow. Carbohydrate stress commonly occurs when fruit demands increase rapidly (e.g. Figure 11.3a) and make the demand greater than the supply (i.e. the control point in Figure 11.3b). Reductions in supply may occur due to weather (low solar radiation, drought stress, high and low temperatures, etc.), edaphic factors (e.g. low levels of nutrients) and of course, competition from other plants including weeds and damage from herbivores and diseases. These factors cause the control point to be moved earlier in time resulting in smaller plants, and reduced yield (Figure 11.3c). Fruits may be shed at a rate proportional to the shortfall and/or the surviving fruits may simply be smaller. Plants cannot greatly increase their capacity to produce photosynthates simply because demands for fruits have increased; hence the maximum yield potential of a crop is set when the demands exceeds the supply.

Herbivores may affect the demand side by removing important sinks such as fruits. This causes the control point to be moved later in time and often results in excessive vegetative growth (Figure 11.3d). This phenomenon is easily demonstrable in plants that produce relatively large fruits (e.g. apple,

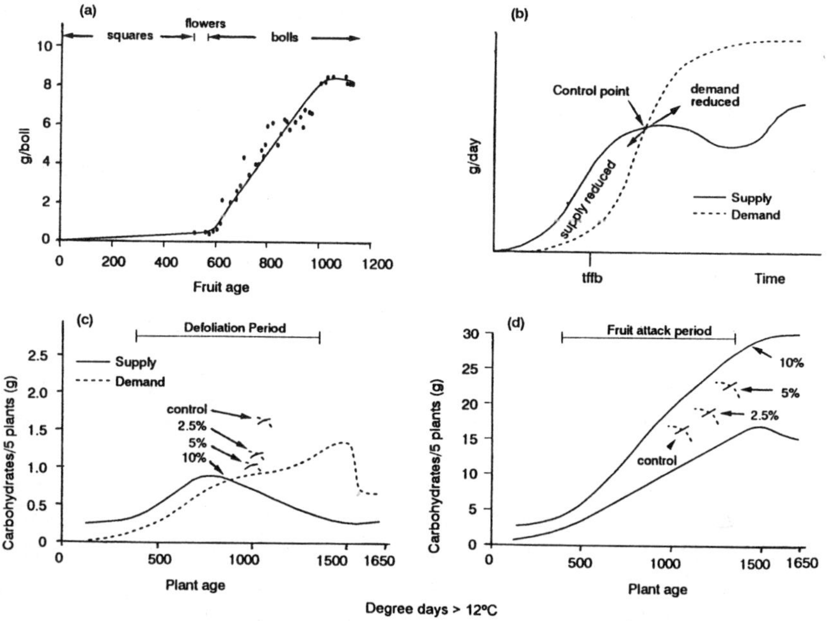

Figure 11.3 Carbohydrate stress in indeterminate cotton. (a) The phenology of fruit growth; (b) the stylized interplay of resource demand and supply indicating the normal time of carbohydrate stress under pest-free conditions (i.e. the control point (•); (c) effects of different percentages of defoliation on the supply side and on the time of stress; (d) the effects of different percentages of fruit loss on reducing demand and on the time of stress. The control point is that expected under pest-free conditions; ttfb, time to first fruiting branch. (After Wang *et al.*, 1977).

common bean, rice), and less so in plants with small fruits where the demand may be small and fruit-related stress may not occur (e.g. cassava). In such cases, carbohydrate stress may occur when the respiration demands of non-photosynthetic parts of a plant become large relative to the capacity of the plant to produce photosynthates (i.e. when the canopy is of fixed or decreasing size), and may lead to senescence and death of the plant. Other pests such as the pink bollworm may attack the standing crop of fruits without affecting either the demand or the supply, and still other pests attack the standing crop of whole plants (e.g. *Verticillium* wilt). We shall examine these relationships below in the sections dealing with the economic threshold.

11.3.1 Growth characteristics of cotton cultivars

Modern crop cultivars have been bred away from wild ancestors by altering mostly the demand side of the equation, leaving the photosynthetic rate per

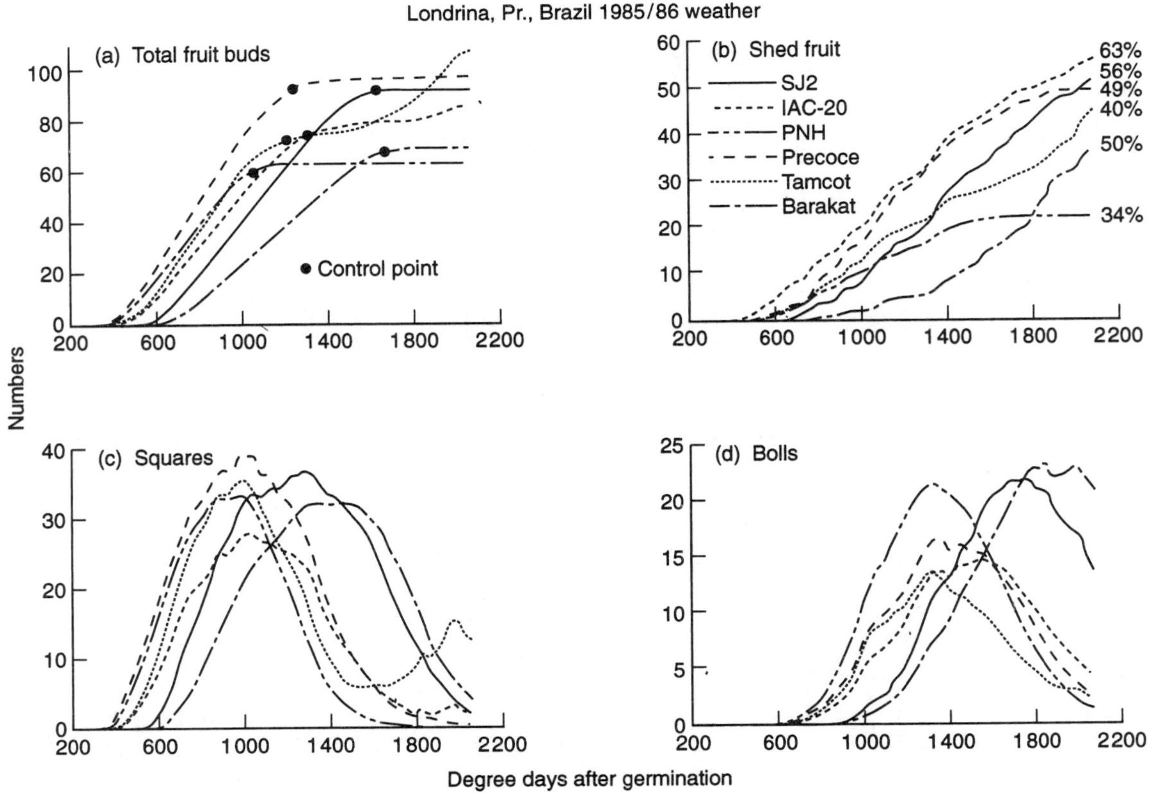

Figure 11.4 Simulated fruit dynamics of six varieties of cotton grown under the same weather and initial conditions showing the approximate time of the control point (•). (a) Cumulative fruit bud production; (b) cumulative fruit bud shedding; (c) the pattern of fruit bud retention; (d) the pattern of boll retention. Weather data are from Londrina, PR, Brazil, 1985–1986. (cf. Gutierrez *et al.*, 1991a).

unit mass of leaf mostly unchanged. During the early season, vegetative growth is emphasized and later in the season fruit growth is emphasized. The growth rates of the different plant parts affect the phenotype, phenology and yield characteristics of a cultivar as well as its ability to avoid and/or compensate for herbivore damage. To compare cultivars plant breeders grow them at the same time and place and protect them from pests. However, the effects of pest dynamics on growth and development of different cultivars have rarely been documented except by researchers interested in IPM (Gutierrez *et al.*, 1975, 1988a,b; Wang *et al.*, 1977; Curry *et al.*, 1980; Villacorta *et al.*, 1984). The growth characteristics of six varieties in two species of cotton (*G. hirsutum* and *G. barbadense*) from different regions of the world were examined to evaluate their responses to boll weevil damage using the cotton model developed by Gutierrez *et al.* (1991a) (Figures 11.4 and 11.5). Among the *G. hirsutum* cultivars compared were some short-season cultivars (PNH-1, Precoce-1 and Tamcot) and long-season cultivars (IAC-20 from Brazil, Acala-SJ2 from California) and *G. barbadense var Barakat* (from the Sudan). The comparisons were made at the same planting density using observed weather, soil water, and nitrogen

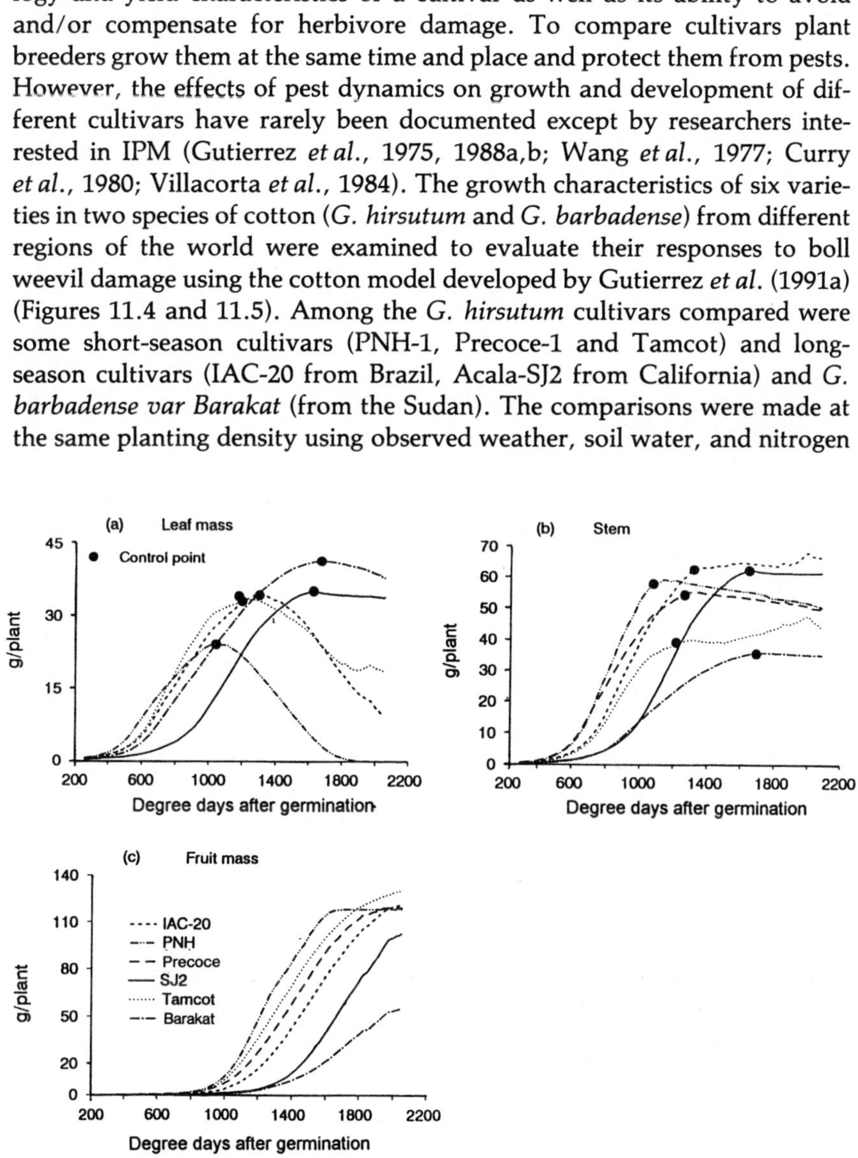

Figure 11.5 Simulated dry biomass dynamics of six varieties of cotton grown under the same weather and initial conditions showing the approximate time of the control point (•). (a) g leaf per plant; (b) g stem per plant; (c) g fruit per plant. Weather data are from Londrina, PR, Brazil, 1985–1986. (cf. Gutierrez *et al.*, 1991a).

Table 11.1 Ranking of attributes that enhance earliness (1–6 or poor to best) in the different cultivars of cotton illustrated in Figures 11.4 and 11.5

	Cultivar					
	SJ-2	*IAC-20*	*PNH*	*Precoce-1*	*Tamcot*	*Barakat*
Characteristics						
Total buds	5	3	5	6	3	1
Time of first fruit	2	4	5	6	4	1
Sum of fruit shed	2	3	6	1	4	5
Peak square	2	3	4	6	5	1
Total bolls	5	2	4	3	1	6
Boll size	5	6	3	3	3	1
Leaf mass	5	4	1	4	2	6
Stem mass	5	6	4	3	2	1
Lint cotton	2	5	5	5	6	1

data from Londrina, PR, Brazil. The results are summarized qualitatively as scores in Table 11.1.

11.4 COMPARISONS OF VARIETIES

The time of stress in cotton is variety dependent, but some generalities are apparent. Cottons which produce fruit early, produce large fruit, or when crowded set the fruit early. Earliness helps plants to avoid pest damage, and some attributes that enhance this have been bred into cultivars (e.g. Tamcot; Bird *et al.*, 1985) that have become integral parts of pest management strategies for pests such as boll weevil (Adkisson *et al.*, 1982). The growth characteristics of several cultivars of cotton are examined below the time of stress ($S/D < 1$) indicated (Figures 11.4 and 11.5).

(a) Fruit dynamics

The physiological time to the first fruiting branch (tffb) is an important component of earliness (Figure 11.4a). The short-season varieties (PNH-1, Precoce-1 and Tamcot) and the long-season variety IAC-20 fruit early, and Acala-SJ2 and *G. barbadense* var Barakat fruit late (Figure 11.4a). The fruiting rates are approximately the same among *G. hirsutum* cottons but slightly lower in *G. barbadense* cottons.

In the absence of pests, shedding of excess fruits in most cultivars is caused by the interplay of factors that affect supply and/or demand, and the quantity shed may be interpreted in terms of earliness (Figure 11.4b). The short-season variety PNH-1 lost ca. 34% of fruits initiated, Tamcot ca. 40%, Precoce-1 ca. 50%, Acala-SJ2 ca. 56% and IAC-20 ca. 63%. Higher early shed rates are correlated with larger fruit size, except in Barakat which produces small fruits yet lost ca. 50% of fruits late in the season.

Earliness is also measured by the time required to reach the peak abundance of fruit buds (i.e. squares). In this regard, Tamcot, Precoce-1 and PNH-1 were earliest followed in sequence by IAC-20, Acala-SJ2 and Barakat (Figure 11.4c). The times and patterns of immature boll formation mirrored the patterns for square earliness: PNH-1, Precoce-1 and Tamcot were the earliest, with IAC-20 being intermediate and Acala-SJ2 and Barakat being much later (Figure 11.4d). The downturns in the boll curves mark the beginning of boll maturation.

(b) Dry matter allocation

In general, long-season cultivars such as IAC-20 and Acala-SJ2 produced more dry matter (100 g of leaf plus stem dry weight with a stem : leaf ratio of 2 : 1 at the time of peak bud production) than did short-season cultivars (Figure 11.5a,b). Precoce-1 produced 87 g of leaf and had a stem : leaf ratio of 1.7 : 1, PNH-1 produced 80 g and the ratio was 2.5 : 1, and that for Tamcot was 76 g and a ratio of 1.1 : 1. The statistics for cultivar Barakat were lower (75 g and ratio 0.9 : 1), and may reflect that it is a different species. Neither plant mass nor stem : leaf ratio was strongly correlated with fruit dry matter.

The pattern of leaf dry matter is hump-shaped in the Brazilian varieties (IAC and PNH) while that for Tamcot, Acala-SJ2 and Barakat show distinctive plateaus (Figure 11.5a). Leaf longevity for the Brazilian varieties and Tamcot were estimated from rain-grown crops, while those for the others were estimated under rain-free desert conditions and irrigation. The different patterns suggest the possibility that rain-induced pathogens affect leaf longevity, but a controlled experiment is required to distinguish among environmental, disease and intrinsic genetic effects.

With regard to earliness of the mature crop (Figure 11.5c), the short-season cultivar Tamcot was the most productive, followed in sequence by PNH-1, IAC-20, Precoce-1, Acala-SJ2 with Barakat being a poor last. Interestingly, predicted IAC-20 yields are among the highest, confirming that before the arrival of boll weevil, it was a good cultivar for southern Brazil.

In nature weather is highly variable and greatly affects the realized yield of cotton. The responses of IAC-17 cottons to non-limiting agronomic conditions and to the weather observed at Londrina, PR, Brazil during 1982 are seen in Figure 11.6. Cultivar characteristics set the basis for yield potential; weather and abiotic and agronomic factors determine how much of the genetic potential is reached, and within that context pests attack either the supply or the demand side of the system.

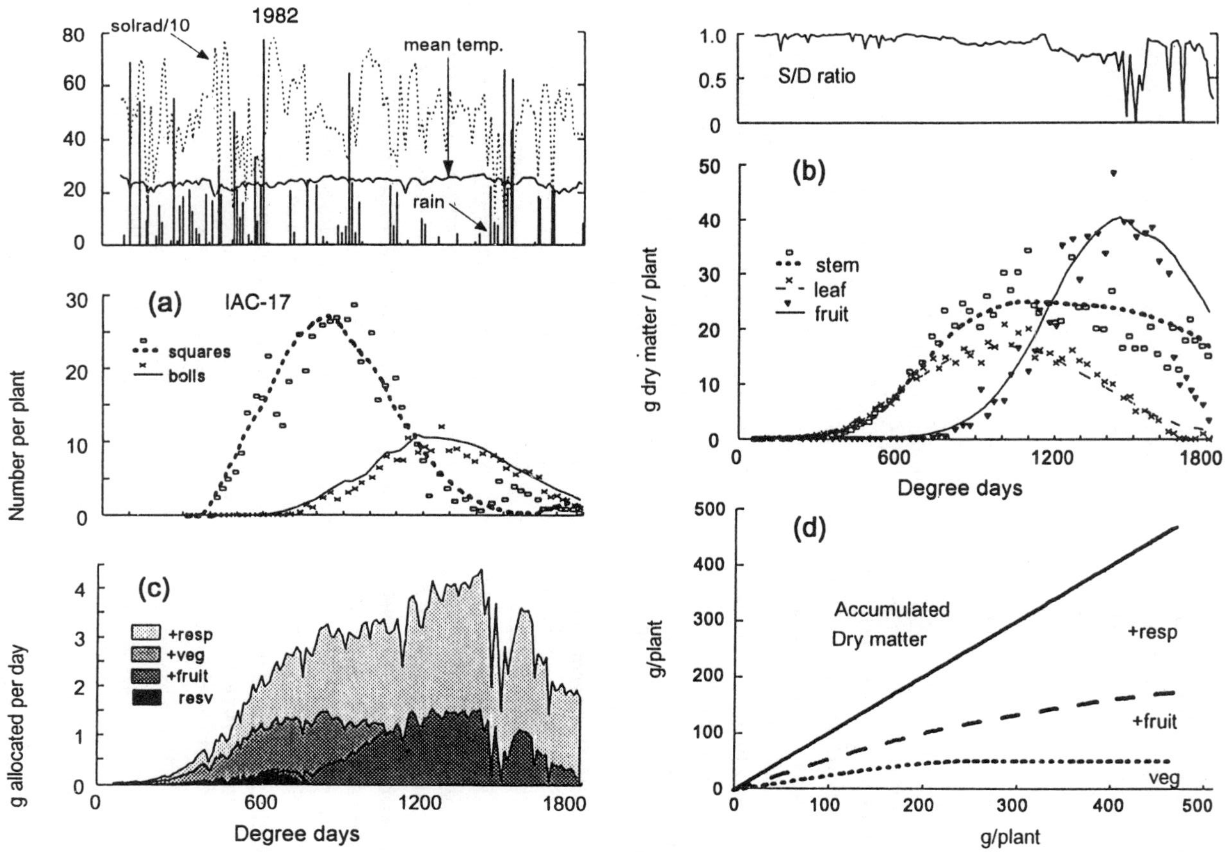

Figure 11.6 Simulated and observed IAC-17 cotton. (a) Fruit dynamics; (b) dry matter dynamics; (c) daily dry matter allocation; (d) cumulative dry matter allocation. (cf. Gutierrez *et al.*, 1984).

11.5 THE ECONOMIC THRESHOLD

The 'holy grail' of the economic threshold for pest damage has been pursued in cotton to a greater extent than in any other crop, but unfortunately with little success. A general review of this subject was made by Pedigo *et al.* (1986), and qualitative notions about plant compensation for pest damage were proposed. Entomologists often use vague thresholds based upon pest numbers to make control decisions (e.g. 10 *Lygus* bugs per 50 sweeps) ignoring the overall dynamics of the biotic and abiotic system, the ability of plants to compensate, and the currency of the biological transactions (i.e. energy or biomass). Approaching this problem using optimality theory has proven difficult as one is forced either to reduce greatly the problem for mathematical tractability, often throwing away essential biology (Gutierrez and Wang, 1984), or to be stymied by the curse of dimensionality (Shoemaker, 1980). A brute force simulation method is also clearly out of the question for even the simplest case of, say, dividing the season into 10 time periods and deciding whether or not to spray as this problem may still result in a *Shoemaker's combination* (2^{10}) of control outcomes (Gutierrez and Wang, 1984).

Headley (1973) set the basis for the definition of the economic threshold. He proposed that yield was some deterministic function of pest damage (D_t) defined in terms of pest numbers (H_t),

$$D_t = bH_t^2 - A, \tag{11.1}$$

where H_t is the pest population at time t squared for compounding effects of damage, b is the per capita pest damage rate, A is a constant to define the damage tolerance level. Headley assumed that H_t grew from a population n time periods before at the rate $(1 + r)$ and that the two densities were related as follows:

$$H_t = H_{t-n}(1 + r)^n \tag{11.2}$$

Substituting this definition for H_t in the damage function (11.1), he got

$$D_t = b[H_{t-n}(1 + r)^n]^2 - A \tag{11.3}$$

Thus if Y is the maximum yield, the observed yield Y^* is

$$Y_t^* = Y_t - c\{b[H_{t-n}(1 + r)^n]^2 - A\} \tag{11.4}$$

with c converting pest damage to units of yield. The yield lost to pests from $t - n$ to t is

$$Y_t - Y_t^* = c\{b[H_{t-n}(1 + r)^n]^2 - A\}. \tag{11.5}$$

The economic threshold is, however, not static but rather it is dynamic and must be assessed at different times during the season to decide when control is warranted. In theory, the economic threshold is defined as the value of yield loss equal (11.6) to the unit cost P_x of the control X (see Regev *et al.*, 1976):

$$P_y \frac{\delta Y}{\delta X(t)} = P_x. \tag{11.6}$$

Unfortunately, the damage we observe at time t has historical roots as well as future implications, and the economic consequences may vary greatly

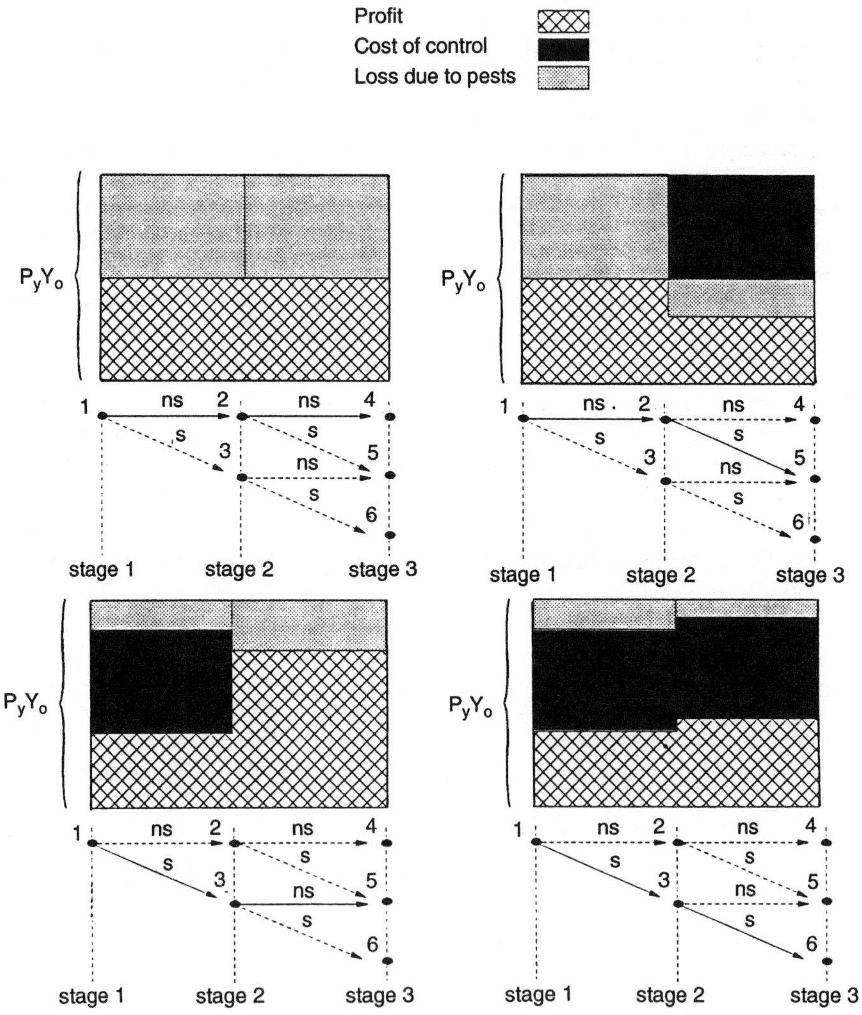

Figure 11.7 Outcomes of different strategies of pest control. See text for details. (cf. Gutierrez and Wang, 1984).

among possible control scenarios. For example, suppose we have a model for yield (Y_0) that includes herbivore (H) damage during a season that has three stages. In this model, pests may be controlled in stages one and two and the crop is harvested in stage three. To illustrate the problem we write the model in terms of profit maximization over the two time period:

$$\text{MAX}[P_y Y_0 - \alpha P_y H e^{-\gamma X_1} - \alpha P_y H e^{-\gamma X_1} e^{-\gamma X_2} - P_x(X_1 + X_2)], \tag{11.7}$$

where P_y is the unit price of Y_0, α is the factor relating the per capita herbivore damage rate to Y, γ_1 and γ_2 are the coefficients relating the effect of X on the survivorship of H and hence pest damage during each time period, and P_x is the cost of each of the two treatment X_1 and X_2 (see Gutierrez and Wang, 1984). The different scenarios and their costs and benefits given arbitrary parameter values are presented in Figure 11.7. The potential revenues are depicted by rectangle $P_y Y_0$, while the costs of damage and the cost of control are sub-components. Although this model is far too simple for practical applications, it helps us visualize the problem by showing that a control treatment during any time period needs to be evaluated in terms of its contribution to end of season profits. If the problem were more complicated, a more efficient approach would be to evaluate the problem in reverse eliminating possible paths to see which maximized net revenues (i.e. dynamic programming) (Bellman, 1957).

Lastly, the economic threshold needs to consider the crop's ability to compensate for pest damage. Pests may attack the demand side or the supply side of the plant's physiology (Figure 11.8), and to evaluate this, more biologically complete models may be required.

11.6 ECONOMICS OF DEMAND-SIDE PESTS

Demand-side pests are those that attack the sink side (i.e. fruit) of the plant's physiology. Fruit bud production rates across cultivars are linear functions of physiological time and plant density, at least until carbohydrate stress

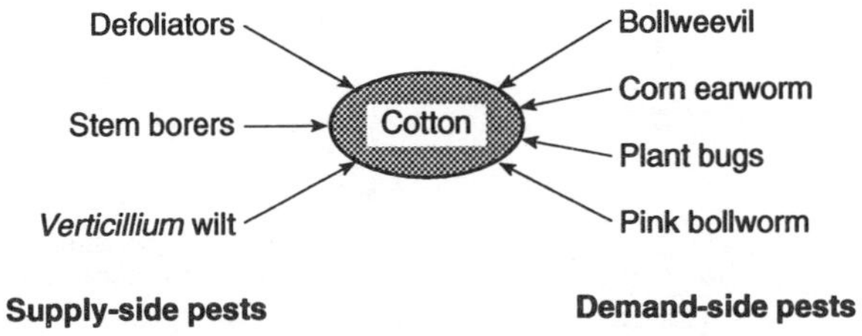

Figure 11.8 Supply- and demand-side pests.

occurs, at which time the rates slow (e.g. Figure 11.9a,b). Plants experiencing high rates of fruit loss produce additional buds because the demand side has been reduced but the supply side remains unaffected. In situations with low fruit loss rates, buds and vegetative growth slow early due to fruit demand induced carbohydrate stress (see Figures 11.4 and 11.5).

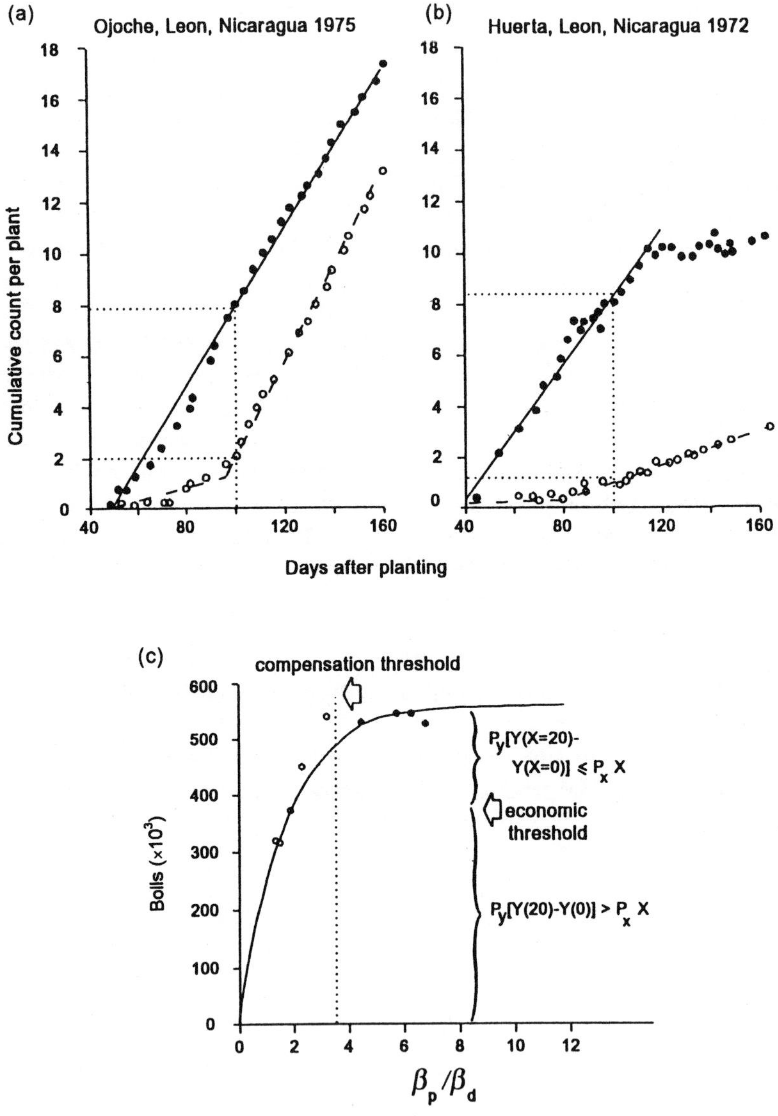

Figure 11.9 (a,b) Observed cumulative cotton fruit bud production (______, β_p) and cumulative buds damaged by boll weevil and bollworm (---------, β_d) at two sites in Nicaragua; (c) yield plotted on the ratio β_p/β_d. (cf. Gutierrez and Daxl, 1984).

Demand-side herbivores such as the cotton boll weevil, tobacco budworm and plant bugs may cause excessive fruit loss causing re-allocation of photosynthates within the plants resulting in root and stem mass three to four times that of pest-free plants. This relationship may be modified by nutrient or water stress which affect the supply side of the equation. It matters little to the plant which pest causes the damage; what matters is its capacity to compensate and the amount of time and energy lost due to abscised fruits. In cotton, *Lygus* bug attacks very small fruits in which little time and energy have been invested; hence the plant is more likely to compensate for this damage unless the rate is high. Boll weevils attack larger fruits that have had proportionately larger investments, and hence the capacity to compensate is less. In the absence of control, the damage rate from boll weevil is likely to be very high. Induced pests such as bollworms attack not only small buds but also large maturing fruits in which considerable investments in time and energy have been made, and often the plant is unable to compensate for such damage.

Data collected by Rainer Daxl and his associates in Nicaragua provided the basis for summarizing the complicated interactions of cotton and two demand-side herbivores: bollworm and cotton boll weevil (e.g. Figure 11.9a,b). Production (β_p) and depletion (β_d) rates of fruits were estimated in the field during different years and for different control practices (see Gutierrez *et al.*, 1982; Gutierrez and Daxl, 1984 for methods). When yields from all trials in the same field but different years were plotted on their respective ratios $1 < \beta = \beta_p/\beta_d \leq \infty$, the data fell on a common concave line having an asymptote near $\beta = 3.5$ (Figure 11.9c). The model suggests that plants compensate for ca. 25–30% of fruit loss before serious reductions in yields occur, i.e. the *compensation point*. Yield declines rapidly after the compensation point, suggesting that irreplaceable damage to the crop has occurred. This simple model explains why *Lygus* bug damage in California cotton did not cause economic damage under most circumstances; its density and per capita damage rate are low and the observed population levels rarely caused damage that exceeded the compensation level which varies with cultivar, location and abiotic factors.

The model predicts that cottons that produce small fruits normally are stressed later in the season, and the compensation point of the function of yield on β is displaced to the left, suggesting a greater ability to compensate.

11.6.1 Using β as an action threshold

Gutierrez *et al.* (1991b) studied the effects of different levels of boll weevil damage on two varieties of cotton: IAC-20 and Tamcot. Cotton crops were simulated at 8 plants/m^2 using Londrina, SP, Brazil 1985/1986 weather, and insecticides were applied when specified values of β (i.e. an action threshold) was exceeded, (Gutierrez *et al.*, 1991a). In the model, only the adult stage of the weevil is readily killed by insecticides as all other life-

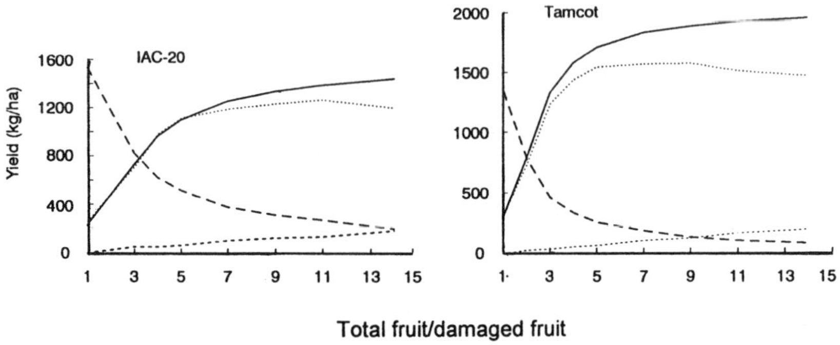

Figure 11.10 The effects of different levels of boll weevil damage on cotton yields (———, kg/ha), profit (· · · · · ·, \$/ha), larvae per boll (– –, × 10^3) and the number of sprays (---------, × 10) at different action thresholds based on the ratio β_p/β_d. Data are from IAC-20 and Tamcot cultivars. (cf. Gutierrez *et al.*, 1991b).

stages are protected inside fruits. An adult mortality rate of 90% on the day of spraying that decays to 0% in 3.5 days and a spraying cost of \$25/ha were assumed. Simulated profits, yields, larvae per boll and numbers of insecticide treatments at different spray threshold are shown in Figure 11.10.

Maximum profits occur at $\beta = 5$ for Tamcot and $\beta = 11$ for IAC-20, suggesting that Tamcot can compensate for recurring fruit damage rates up to 20%, while IAC-20 can compensate for only 9%. This is not surprising as Tamcot was bred to avoid and tolerate boll weevil damage and because the loss per unit fruit is smaller than in IAC-20. These results are for a specific location and season, but the results for other seasons are qualitatively similar. Compared with IAC-20, maximum profits for Tamcot are about 20% higher, require half the number of sprays (6 versus 12), and bolls are infested with half the number of weevil larvae.

The effect of over-using pesticides on profits is negligible, as the cost of each application is trivial compared with the value of the crop, and the external costs of induced pest problems and the social costs are not included. However, if there are marginal increases of 1, 5, 10 and 15% per application, the effects on profit are dramatic resulting in a fairly robust optimum at seven sprays across all penalty rates (see Gutierrez *et al.*, 1991a). In practice, the treatments could be lowered in most cases by cultural methods such as trap cropping and fall diapause control which reduces initial numbers of weevil adults. Adoption of these additional cultural control methods is an important aspect for ecologically sound integrated crop production and pest management. Lastly, these are not optimal soultions as the appropriate β is likely to vary over time and with the state of the crop's development.

11.6.2 Marginal analysis of factors affecting yields due to demand-side pests

The interactions of plant density (ρ), boll weevil larvae per boll (ω) and variety $(\upsilon) = 0, 1$, i.e. dummy variables for Tamcot and IAC-20 respectively, on yield (*Y*) were analysed using multiple regression. It should be noted that the assumption of normality required for such analyses is likely not met by our simulated data; hence the purpose of the analysis is heuristic and not predictive. The regression equation includes only those variables giving significant *t*-values for slopes. Because of the concave nature of the yield response, the yields were log transformed:

$$\log_e Y = 7.0 + 0.075\,\rho - 0.485\omega - 0.505\omega\upsilon,\ R^2 = 0.84,\ n = 41. \quad (11.8)$$

Increasing planting density (ρ) had a positive effect on yield, but increasing boll weevil larval density (ω) as expected had a negative effect as did the interaction of cultivar and weevil density. Taking the derivative of the antilog of the regression with respect to weevil density yields

$$\delta Y/\delta\omega = (0.0.485 - 0.505\upsilon)e^{7.0 + 0.075\rho - 0.485\omega - 0.505\upsilon\omega} \quad (11.9)$$

Substituting for planting density and variety, the relative benefits of the weevil tolerant varieties may be estimated.

$$\delta Y/\delta\omega = \begin{cases} -702.7\,\mathrm{kg\,ha^{-1}} \text{ for } X_3 = \mathrm{IAC} - 20 \\ -596.7\,\mathrm{kg\,ha^{-1}} \text{ for } X_3 = \mathrm{Tamcot} \end{cases} \quad (11.10)$$

On average, the model predicts a 106 kg/ha greater loss in IAC-20 $(\upsilon = 1)$ per unit of boll weevil per boll at, say, 8 plant/m^2 $(\rho = 8)$ compared with Tamcot $(\upsilon = 0)$.

11.7 ECONOMICS OF SUPPLY-SIDE PESTS

Defoliating pests in general reduce photosynthesis and cause stress to occur earlier than in healthy plants. Pests such as noctuid moths eat leaves and tetranychid mites and foliar diseases reduce the active photosynthetic surface, but all of them may reduce the photosynthetic capacity of plants. Pests such as stem borers and plant diseases such as *Verticillium* wilt destroy or block the vascular system of a plant and affect its photosynthetic rate. Other pests such as thrips may damage the terminal and introduce delays in development, but they also kill patches of leaf cells and reduce photosynthesis. If the damage is severe enough the plant may die.

11.7.1 Defoliators

Different species often attack different ages of leaves, and hence the same damage rate may have very different effects on the overall photosynthetic rate (i.e. the supply side). Defoliation of the upper portions of the canopy may cause older intact leaves to maintain and possibly increase their photosynthetic activity for a longer period of time. Attack on older leaves may reduce respiration rates without affecting photosynthesis. In contrast, mites kill cells, reducing the photosynthetically active but not the total leaf area, reducing the amount of light falling on photosynthetically active leaves.

A model based on the biology of *S. exigua* (Hogg and Gutierrez, 1980) was used to simulate the effects of defoliators that prefer to feed on the top, middle (*T. ni*) and bottom (*S. exigua*) of the plant. A mortality rate of 30% per day of all immature stages of the pest (A.P. Gutierrez, unpublished data) and different initial adult densities starting at 0.01/plant/day were assumed. Light-trap data indicated that populations of adult moths increase five-fold during the season (Hogg and Gutierrez, 1980), and this effect was included in the model.

In the multiple regression analysis of the simulation data only those variables yielding significant *t*-values for slopes were included. The final model is

$$Y = \text{bales/acre} = 0.164 + 0.479A - 0.88I + 0.149L - 0.157C,\ R^2 = 0.88,\ n = 82, \quad (11.11)$$

where A is leaf age class ($A = 1,2,3$), I is the initial immigration rate of adult moths/day/plant, L is cumulative g leaf mass/plant and C is cumulative g leaves consumed/plant. The model predicts that yields decrease monotonically with increasing immigration rates (i.e. pest pressure) and with decreasing age of leaves attacked (Figure 11.11). The

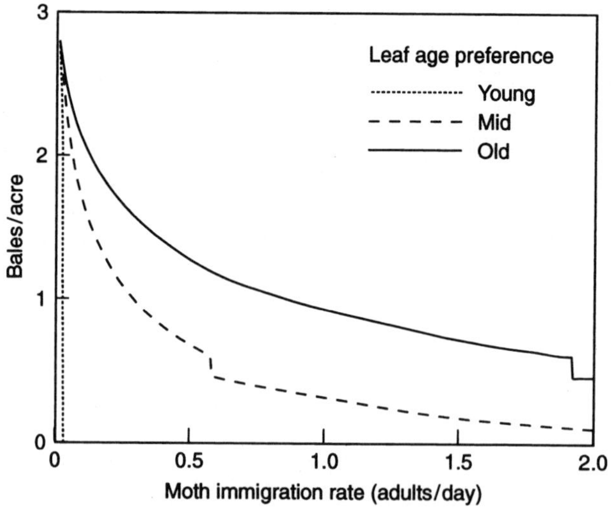

Figure 11.11 The simulated effects of defoliation on yields by different levels of adult immigration and preference for different aged leaves.

death of young leaves severely affects future productivity, but the death of old leaves affects photosynthetic productivity little and may produce a beneficial net gain by reducing respiration costs.

11.7.2 Stem borers

Stem borers such as *E. braziliensis* are early-season pests that kill whole plants and stunt the growth of others. The degree of stunting depends on when the plants were attacked. Dos Santos *et al.* (1989) examined the effects on yield of treating seeds with insecticide (T_1 vs T_2), planting the crop at four different times (t_1–t_4), and planting density (see Figure 11.12a).

Highly significant direct effects of the seed treatment on yield (428 kg/ha) and plant survival were observed. In the range 2–8 plants/m^2, yields increased 198 kg/ha per surviving plant. Delaying planting up to 3 weeks increased yields, but delaying it 1 week longer reduced yields almost as much as the pest itself. Plant density is an important consideration for managing this and other pests that kill whole plants, as some compensation for yield occurs when healthy plants grow into the area vacated by dead or infested neighbours (Figure 11.12a,b).

11.7.3 Diseases

The infective agent of the fungal pathogens such as *Verticillium* wilt are called propagules, and are released into the soil when plants die and decay.

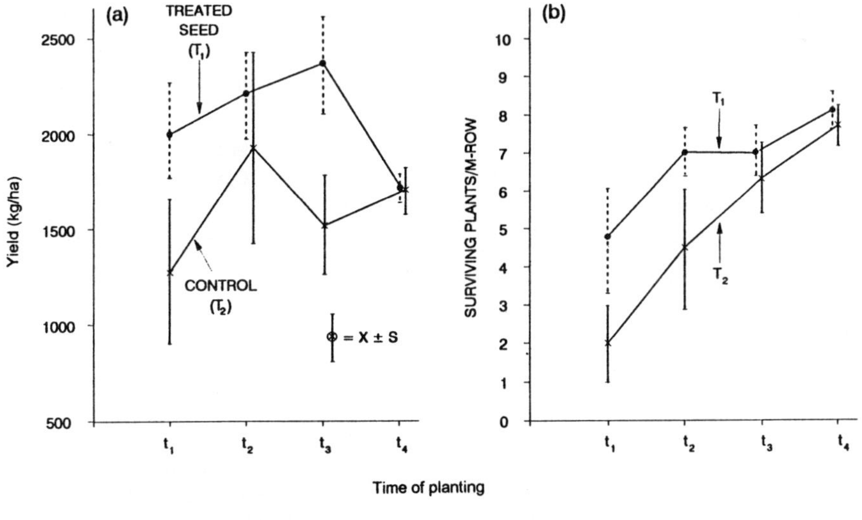

Figure 11.12 The effects of seed treatment and planting time in IAC cotton against stem borer damage on (a) yield and (b) survivorship of plants. (cf. Dos Santos *et al.*, 1989).

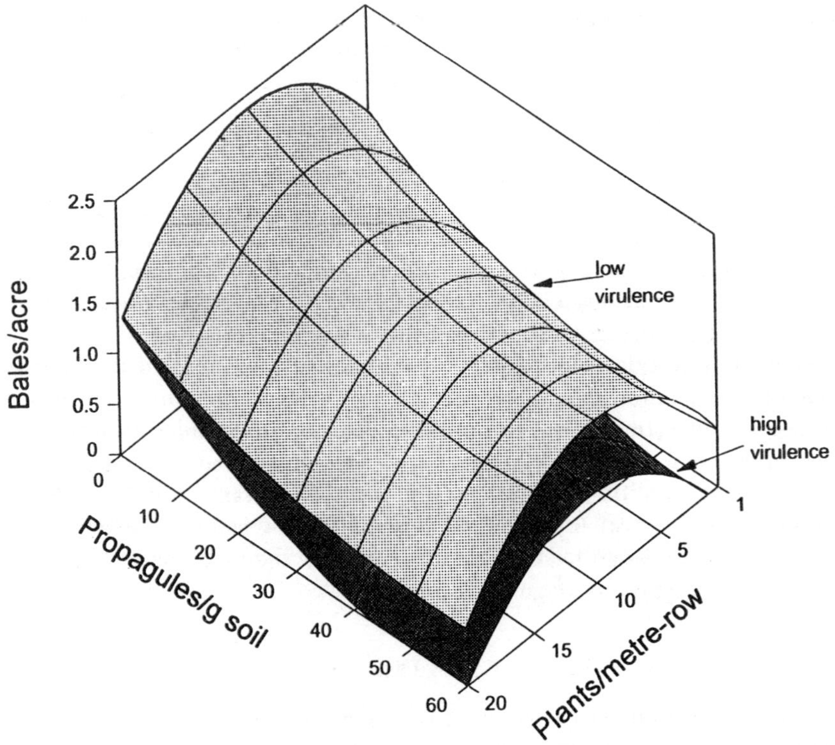

Figure 11.13 The simulated effects of *Verticillium* wilt virulence and propagules density and planting density on cotton yields. (cf. Gutierrez *et al.*, 1983).

This pathogen blocks the plant's vascular system and restrict the flow of nutrients and water, and in this manner slows photosynthesis. Propagules infect plants when roots growing through the soil come in contact with them, hence the infection rate of plants depends on propagule density and virulence. The number of plants infected and killed by *Verticillium* wilt increases through the season, with those infected early being more likely to be killed or severely stunted than those infected late in the season (DeVay *et al.*, 1974). The degree to which a healthy plant grows into the area occupied by a diseased neighbour depends upon the degree of stunting of the diseased plant and the time in the season when the interaction begins. On an area basis, this encroachment allows the crop to partially compensate for lost yield (see section on stem borers). The degree of compensation that is possible depends on the virulence of the pathogen, inoculum density, planting density and the length of the season (Figure 11.3) (Gutierrez *et al.*, 1983). Control of *Verticillium* wilt normally occurs between seasons, and may include crop rotation, solarization of the soil, crop free periods and, as a last resort, fumigation (Regev *et al.*, 1990).

11.7.4 Pests that deplete the standing crop of fruits

Herbivores such as pink bollworm attack buds and large fruits, but their activities do not alter plant growth patterns. This occurs because the larvae feed on non-vital parts within buds (i.e. excess pollen), and attacked buds are abscised at approximately the same rate as healthy ones (Westphal *et al.*, 1979). Surviving buds form viable flowers and may produce bolls which are the preferred host stage of this pest. From a physiological perspective this pest does not perceptibly alter either the supply or the demand side of the equation, but rather it depletes the standing crop of bolls after nearly full investments in time and energy have been made, making compensation unlikely. Yield losses accrue via the destruction of seed and reduced lint quality (Brazzel and Gaines, 1956; Gutierrez *et al.*, 1977a; Stone and Gutierrez, 1986a,b), and are characteristically a dose–response function of larvae per boll (Brazzel and Gaines, 1956). The measures needed to control this pest were reviewed above (Burrows *et al.*, 1984).

11.7.5 Pests that damage the terminal of the plant

The economic damage caused by thrips has been seriously questioned (van den Bosch, 1978), as it is generally thought that the plant can compensate for this damage, unless in some rare circumstance the damage rate is very large (e.g. cowpea in West Africa; M. Tamo personal communication). Data from Brazil (W.J. Dos Santos, unpublished results) showed no significant effect of seed treatment for thrips control in cotton, with plant height (size) and replicates accounting for 73% of the variation in the data and thrips symptoms adding a further 4%. No other variable (treatment, or thrips density) was significant. Hence, control measures for thrips in cotton in Brazil (and California) are difficult to justify.

Other pests such as early season defoliators (e.g. beet armyworm) also kill the terminal of the plant and introduce developmental delays that ultimately may affect yields. Again, this damage is not expected to be large, and may not warrant chemical control because it may induce other pests which may cause greater damage (i.e. pest resurgence and secondary pest outbreak) (Gutierrez *et al.*, 1975; van den Bosch, 1978).

11.8 ABIOTIC EFFECTS ON COMPENSATION

Adverse edaphic and weather factors affect compensation by lowering overall plant productivity (i.e. the supply side of the ratio). For example, long periods of cloudy weather, drought and shortages of nitrogen reduce

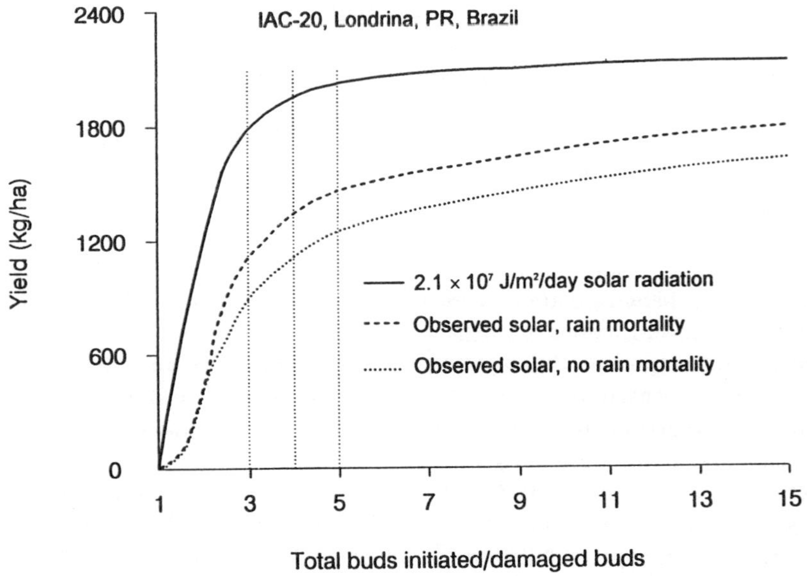

Figure 11.14 The effects of abiotic factors on the capacity of cotton to compensate for different levels of fruit damage as measured by the ratio of β_p/β_d. (cf. Gutierrez *et al.*, 1991b).

photosynthetic rates, induce carbohydrate stress and increase fruit shedding (Gutierrez *et al.*, 1991a,b). Rain is thought to reduce the successful fertilization of flowers causing them to abort, and thus affect the demand side of the ratio. Here we estimate the simulataneous effects on yield of variable solar radition and rainfall compounded by different levels of attack by, say, a demand-side pest such as the boll weevil via simulation using non-limiting conditions (Figure 11.14). A constant high solar radiation level of 2.1×10^7 J/m/day and non-limiting water are used as the standard and the effects of observed solar radiation and rainfall included sequentially to see their effects on yield. The first factor added is observed variable solar radiation, and the next is the effect of rainfall on pollination of flowers. Under non-limiting conditions, the response of different varieties of cotton to pest damage are similar, but the yields are different. Adding the observed variable solar radiation of Southern Brazil causes dramatic reductions in yield, and adding rain and its effect on flower mortality lowers yields by about a further 10%. This analysis points out an obvious result: abiotic factors determine the potential dynamics of the system given cultivar growth characteristics, and biotic factors such as pests operate within those constraints.

11.9 DISCUSSION

In pest-free cotton, vegetative growth increases exponentially until fruits begin rapid growth and the total plant demand for photosynthates exceeds the supply (i.e. the control point). The supply of photosynthates is that produced daily plus a small fraction of available reserves, and the demand is the total required to meet all growth demands. Vegetative growth slows or stops when fruit growth-induced stress occurs causing photosynthates to be allocated almost exclusively to fruits. This occurs when the plant's leaf area and photosynthetic rate are at a maximum. The imbalance is in part corrected by shedding young squares and young bolls that are not likely to complete development while at the same time minimizing the loss of biomass. To compensate for herbivore damage and fruit shed due to stress, fruits are initiated in considerable excess, and it is this excess reproductive capacity that is the basis of yield compensation in cotton (and in other organisms). It is upon this backdrop that herbivore damage and plant compensation must be evaluated.

Compensation may be adversely influenced by any factor that decreases the supply or increases the demand. For example, changes in supply may be due to cloudy weather, edaphic factors, herbivore damage and other factors. Among the factors that increase the demand side of the interaction may be those that increase respiration (e.g. temperature), varietal characteristics such as high stem to leaf ratios, high varietal fruit demand rates due to high retention and/or growth rates and interactions.

In assessing the economic threshold, it is obvious that we must consider the plant's capacity to compensate for pest damage. In cotton, up to 50% of all fruit initiated may be shed, even in the absence of pests, thus if A is the probability at time t that a fruit will be shed due to plant stress and B is the probability that it will be shed due to herbivore damage, then $A + B - A \cap B$ is the combined probability that the fruit will be shed. If over the season, A is large relative to B (i.e. as in the case of *Lygus* bugs), then $A \cap B$ will be large relative to B, suggesting that fruit buds attacked by pests are likely to be shed anyway.

11.9.1 Overview of models and decision support systems

Models are useful tools for understanding plant–herbivore interactions. In agriculture they help sketch the nature of the problem, are useful for estimating losses due to various factors, and suggest directions.that control measures should take. In general, model results should be viewed with extreme caution to avoid model-driven disasters (*sensu* C.S. Holling), because under the best of circumstances models are incomplete, they may be formulated wrongly, and not all of the initial conditions are known. However, models are the only tools that let us address questions of pest damage from a holistic perspective. To avoid inappropriate uses of models while taking advantage of their synthetic powers, many modellers have turned to the field of decision support systems (knowledge-based or expert systems). An expert system addresses specific problems like, 'When should I plant cotton?' To answer such questions the expert system might run a model or a set of models as necessary, but it explicitly controls how and when a model is accessed and how its output is interpreted. Several decision support systems for cotton have been developed since the early 1980s including SIRITAC in Australia (Hearn *et al.*, 1981), CALEX/COTTON in California (Plant, 1989; Goodell *et al.*, 1990), COTFLEX (Stone and Toman, 1989) and TEXCIM (Sterling *et al.*, 1993) in Texas, and GOSSYM/COMAX (McKinion and Lemmon, 1985; Lemmon, 1986) in Mississippi.

Models alone or as components of decision support systems have shortcomings because they will always be incomplete. A major shortcoming is that the detrimental effects of insecticide-induced resurgence of primary and secondary pests (see van den Bosch, 1978) are rarely included in models except in intuitive ways (Sterling *et al.*, 1993). Despite these drawbacks, models have been used to capture in a realistic manner the relevant issues of many cotton/pest interactions as modified by abiotic factors.

REFERENCES

Adkisson, P.L., Niles, G.A., Walker, J.K., Bird, L.S. and Scott, H.B. (1982) Controlling cotton's insect pests: a new system. *Science*, **216**, 19–22.

Baker, D.N., Hesketh, J.D., and Duncan, W.G. (1972) Simulation of growth and yield in cotton. I. Gross photosynthesis, respiration and growth. *Crop Science*, **11**, 431–5.

Baker, D.N., Lambert, J.R. and McKinion, J.M. (1983) GOSSYM: a simulator of cotton growth and yield. *South Carolina Agricultural Experiment Station Technical Bulletin* No. 1089.

Bellman, R. (1957) *Dynamic Programming*, Princeton University Press, Princeton, New Jersey.

Bird, L.S., El-Zik, K.M and Thaxton, P.M. (1985) Registration of (TAMCOT CAB-CS) Upland cotton. *Crop Science*, **26**, 384–5.

Brazzel, J.R. and Gaines, J.C. (1956) The effects of pink bollworm infestations on yield and quality of cotton. *Journal of Economic Entomology*, **49**, 852–4.

Burke, H.R. (1976) Bionomics of the anthonomine weevils. *Annual Review of Entomology*, **21**, 283–303.

Burrows, T.M., Sevacherian, V., Browning, H. and Baritelle, J. (1982) The history and cost of the pink bollworm in the Imperial Valley. *Bulletin of the Entomological Society of America*, **28**, 286–90.

Burrows, T.M., Sevacherian, V., Moffitt, L.J. and Baritelle, J.L. (1984) Economics of pest control alternatives for Imperial Valley cotton. *California Agriculture*, **38**, 15–16.

Byerly, K.F., Gutierrez, A.P., Jones, R.E. and Luck, R.F. (1978) Comparison of sampling methods for some arthropod populations in cotton. *Hilgardia*, **46**, 257–82.

Cave, R.D. and Gutierrez, A.P. (1983) *Lygus hesperus* field life table studies in cot-

ton and alfalfa (Heteroptera: Miridae). *Canadian Entomologist*, **115**, 649–54.
Curry, G.L. and Cate, J.R. (1984) Strategies for cotton-boll weevil management in Texas, in *Pest and Pathogen Control; Strategic, Tactical, and Policy Models*, (ed. G.R. Conway), John Wiley and Sons, Ltd., Chichester, pp. 169–83.
Curry, G.L., Sharpe, P.J.H., DeMichele, D.W. and Cate, J.R. (1980) Towards a management model of the cotton-boll weevil ecosystem. *Journal of Environmental Management*, **11**, 187–223.
DeBach, P. (1975) *Biological Control by Natural Enemies*, Cambridge University Press, London.
DeVay, J.E., Forrester, L.L., Garber, R.H. and Butterfield, E.J. (1974) Characteristics and concentration of propagules of *Verticillium dahliae* in air-dried field soils in relation to the prevalence of *Verticillium* wilt in cotton. *Phytopathology*, **64**, 22–9.
DeWit, C.T. and Goudriaan, J. (1978) *Simulation of Ecological Processes*, 2nd edn, PUDOC Publishers, The Netherlands.
Dos Santos, W.J. (1989) Recomendacoes tecnicas para a convivencia com o bicudo do algodoeiro (*Anthonomus grandis* Boheman 1843), no estado do Parana. *IAPAR Circular Tecnica*, Londrina, PR, Brazil.
Dos Santos, W.J., Gutierrez, A.P. and Pizzamiglio, M.A. (1989) Evaluating the economic damage caused by the cotton stem borer *Eutinobothrus brasiliensis* (Hambleton 1937) in cotton in southern Brazil. *PAB*, **24**, 337–45.
Ephrath, J.E. and Marani, A. (1993) Simulation of the effect of drought stress on the rate of photosynthesis in cotton. *Agricultural Systems*, **42**, 327–42.
Falcon, L.A., van den Bosch, R., Ferris, C.A., Strombert, L.K., Etzel, L.K., Stinner, R.E. and Leigh, T.F. (1968) A comparison of season-long cotton pest-control programs in California during 1966. *Journal of Economic Entomology*, **61**, 633–42.
Falcon, L.A., van den Bosch, R., Gallegher, J. and Davidson, A. (1971) Investigation of the pest status of *Lygus hesperus* in cotton in Central California. *Journal of Economic Entomology*, **64**, 56–61.
Falcon, L.A., Gutierrez, A.P. and Mueller, D. (1986) Exotic pest profile: boll weevil. *University of California Statewide IPM Special Report*.
Frisbie, R.F., El-Zik, K.M. and Wilson, L.T. (1989) *Integrated Pest Management Systems and Cotton Production*, John Wiley and Sons, Chichester.
Fryxell, P.A. (1979) *The Natural History of the Cotton Tribe*, Texas A and M University Press, College Station.
Gaston, L.K., Kaae, R.S., Shorey, H.H. and Sellers, D. (1977) Controlling the pink bollworm by disrupting sex pheromone communication between adult moths. *Science*, **16**, 904–5.
Goodell, P.B., Plant, R.E., Kerby, T.A. *et al.* (1990) Calex/Cotton: an integrated expert system for crop production and management in California cotton. *California Agriculture*, **44**, 18–21.
Gutierrez, A.P. (1992) The physiological basis of ratio-dependent predator–prey theory: the metabolic pool model as a paradigm. *Ecology*, **73**, 1552–63.
Gutierrez, A.P. and Baumgaertner, J.U. (1984) Multitrophic level models of predator–prey-energetics, I. Age specific energetics models – pea aphid *Acyrthosiphon pisum* (Harris) (Homoptera, Aphididae) as an example. *Canadian Entomology*, **116**, 924–32.
Gutierrez, A.P. and Curry, G.L. (1989) Conceptual framework for studying crop–pest systems, in *Integrated Pest Management Systems and Cotton Production*, (eds R.E. Frisbie, K.M. El-Zik and L.T. Wilson), John Wiley and Sons, New York, pp. 37–64.
Gutierrez, A.P. and Daxl, R. (1984) Economic threshold for cotton pests in Nicaragua, ecological and evolutionary perspectives, in *Pests and Pathogens*

Control, Strategic, Tactical and Policy Models, (ed. G.R. Conway), John Wiley and Sons, New York, pp. 184–205.
Gutierrez, A.P. and Wang, Y.H. (1976) Applied population ecology: models for crop production and pest management, in *Pest Management*, (eds G.A. Norton and C.S. Holling), International Institute for Applied Systems Analysis Proceeding, Pergamon Press, Oxford, pp. 255–80.
Gutierrez, A.P. and Wang, Y.H. (1984) Models for managing the economic impact of pest populations in agricultural crops, in *Ecological Entomology*, (ed. C.B. Huffaker), John Wiley and Sons, New York, pp. 729–61.
Gutierrez, A.P. and Wilson, L.T. (1989) Development and use of pest models, in *Integrated Pest Management Systems and Cotton Production*, (eds R.E. Frisbie, K.M. El-Zik and L.T. Wilson), John Wiley and Sons, New York, pp. 65–83.
Gutierrez, A.P., Falcon, L.A., Loew, W., Leipiz, P.A. and van den Bosch, R. (1975) An analysis of cotton production in California, a model for Acala cotton and the effects of defoliators on yield. *Environmental Entomology*, **4**, 125–36.
Gutierrez, A.P., Butler, G.D. Jr, Wang, Y.H. and Westphal, D.F. (1977a) The interaction of pink bollworm (Lepidoptera: Gelichiidae), cotton, and weather: a detailed model. *Canadian Entomologist*, **4**, 125–36.
Gutierrez, A.P., Leigh, T.F., Wang, Y.H. and Cave, R.D. (1977b) An analysis of cotton production in California: *Lygus hesperus* (Heteroptera: Miridae) injury – an evaluation. *Canadian Entomologist*, **109**, 1375–86.
Gutierrez, A.P., Wang, Y.H. and Regev, U. (1979) An optimization model for *Lygus hesperus* (Heteroptera: Miridae) damage in cotton: the economic threshold revisited. *Canadian Entomologist*, **111**, 41–54.
Gutierrez, A.P., Baumgärtner, J.U. and Hagen, K.S. (1981) A conceptual model for growth development and reproduction in the ladybird beetle *Hippodamia convergens* G.-M. (Coccinellidae, Coleoptera). *Canadian Entomologist*, **113**, 21–33.
Gutierrez, A.P., Daxl, R., Leon Quant, G. and Falcon, L.A. (1982) Estimating the economic threshold for bollworm (*Heliothis zea* Boddie) and bollweevil (*Anthonomus grandis* Boh.) damage in Nicaraguan cotton (*Gossypium hirsutum* L.). *Environmental Entomology*, **10**, 873–9.
Gutierrez, A.P., DeVay, J.E., Pullman, G.S. and Friebertshauser, G.E. (1983) A model of *Verticillium* wilt in relation to cotton growth and development. *Phytopathology*, **75**, 89–95.
Gutierrez, A.P., Pizzamiglio, M.A., Dos Santos, W.J., Tennyson, R. and Villacorta, A.M. (1984) A general distributed delay time varying life table plant population model, cotton (*Gossypium hirsutum* L.) growth and development as an example. *Ecological Modelling*, **26**, 231–149.
Gutierrez, A.P., Schulthess, F., Wilson, L.T., Villacorta, A.M. Ellis, C.K. and Baumgärtner, J.U. (1987) Energy acquisition and allocation in plants and insects, a hypothesis for the possible role of hormones in insect feeding patterns. *Canadian Entomologist*, **119**, 109–29.
Gutierrez, A.P., Wermelinger, B., Schulthess, F., Baumgaertner, J.U., Herren, H.R., Ellis, C.K. and Yaninek, S.J. (1988a) Analysis of biological control of cassava pests in Africa: I. Simulation of carbon, nitrogen and water dynamics in cassava. *Journal of Applied Ecology*, **25**, 901–20.
Gutierrez, A.P., Neuenschwander, P., Schulthess, F., Wermelinger, B., Herren, H.R., Baumgärtner, J.U. and Ellis, C.K. (1988b) Analysis of the biological control of cassava pests in West Africa, II. The interaction of cassava and cassava mealybug. *Journal of Applied Ecology*, **25**, 921–40.
Gutierrez, A.P., Dos Santos, W.J., Pizzamiglio, M.A., Villacorta, A.M., Ellis, C.K., Fernandes, C.A.P. and Tutida, I. (1991a) Modelling the interaction of

cotton and the cotton bollweevil (*Anthonomus grandis*) in Brazil. *Journal of Applied Ecology*, **28**, 398–418.

Gutierrez, A.P., Dos Santos, W.J., Villacorta, A.M., Pizzamiglio, M.A., Ellis, C.K., Carvalho, L.H. and Stone, N.D. (1991b) Modelling the interaction of cotton and the cotton boll weevil. I. A comparison of growth and development of cotton varieties. *Journal of Applied Ecology*, **28**, 371–97.

Headley, J.C. (1973) The economics of pest management, in *Introduction to Pest Management*, (eds R.L. Metcalf and W.H. Luckmann), John Wiley and Sons, Inc., New York, pp. 69–91.

Hearn, A.B., Room, P.M., Thomson, N.J. and Wilson, L.T. (1981) Computer-based cotton pest management in Australia. *Field Crops Research*, **4**, 321–2.

Hogg, D.B. and Gutierrez, A.P. (1980) A model of the flight phenology of the beet armyworm (Lepidoptera: Noctuidae) in Central California. *Hilgardia*, **48**, 1–36.

Hokkanen, H. and Pimentel, D. (1984) New approaches for selecting biological control agents. *Canadian Entomologist*, **116**, 1109–21.

Kohel, R.J. and Lewis, C.F. (1984) *Cotton*, Agronomy Monograph, vol. 24, Soil Science Society, Inc., Madison, Wisconsin, p. 605.

Lemmon, H. (1986) Comax: an expert system for cotton crop management. *Science*, **233**, 29–33.

Matthews, G.A. (1989) *Cotton Insect Pests and their Management*, Longman Scientific and Technical, Harlow, Essex.

McKinion, J.M. and Lemmon, H. (1985) Expert systems in agriculture. *Computers and Electronics in Agriculture*, **1**, 31–40.

Pedigo, L.P., Hutchins, S.H. and Higley, L.G. (1986) Economic injury levels in theory and practice. *Annual Review of Entomology*, **31**, 341–68.

Plant, R.E. (1989) An integrated expert decision support system for agricultural management, *Agricultural Systems*, **29**, 49–66.

Regev, U. (1984) Man's addiction to pesticides, in *Strategies, Tactics and Policy Models*, (ed. G.R. Conway), Wiley-Interscience, Chichester.

Regev, U., Shalit, H. and Gutierrez, A.P. (1976) Economic conflicts in plant protection: the problems of pesticide resistance, theory and application to the Egyptian alfalfa weevil. *Proceeding of the Conference on Pest Management*, (eds G.A. Norton and C.S. Holling), IIASA, Laxenburg, Austria.

Regev, U., Gutierrez, A.P., Devay, J.E. and Ellis, C.K. (1990) Optimal strategies for management of verticillium wilt. *Agricultural Systems*, **33**, 139–52.

Shoemaker, C.A. (1980) The role of systems analysis in integrated pest management, in *New Technology of Pest Control*, (ed. C.B. Huffaker), Wiley, Chichester, pp. 24–49.

Sterling, W.L., Hartstack A.W. Jr, Dean, D.A., Shahed, S. and Burudgunte, R. (1993) *Texcim for Windows: the Texas Cotton Insect Model*. Texas Agricultural Experiment Station, College Station, Miscellaneous Publication #MP-1646 (revised).

Stern, V.M., van den Bosch, R. and Leigh, T.F. (1964) Strip cutting alfalfa for lygus bug control. *California Agriculture*, **18**(4), 4–6.

Stone, N.D. and Gutierrez, A.P. (1986a) Pink bollworm control in Southwestern desert cotton. I. A field oriented simulation of pink bollworm in south western desert cotton. *Hilgardia*, **54**, 1–24.

Stone, N.D. and Gutierrez, A.P. (1986b) Pink bollworm control in Southwestern desert cotton. II. A strategic management model. *Hilgardia*, **54**, 25–41.

Stone, N.D. and Toman, T.W. (1989) A dynamically linked expert-database system for decision support in Texas cotton production. *Computers and Electronics in Agriculture*, **4**, 139–48.

van den Bosch, R. (1978) *The Pesticide Conspiracy*, Doubleday and Company, Inc., New York.

Villacorta, A.M., Gutierrez, A.P., dos Santos, W.J. and Pizzamiglio, M.A. (1984) Analise do crescimento e desenvolvimento do algodoeiro no Parana: un modelo di simulaçao para a variedadi IAC 17. *Pesg. Agropec. Bras.*, **20**, 115–28.

Wang, Y.H., Gutierrez, A.P., Oster, G. and Daxl, R. (1977) A population model for cotton growth and development: coupling cotton–herbivore interaction. *Canadian Entomologist*, **109**, 1359–74.

Westphal, D.F., Gutierrez, A.P. and Butler, G.D. Jr (1979) Some interactions of the pink bollworm and cotton fruiting structures. *Hilgardia*, **47**, 177–90.

CHAPTER 12

Integrated pest management in protected crops

J.C. van Lenteren

12.1 INTRODUCTION

The total world area covered by greenhouses is about 200 000 ha. Developments in integrated pest management (IPM) in this cropping system have been unexpectedly fast and illustrate the great potential of alternatives to chemical methods. Greenhouses offer an excellent opportunity to grow high-quality products in large quantities on a small surface area. For example, in the Netherlands only 0.5% of the area in use for agriculture is covered with glasshouses (9300 ha). On this small area, 17% of the total value of agricultural production is realized. Few specialists in biological control anticipated being able to employ natural enemies in greenhouses because growing vegetables and ornamentals in this protected situation is very expensive and pest damage is not tolerated.

Greenhouse production requires well-trained, intelligent growers who cannot afford to risk any damage from insects for ideological reasons such as reduced environmental side effects compared with chemical control. If chemical control works better, they will certainly use it. In tomatoes, for example, pest control represents less than 2% of the total overall cost of production (Figure 12.1), so this is not a limiting factor (Oskam *et al.*, 1992). Yet despite the serious constraint that chemical control is comparatively simple and inexpensive, adoption of IPM has been remarkably quick in greenhouses. The main reason for use of biological control methods in the 1960s was the occurrence of resistance to pesticides in several key pests in greenhouses. Nowadays, other important stimuli include demands by policy makers for a reduction in usage of pesticides and consumers requiring production of residue-free food and flowers.

Integrated Pest Management Edited by David Dent. Published in 1995 by Chapman & Hall, London. ISBN 0 412 57370 9

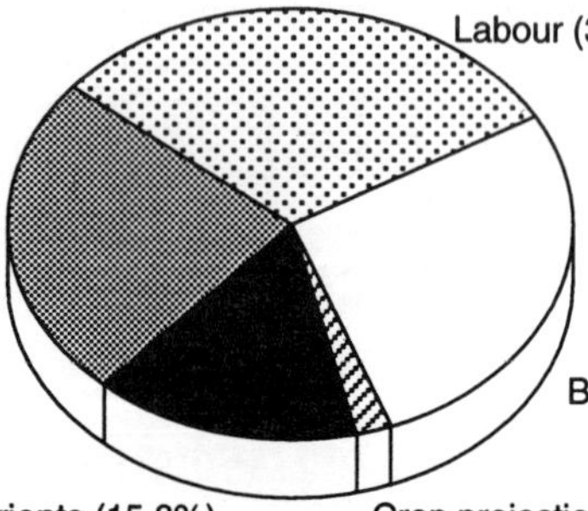

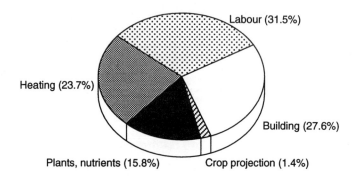

Figure 12.1 Percentage spent on pest control of the overall costs of producing tomatoes. The total costs of producing tomatoes in The Netherlands are about 25 US$ per square metre per season.

12.2 INITIATION OF IPM

IPM has been defined in many ways. The Food and Agriculture Organization (FAO) of the United Nations agreed on the following description: 'a pest population management system that utilises all suitable techniques in a compatible manner to reduce pest populations and maintains them at levels below those causing economic injury' (Smith and Reynolds, 1966). I consider the word pest to embrace animal pests, diseases and weeds in this paper, in accordance with FAO. Many IPM researchers consider the FAO definition too meagre, and it can easily be misused to defend slightly adapted conservative pest control programmes which are almost completely built on conventional chemical control. They opt for a description which contains philosophical and ecological elements in addition to the more technical aspects. An example of such a definition is the one developed by P. Gruys (personal communication 1976): a durable, environmentally and economically justifiable system in which damage caused by pests, diseases and weeds is prevented through the use of natural factors which limit the population growth of these organisms, if needed supplemented with appropriate control measures.

IPM is not a technology of the last 50 years. A number of methods to prevent or reduce pests has been in use since the evolution of agriculture (Table 12.1). The new aspects are: (i) that the IPM technology was developed in reaction to non-critical and superfluous application of chemical control; and (ii) the introduction of the concept of economic injury level. A first wave of IPM research, initiated by scientists who were concerned about the environmental effects of pesticides, took place between 1950 and 1970. Presently we experience a second wave of research interest, which is

Table 12.1 Methods to prevent or reduce development of pests

Prevention
• prevent introduction of new pests (inspection and quarantine)
• start with clean seed and plant material (thermal disinfection)
• start with pest-free soil (steam sterilization and solarization)
• prevent introduction from neighbouring crops
Reduction
• apply cultural control (crop rotation)
• use plants which are (partly) resistant to pests
• apply one of the following control methods:
– mechanical control (mechanical destruction of pest organisms)
– physical control (heating)
– control with attractants, repellents and anti-feedants
– control with pheromones
– control with hormones
– genetic control
– biological control
– (selective) chemical control
Control based on sampling and spray thresholds: guided or supervised control
Control based on the integration of methods which cause the least disruption of ecosystems: integrated control/IPM

now supported much more widely: policy makers, extension specialists and farmers have realized after a period of euphoria that there are limits to chemical pest control and that durable and safe production of food is possible only if alternatives for pesticides will become available.

IPM has received widespread acclaim since the 1950s as the only rational approach to providing long-term solutions to pest problems, but the rate of adoption of IPM by farmers has been slow during the first wave of IPM, because of problems with the transfer of IPM technology (Wearing, 1988). Now, IPM is strongly supported by governments in Europe (van Lenteren *et al.*, 1992a) and Asia (Ooi *et al.*, 1992), and is applied to vast areas. IPM in protected crops materialized in the Netherlands and the UK, and has since largely been a European affair.

12.3 THE GREENHOUSE ENVIRONMENT

In temperate zones, differences between greenhouse and field environments may partly explain the success of IPM in greenhouses. Greenhouses are relatively isolated units, particularly during the cold season. Before the start of a cropping period, usually during the winter, the greenhouse can be cleansed of pest organisms and subsequently kept free for several months. Later in the season, isolation prevents massive immigration of pest organisms. Furthermore, a limited number of pest species occurs in greenhouses, partly because of isolation and partly because not all pests specific to a certain crop have been imported into countries with greenhouses. Many greenhouse pests cannot survive in the field in winter or develop very slowly. This makes biological control easier because the natural enemies of only a few pest species have to be introduced. In addition, cultivars resistant to a number of diseases (viruses and fungi) had been developed already for the most important vegetable crops. As a result, chemical control of fungi - which may lead to high mortalities of natural enemies - does not have to be applied frequently. During the past 20 years growing crops on inert media instead of in the soil, has greatly decreased soil diseases and nematode problems.

Another factor easing implementation of IPM in protected crops is that cultural measures and pest management programmes can be organized for each separate greenhouse unit. Interference with pest management in neighbouring greenhouses is limited. The influence of pesticide drifts on natural enemies, which is a common problem in field crops, does not play a very important role here.

On the other hand, pest control is complicated by the virtually year-round culture of crops and by continuous heating during cold periods. These conditions provide excellent opportunities for the survival and development of a pest once it has invaded the greenhouse. Some field pests, e.g. spider mites, have adapted to the greenhouse climate by no longer reacting to diaspause-inducing factors. As a result, rates of population growth are often much higher than in the field. These complications do not, however, create specific problems for biological control. The greenhouse climate is managed within certain ranges, and this makes prediction of the population development of pest and natural enemies easier and more reliable than in field situations. The moment of introduction of natural enemies, the number and spacing of releases can be fine-tuned, resulting in season-long economic control.

The situation is more complicated in warm climates (van Lenteren *et al.*, 1992b). In the Mediterranean for example, many greenhouse frames are constructed of wood which harbour pests and diseases and are very difficult to clean. Growers in warmer climates are often less specialized than those of temperate zones and grow a diversity of crops on one holding, while at the same time some of the crops may also be present in the field. Most crops are grown in the soil which can lead to nematode and fungal problems. Often, little attention is paid to farm hygiene. Climate control is limited to opening and closing of the greenhouse, the use of shade screens or white-washing of the plastic. The mild climate outside enables pests to develop year-round and pest pressure is, therefore, very high. Ventilation leads to continuous migration of organisms into and out of the greenhouse; the pest and disease spectrum is much broader in this situation. On the other hand, numerous natural enemies which occur in the field can invade the greenhouse and exert natural control free of charge.

12.4 EMERGENCE OF IPM IN GREENHOUSES

One of the building blocks of IPM, biological control of pests, has been practised in greenhouses for more than 60 years. At the end of the 1920s black pupae were found in the UK among the normally white scales of the greenhouse whitefly, *Trialeurodes vaporariorum*. From the black pupae, parasites emerged that were identified as *Encarsia formosa* (Speyer, 1927). Within a few years a research station in England was supplying 1.5 million

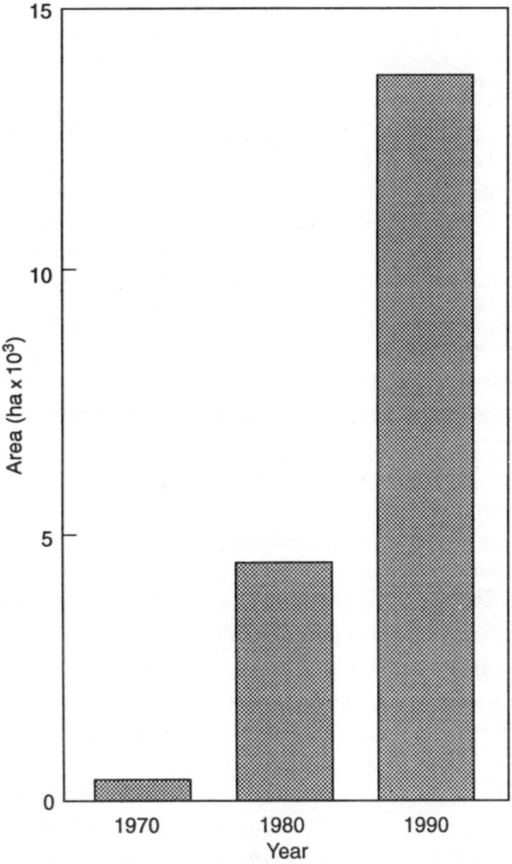

Figure 12.2 World-wide use of IPM in greenhouses.

of these parasites annually to about 800 nurseries in Britain. During the 1930s *E. formosa* was shipped to some other European countries, Canada, Australia and New Zealand.

After World War II, the distribution of *E. formosa* was discontinued because newly introduced insecticides provided convenient and efficient control on most greenhouse crops. A few years later, however, the first signs of resistance to a number of pesticides were observed in spider mites (*Tetranychus urticae*). Dosse (1959) received predatory mites from South

Table 12.2 Commercially produced natural enemies for control of greenhouse pests. (After van Lenteren *et al.*, 1992b, with additions)

Natural enemy	*Target pest*	*In use since*
Phytoseiulus persimilis	*Tetranychus urticae*	1986
Encarsia formosa	*Trialeurodes vaporariorum*	1970 (1926)
	Bemisia tabaci	1988
Opius pallipes	*Liriomyza bryoniae*	1980–1988*
Amblyseius barkeri	*Thrips tabaci*	1981–1990*
	Frankliniella occidentalis	1986–1990*
Dacnusa sibirica	*Liriomyza bryoniae*	1981
	Liriomyza trifolii	1984
	Liriomyza huidobrensis	1990
Diglyphus isaea	*Liriomyza bryoniae*	1984
	Liriomyza trifolii	1984
	Liriomyza huidobrensis	1990
Bacillus thuringiensis	Lepidoptera	1983
Heterorhabditis spp.	*Otiorrhynchus sulcatus*	1984
Steinernema spp.	Sciaridae	1984
Amblyseius cucumeris	*Thrips tabaci*	1985
	Frankliniella occidentalis	1986
Chrysoperla carnea	Aphids	1987
Aphidoletes aphidimyza	Aphids	1989
Aphidius matricariae	*Myzus persicae*	1990
Orius spp. (*c.* 5)	*Frankliniella occidentalis*	1991
Verticillium lecanii	Aphids	1992
Aphidius colemani	*Aphis gossypii*	1992
Aphelinus abdominalis	*Macrosiphum euphorbiae*	1992
Trichogramma spp.	Lepidoptera	1992
Leptomastix dactylopii	*Planococcus citri*	1992
Cryptolaemus montrouzieri		
Anthocorus nemorum	Thrips	1992
Metaphycus helvolus	Scales	1992
Trichoderma harzianum	*Fusarium* spp.	1992
Amblys. cucumeris/degenerans, non-diapausing strains	Thrips	1993
Eretmocerus californicus	*Bemisia tabaci*	1993
Metaseiulus occidentalis	*Tetranychus urticae*	1993
Hippodamia convergens	Aphids	1993
NPV-virus *Spodoptera*	*Spodoptera exigua*	1993

*Use terminated, other natural enemy available.

America and Bravenboer (1963) observed that these predators, *Phytoseiulus persimilis*, were effective in reducing the number of spider mites. Small-scale commercial production and application of biological control in greenhouses was started in 1968 with the use of the predatory mite *P. persimilis*. Interest in whitefly parasites increased at the start of the 1970s when enormous outbreaks of this pest took place in Europe. The knowledge of the availability of an efficient parasite paved the way for the development of a control programme. The revival of biological control in the 1970s became an established fact and since then the area of greenhouses under IPM has increased steadily (Figure 12.2).

12.5 THE PRESENT SITUATION

The natural enemies which have been selected, tested and introduced in programmes for commercial integrated pest control in protected crops in Europe are listed in Table 12.2. Biological control of the two key pests in greenhouses, whitefly (*Trialeurodes vaporariorum*) and spider mite (*Tetranychus urticae*), is now applied in more than 20 out of a total of 35 countries having a greenhouse industry. The increase in use of these natural enemies is shown in Figures 12.3 and 12.4. Details of the developments in this field can best be traced in the Proceedings of the Working Group on Integrated Control in Glasshouses of the International Organisation for Biological Control of Noxious Animals and Plants (Bulletins of the IOBC/WPRS from 1970 onwards; see e.g. van Lenteren, 1993 and Wardlow and van Lenteren,

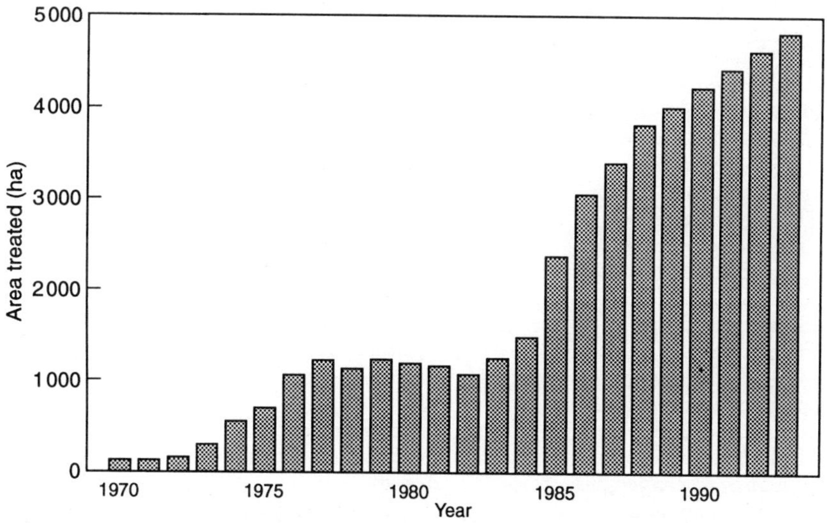

Figure 12.3 World-wide use (ha) of the parasite *Encarsia formosa*.

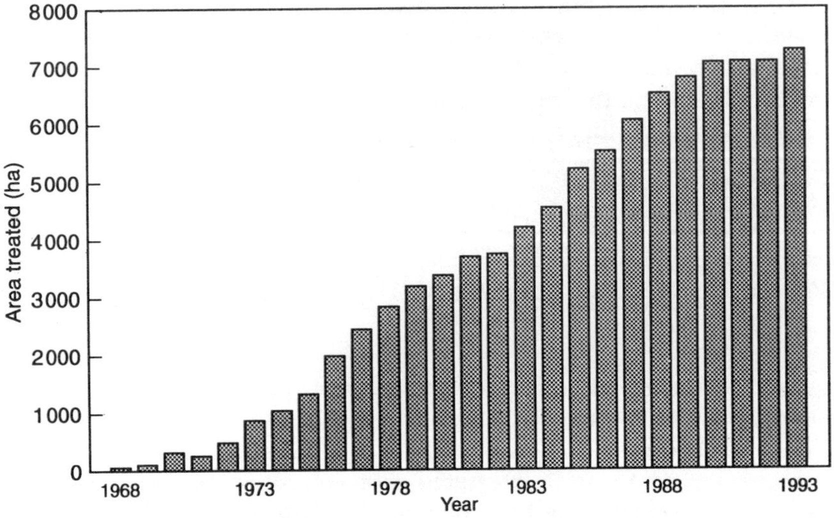

Figure 12.4 World-wide use (ha) of the predatory mite *Phytoseiulus persimilis*.

1993). Various natural enemies are being tested for use in protected crops (Table 12.3). The number of companies that produce natural enemies for use in greenhouses has increased from one in 1968 to more than 50 in 1993 (Figure 12.5). Europe has the largest greenhouse area under IPM, followed by the former USSR (Figure 12.6b). These two areas are proportionally over-

Table 12.3 Natural enemies in testing phase

Natural enemy	*Target pest*
Parasites, pathogens, predators	Thrips spp.
Parasites, pathogens, predators	Aphids
Parasites	Leafminers
Pathogens, nematodes	Soil pests, *Otiorrhynchus sulcatus* Sciaridae
Pathogens, *Aschersonia aleyrodis* and *Verticillium lecanii*	Whiteflies, *Bemisia tabaci*
Pathogens, *Metarrhizium anisopliae*	*Otiorrhynchus sulcatus*
Pathogens, *Bacillus thuringiensis*	Lepidoptera
Pathogens, *Bacillus thuringiensis* var. *israelensis*	Sciaridae
Mycoparasites/antagonists	Leaf pathogens
Mycoparasites/antagonists, *Streptomyces* and *Trichoderma* spp.	Soil pathogens

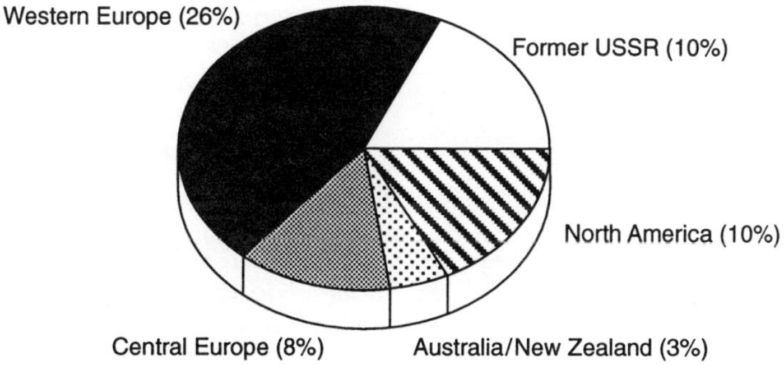

Figure 12.5 Number of producers of natural enemies for greenhouse biological control per region world-wide.

represented in the application of IPM, while all other areas are under-represented, particularly Asia and North Africa (Figure 12.6a,b).

Most of the natural enemies mentioned above are employed in IPM programmes, with differences in use of insecticides and natural enemies per crop and per country. Work of the IOBC/WPRS Working Group 'Pesticides and Beneficial Arthropods', has been elementary in selecting for those pesticides which interfere as little as possible with natural enemy activity (Hassan, 1992). If selective insecticides are not available, there are still other alternatives: (i) apply chemicals at a time when natural enemies are not seriously harmed (separation of application in time, or selective timing); and (ii) spray only the most seriously infested zones on individual plants or groups of plants (separation of application in space, or selective spraying). Careful guidance of the growers by producers of natural enemies and extension service personnel on the use of pesticides in combination with natural enemies is essential.

Another working group of the IOBC/WPRS, 'Breeding for Resistance to Insects and Mites', concentrates on the development of hostplant resistance. For those crops where the population development of the phytophagous insects is so fast that the parasite or predator cannot keep up with the pest, plant breeders search for partially resistant varieties so that population development of the pest is reduced. The first results of this were obtained with partial resistance in tomato and cucumber against whitefly and spider mite, respectively (de Ponti, 1982). An unexpected development from plant breeding research was to change morphological features of the hostplant in order to facilitate the searching for hosts by natural enemies. In cucumber, the number of hairs per unit of leaf area has been reduced through a breeding programme, leading to improved parasitization of whitefly by *E. formosa* (van Lenteren and de Ponti, 1990). When it was found that plants

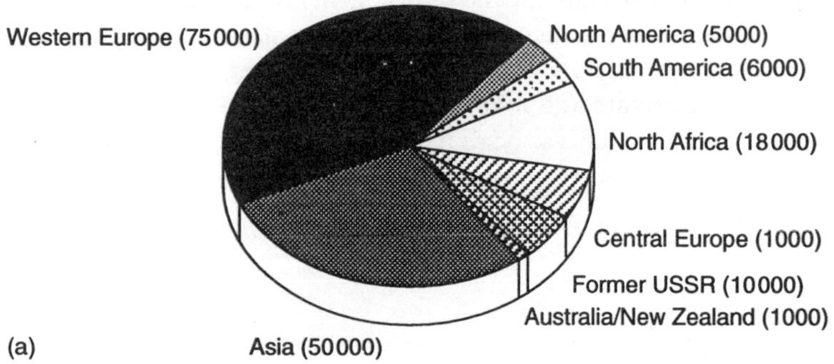

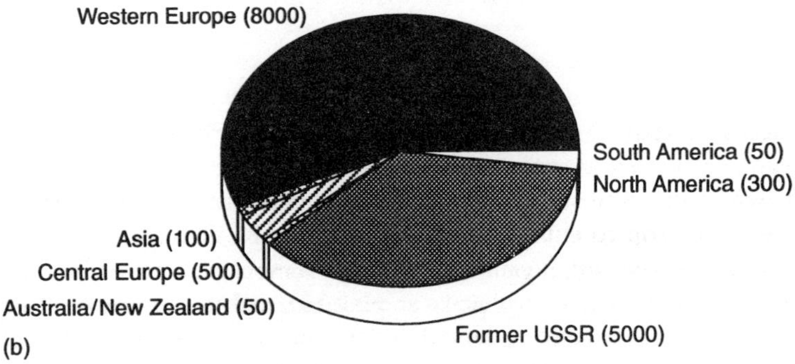

Figure 12.6 (a) World-wide greenhouse area in hectares per region; (b) areas (ha) under IPM.

may attract natural enemies after being attacked by a pest insect (Dicke *et al.*, 1990), interest in identifying variability in host-induced infochemicals increased. Plant breeders may select for those cultivars that best attract natural enemies after attack by a pest.

IPM can sometimes be stimulated by the use of insecticide-resistant strains of natural enemies. In greenhouse IPM, an organophosphorus-resistant strain of *P. persimilis* is used. Climate management to improve the performance of natural enemies and/or to decrease development of pests and diseases is already part of greenhouse IPM programmes. IPM was limited mainly to the control of insects until a few years ago. In North-west Europe as well as in Mediterranean countries, disease problems are considerable, particularly in tomatoes, cucumbers and cut flowers (Gullino, 1992; Fokkema, 1993). A number of fungicides integrate well with the use of natural enemies, but here problems of resistance are also strongly increasing. The lack of biological control agents of diseases is of major concern. Although use of fungicides remains substantial for foliar pathogens, disease management is now evolving towards strategies which rely on the use of resistant cultivars and manipulation of the environment. During the past decade several initiatives have led to research in non-chemical control, such as the effect of soil solarization on nematodes and fungi, the role suppressive soils may play in reduction of fungi and the potential use of antagonistic leaf fungi (Garibaldi and Gullino, 1991).

12.6 EXAMPLES OF IPM PROGRAMMES

Some examples of IPM programmes are given below. Chemicals used in such programmes vary among countries, depending on availability and registration. Therefore, names of chemicals are not given.

12.6.1 Tomato

The success in tomato crops is related to the rather simple pest and disease spectrum of this crop (van Lenteren and Woets, 1988). When tomatoes are grown in soil, soil sterilization by steaming is used shortly before planting the main crop to eliminate soil-borne diseases such as tomato mosaic virus (TMV), *Fusarium*, *Verticillium* and pests such as *Laconobia oleracea* (tomato moth) and three *Liriomyza* spp. (leafminers). Previously, cultivars lacking TMV resistance were inoculated as young plants with a mild strain of the TMV virus to make them less susceptible. Now, TMV-resistant cultivars are available. Furthermore, many tomato cultivars in West Europe are resistant to *Cladosporium* and *Fusarium*. Some cultivars are also tolerant to *Verticillium* and root-knot nematodes. Problems with soil diseases can also be strongly reduced by growing the crop in inert media, which has become common practice in West Europe. In tomatoes, therefore, only foliage pests and *Botrytis cinerea* require direct control measures.

The few pest organisms that 'overwinter' in greenhouses and survive soil sterilization are the greenhouse red spider mite (*T. urticae*) and the tomato looper (*Chrysodeixis chalcites*). Transferring young plants free of the other pest organisms into the greenhouse is important to prevent early pest development. For the last 10 years, the bulk of greenhouse tomatoes have been grown on rockwool systems, which makes soil sterilization redundant. As a result more organisms 'overwinter', such as *Liriomyza* spp. and their natural enemies, and *L. oleracea*. A recent development is the use of bumble bees for pollination, which gave an extra stimulus to the exploitation of biological control. Table 12.4 illustrates the tomato IPM programme.

Table 12.4 Commercially applied IPM programme for tomato crops

Pests and diseases	*Commercial IPM programme*
Trialeurodes vaporariorum	*Encarsia formosa*
Tetranychus urticae	Chemical control, *Phytoseiulus persimilis*
Liriomyza bryoniae, L. trifolii and *L. huidobrensis*	*Dacnusa sibirica, Diglyphus isaea* and natural control
Noctuid spp.	*Bacillus thuringiensis*
Aphid spp.	*Aphidoletes aphidimyza, Aphidius matricariae, Aphidius colemani* and natural control
Botrytis cinerea	Fungicides
Oidium lycopersicum	Fungicides
Bombus terrestris	Pollination

12.6.2 Cucumber

The situation is somewhat more complicated in cucumber crops than in tomato, because more pests and diseases occur. In addition to the pests mentioned under tomato, thrips and aphids create serious problems and fungal diseases are more common. Several aphid species can be controlled with the selective pesticide pirimicarb, but *Aphis gossypii*, which is resistant to pirimicarb, occurs frequently in cucumber. An effective natural enemy for control of cotton aphid, *Aphidius colemani*, was found recently. Biological control of thrips (*Frankliniella occidentalis* and *Thrips tabaci*) with *Amblyseius* spp. is still problematic, but presently *Orius* spp. are used with reasonable results. Chemical control of thrips on the plants with organophosphorus compounds precludes IPM, but the application of polybutenes mixed with deltamethrin, which can be applied topically on the plastic sheeting on the soil and which catches and kills thrips when leaving the plants, can be combined with natural and biological control of other pests. The IPM programme for cucumber is given in Table 12.5.

12.6.3 Sweet peppers

IPM in sweet pepper crops has long been problematical because natural enemies of thrips were not available. For some 5 years, predatory mites of the genus *Amblyseius* have shown to be able to control thrips, leading to a rapid increase in the use of the IPM programme as presented in Table 12.6.

Table 12.5 Commercially applied IPM programme for cucumber crops

Pests and diseases	*Commercial IPM programme*
Trialeurodes vaporariorum	*Encarsia formosa* and chemical control
Tetranychus urticae	*Phytoseiulus persimilis*
Thrips tabaci	*Amblyseius* spp. *Orius* spp.
Frankliniella occidentalis	*Amblyseius* spp. *Orius* spp.
Aphis gossypii	*Aphidius colemani*
Other aphid spp.	*Aphidius matricariae, Aphidoletes aphidimyza,* natural and chemical control
Noctuid spp.	*Bacillus thuringiensis*
Botrytis cinerea	Fungicides
Didymella bryoniae	Fungicides
Sphaerotheca fuliginea	Fungicides

Table 12.6 Commercially applied IPM programme for sweet pepper crops

Pests and diseases	*Commercial IPM programme*
Tetranychus urticae	*Phytoseiulus persimilis* and chemical control
Thrips spp.	*Amblyseius cucumeris, Orius insidiosus*
Myzus persicae, Aphis gossypii	*Aphidius matricariae, A. colemani, Aphidoletes aphidimyza*
Macrosiphum euphorbiae	*Aphelinus abdominalis, Aphidoletes aphidimyza*
All aphid spp.	Also natural control and chemical control
Noctuid spp.	*Bacillus thuringiensis* and chemical control
Liriomyza spp.	Natural control and see table for tomato
Tarsonemid mites	Chemical control
Mirid bugs	Chemical control
Botrytis cinerea	Fungicides
Leveilluta taunica	Fungicides
Rhizoctonia solani	Fungicides

12.6.4 Egg plants (aubergines)

IPM has been implemented in egg plants only recently. The main causes for this slow development were: (i) the rich pest spectrum; and (ii) the very rapid development of several pests demanding frequent introductions of natural enemies. In combination those factors led to costs for biological control that were prohibitive. After the discovery that bumble bees were very efficient and cheap pollinators, IPM gained quick acceptance in egg plant (Table 12.7). Bumble bees can only be used in combination with non-chemical pest control. The reduced costs of pollination make it possible for the grower to compensate for the relatively high costs of biological control in this crop. Generally, natural enemies have to be released more often and/or in greater numbers in egg plant.

Table 12.7 Commercially applied IPM programme for egg plant crops

Pests and diseases	*Commercial IPM programme*
Tetranychus urticae	*Phytoseiulus persimilis*
Trialeurodes vaporariorum	*Encarsia formosa* and chemical control
Thrips spp.	See Table 12.5 for cucumber
Aphid spp.	See Table 12.5 for cucumber
Leafminer spp.	See Table 12.4 for tomato
Noctuid spp.	*Bacillus thuringiensis*
Bombus terrestris	Pollination

12.6.5 Flowers

Development of IPM for ornamentals is much more complicated than for vegetables. The first problem is that many more different species and cultivars of ornamentals are grown than of vegetables. In the Netherlands for example, 110 species of cut flowers and 300 species of pot plants are cultivated, and some 150 cultivars of chrysanthemum are produced (Fransen, 1993). Other problems for implementation of IPM in ornamentals are that: (i) more pesticides are available than for vegetables; (ii) the whole plant is marketed, instead of only the fruits, so no leaf damage is allowed; and (iii) a zero-tolerance is applied to export material.

Although IPM programmes are available for flower crops, they are applied on a very limited scale and usually only for non-export flowers. Hesselein *et al.* (1993) and Wardlow (1993) give examples of IPM programmes for chrysanthemums produced for the home market. Fransen (1993) provides information on commercial use of biological control in cut flowers (e.g. bouvardia, chrysanthemum, cymbidium and gerbera) and ornamentals (e.g. ficus, palms and poinsettia), totalling some 65 ha in 1992. She also summarized the natural enemies which are being considered for biological control of pests in several ornamental crops (Fransen, 1992). An example of an IPM programme for gerbera is given in Table 12.8. Worldwide, only a few hundreds of hectares of ornamentals are under IPM.

In experimental glasshouses of institutes, botanical gardens, zoos and in office buildings ('interior landscapes'), where somewhat higher pest levels are tolerated than on crops and certain pesticides cannot be used, quite complicated IPM programmes are used, for example, in the USA, Canada, the UK and The Netherlands.

12.7 NEW ASPECTS OF IPM IN PROTECTED CROPS

A number of recent developments in IPM will be summarized in this section.

Table 12.8 Commercially applied IPM programme for gerbera crops

Pests and diseases	*Commercial IPM programme*
Trialeurodes vaporariorum	*Encarsia formosa, Verticillium lecanii*
Bemisia tabaci	*Encarsia formosa, Verticillium lecanii*
Tetranychus urticae	*Phytoseiulus persimilis* and chemical control
Frankliniella occidentalis	*Amblyseius cucumeris, Orius insidiosus* and chemical control
Aphid spp.	*Aphidius colemani, Aphidius matricariae, Aphidoletes aphidimyza*, natural and chemical control
Leafminer spp.	*Dacnusa sibirica, Diglyphus isaea* and natural control
Noctuid spp.	*Bacillus thuringiensis* and chemical control
Spodoptera exigua	NPV-virus *Spodoptera*
Mirid bugs	Chemical control
Mildew spp.	Chemical control (sulphur)
Botrytis cinerea	Fungicides

12.7.1 Disease control

Although resistant cultivars to soil and foliar diseases are available for many vegetables, resistance breeding in ornamental crops is less advanced. By using soil solarization, a technique based on solar heating of the soil by covering it with transparent plastic film, soil-borne pathogens, weeds and other pests are being controlled effectively and economically in many sub-tropical countries (Gullino, 1992). Soil sterilization by solarization has important advantages over chemical soil disinfection; it is cheaper and has a low hazard to the user and his environment.

Soil suppressiveness could provide another good opportunity for control of soil-borne diseases, but as yet has not advanced to practical use. Organisms isolated from suppressive soils can already be used against some soil-borne pathogens, such as the antagonist bacterium *Streptomyces griseoviridis* (Mycostop), which is registered in several European countries for use against *Fusarium* wilt and seedling blight both in flowers and in cucurbits. Its mode of action is based on antibiotic effects. The microbe colonizes the rhizosphere before the pathogens, and secretes antibiotic substances which inhibit the growth of fungal pathogens (Lahdenperä *et al.*, 1990). It can be applied as seed dressing or in the soil.

In future, use of antagonistic *Fusarium* spp. and fluorescent pseudomonads active against *Fusarium oxysporum* will permit biological control of *Fusarium* wilts. Also the use of *Trichoderma* spp. as seed dressing or soil treatment will provide control of damping off (*Pythium* spp) and root rot (*Phytophthora* spp., *Rhizoctonia solani*) (Gullino, 1992). In tomatoes, crown and root rot (*Fusarium oxysporum*) can be controlled with *Trichoderma harzianum* (van Steekelenburg, 1992).

Foliar diseases can often be reduced by proper climate regulation, and use of computers to control temperature, light, humidity, water, ventilation, carbon dioxide and nutrition has resulted in improved disease management. Manipulation of the interactions of temperature and humidity is most important but also very costly to achieve. It is easier to carry out in modern glasshouses than in plastic houses or tunnels which are mainly used in the sultropics, resulting in disease-prone situations in the plastic structures. Therefore, growers rely heavily on fungicides in such situations. Widespread resistance of foliar pathogens to fungicides has resulted in demands for alternatives (Gullino, 1992), but no biological control agents are registered. *Trichoderma* spp.-based products are currently under development.

12.7.2 Open rearing units or 'banker plants' with natural enemies

When natural enemies need to be continuously available over long periods, the provision of banker plants may be more economic than frequent inundative releases. The idea behind the banker plant system is that non-crop plants are placed in the greenhouse with a host insect (not a pest of the target crop) that serves as food and for reproduction of parasites or predators. In greenhouse IPM, the banker plant system is used in several crops:

1. In sweet pepper, to control aphids, particularly *Myzus persicae*. The banker plants are broad bean plants (*Vicia faba*), infested with vetch aphids (*Megoura viciae*) and the predator *Aphidoletes aphidimyza*. Early placement of the banker plants in the crop guarantees the presence of a large predator population when the pest occurs, and control over a long period. Additionally, no time consuming checks of the crop are needed (Stengard Hansen, 1983).
2. In cucumbers, to control *Aphis gossypii*. Banker plants, comprising wheat or barley seedlings infested with bird-cherry aphids, *Rhopalosiphum padi*, enhance the early establishment of *Aphidius colemani* and *Aphidoletes aphidimyza* in advance of infestation of the cucumbers with *A. gossypii*. This system gives effective and prolonged control of the cotton aphid (Bennison and Corless, 1993).
3. In lettuce, to control leafminers. The banker plants are *Ranunculus asiaticus* with the leafminer *Phytomyza caulinaris* as alternative host for the parasites *Dacnusa sibirica* and *Diglyphus isaea*. These parasites control *Liriomyza huidobrensis* in lettuce (van der Linden, 1993)

12.7.3 Multiple introductions of natural enemies

Biological control in glasshouses was initially based on seasonal inoculative introductions of natural enemies shortly after the first pest insects had been observed with the aim to obtain season-long control (van Lenteren and Woets, 1988). In several countries, multiple inundative introductions are made, whether pests are seen or not (Wardlow and O'Neill, 1992). This is possible because of diminished costs of natural enemies. Such programmes can be compared with standard prophylactic chemical control, remove the need to monitor for pests, and hence reduce uncertainty.

Diapause manipulation of natural enemies

In temperate zones, some of the indigenous natural enemies used in greenhouses enter diapause in response to photoperiod, and may therefore be inactive during parts of the year when the pests they are targeted against are active. There are several possibilities for extending the seasonal limits of natural enemies:

1. The provision of low-intensity lights to create a long photoperiod can be used to keep natural enemies from diapausing in winter, e.g. with *Aphidoletes aphidimyza*, thus resulting in good aphid control (Gilkeson and Hill, 1986).
2. The selection for non-diaspausing strains of natural enemies, e.g. in the predatory mite *Amblyseius cucumeris* to obtain sufficient thrips control in winter (Morewood and Gilkeson, 1991; van Houten and van Stratum, 1993).
3. The collection of a geographical strain of the natural enemy with a different critical photoperiod, e.g. the predator *Orius tristicolor*, populations from California had much shorter critical daylengths than populations of Canada and use of Californian predators can be a useful approach to extend the seasonal limits of thrips control (Gillespie and Quiring, 1993).

12.7.4 Climate manipulation

Early use of entomopathogenic fungi often gave improper results because the humidity appeared to be too low for proper germination of the fungi (Fransen, 1990). New work has shown that cycling of 2 nights' elevated humidity with 2 nights' ambient conditions was adequate for reliable control of several insect pests without resulting in increased problems with plant pathogenic fungi (Helyer, 1993).

The pepper weevil, *Anthonomus eugenii* (Coleoptera: Curculionidae) is a recent, accidental introduction into Canada. It could be eradicated locally by maintaining growing temperatures between crops, instead of chilling the greenhouse. At high temperatures and without food, the adults died quickly (Costello and Gillespie, 1993).

Climate manipulation is essential for proper disease management. With monitoring of within-crop relative humidity, coupled with an automatic response in heating and/or ventilation, *Botrytis* problems can be prevented or reduced. Many long-season crops of heated tomatoes are now grown without a single fungicide application for control of diseases. The development of permeable thermal screens has reduced the risk of long periods with high humidity and thus, the incidence of white rust (*Puccinia horiana*) in chrysanthemums. Proper management of the relative humidity reduces or prevents attack of cucumber by *Didymella bryoniae* and roses by *Botryosphaeria rhodina*. Irrigation at, or below soil level rather than from overhead systems prevents spread of rusts (*Puccinia* and *Uromyces*) in cut flowers.

12.7.5 Use of pheromones

Pheromones were tested for monitoring, mass trapping and confusion at the start of the 1980s for control of the tortricid *Clepsis spectrana* (van den Bos, 1983), and more recently with the pheromone of *Chrysodeixis chalcites*. Results were disappointing in most of these experiments, suggesting that the artificial greenhouse situation with very different air current patterns does result in unnatural pheromone distribution patterns to which the target insects cannot react properly. In particular, the long-distance reactions are disrupted.

12.7.6 Hostplant resistance

Selecting plants resistant to diseases is a method commonly used by private breeding companies, particularly for diseases of vegetables. Breeding for hostplant resistance against insects and mites is a rather recent approach in greenhouse crops. Tomatoes accessions resistant to whiteflies (de Ponti *et al.*, 1990) and cucumbers accessions resistant to spider mites (de Ponti, 1982) are available. Fairly high levels of resistance to *Frankliniella occidentalis* have been found in cucumber accessions (Mollema *et al.*, 1993). In chrysanthemums resistance to leafminers (van Dijk *et al.*, 1993) and *Frankliniella occidentalis* (de Jager and Butöt, 1993) is under study.

12.7.7 Use of bumble bees for pollination

The quick rise in use of bumble bees for pollination strongly stimulated application of biological control as chemical pesticides could not be integrated with the use of bees. Since bumble bees save labour compared with hand pollination, growers are willing to pay more for biological control. In situations where costs for biological control were a limiting factor, it has now become an interesting option.

12.8 HOW IMPLEMENTATION OF IPM HAS BEEN REALIZED

IPM was not recognized as such until 1950. Organic pesticides were hardly available before that period and many different control techniques were combined. Cultural control, hostplant resistance and biological control were important aspects of the overall activities to reduce pests and diseases. Interest in integrated control developed shortly after the appearance of the synthetic pesticide resistance and the recognition of unwanted side effects. This led to the formation of the International Organisation for Biological Control (IOBC). The European section of the IOBC has been the driving force behind a change of thinking in crop protection since, and has coordinated many cooperative IPM projects (for an overview see van Lenteren *et al.*, 1992a), including those in greenhouses.

The successful IPM programmes for greenhouses have a number of characteristics in common, including:

1. Their use was promoted only after a complete IPM programme had been developed covering all aspects of pest and disease control for a crop.
2. An intensive support of the IPM programme by the extension service during the first few years.
3. The total costs of crop protection in the IPM programme were not higher than in the chemical control programme.
4. Resources for non-chemical control methods (like natural enemies, resistant plant material) were as easily available, as reliable, as constant in quality and as well guided as chemical agents.

Technically, implementation of IPM is not different from that of chemical control. At the introduction of the first IPM programme for a new crop, special attention should be paid to extension. The degree of knowledge makes acceptance of complicated IPM programmes initially difficult for the farmer. IPM methods demand a different attitude based on the principle of introducing a natural enemy or pesticide only when the pest insect is present and expected to lead to economic loss. A misconception is that such a practice is adopted readily if it is superior to current ones. Only when the IPM method is *perceived* to be better than conventional methods will it be adopted by growers (Wilson, 1985). All participants in an IPM programme must be receptive to new developments and willing to implement them. Quite often IPM research was done in isolation at university departments and thus, a great deal of IPM work remained 'ivory tower' research. When growers, extension workers and researchers agree that the IPM programme is as cheap as chemical control and that production and delivery of alternative control methods is reliable, IPM can be applied in a similar way as chemical control and becomes a normal commercial affair.

Before we, in The Netherlands, start developing IPM for a new crop the following is considered: can we develop a complete IPM programme in which chemical pesticides do not interfere negatively with non-chemical means of control? If the answer to this question is positive, a meeting is organized with researchers, extension workers, producers of natural enemies and representatives of growers' organizations, and questions are discussed on how to attack the new pest or disease problem and in what way to divide the work.

Usually the research starts at a university department with a literature search, followed by a collection of resistant plant material, different natural enemies, etc. Elements of an IPM programme are first tested in the laboratory or in small experimental greenhouses, followed by tests at local research stations, and finally experiments with the complete IPM programme are run in commercial greenhouses. All participants in a project work together with the extension service, on an extension programme. Progress reports are published regularly in growers' journals. Our growers have learned to rely on IPM and now ask for new natural enemies before we can provide them with the necessary information. This enthusiasm might, however, create a new problem: too early a release of a natural enemy can result in a bad control effect and thus in a negative advertisement for IPM!

12.8.1 Role of the extension service

The situation with extension in IPM is very complex, since each country tends to have a different system. Farmers obtain information on crop production and crop protection from many different sources. The most important (as listed by Dutch growers) for the greenhouse situation are in order of descending importance:

1. Specialized growers' journals
2. Growers' own voluntary study groups
3. Specialized agricultural exhibitions
4. Regional research stations
5. Private and state extension service.

Information related to IPM comes mainly from categories 5, 2 and 1. The producers of resistant plant material, natural enemies and other alternative control agents play an important role under category 5. The universities and the research institutes of the Ministry of Agriculture are missing from this list. Their information is usually provided to the state extension service. Further, they use the specialized growers' journals to publish their information. The high ranking of these journals indicates the value of this source.

An implication of introduction of IPM is a 2-year period of intensive attention to the growers by the extension service. Also the teaching of crop protection has had to be drastically changed at all levels (from vocational schools to university) during the last two decades. Essentially purely technical information was taught on how to spray and with what chemicals. This has now partly been replaced with information on other forms of pest control especially IPM. Students currently learn much more about pests *and* beneficial organisms.

The importance of IPM specialists in the extension service in order to realize implementation in greenhouses is well documented in the literature (Wearing, 1988; Wardlow, 1992; Beck *et al.*, 1993; Murphy and Broadbent, 1993). Besides, information to individual growers, group meetings of farmers to instruct, discuss and further evolve IPM, based on their experiences seem to be particularly effective. Although most countries have some kind of extension service, adequate expertise in IPM is often lacking.

12.9 FACTORS LIMITING THE INTRODUCTION OF IPM

During the past three decades many countries have invested public money for the development of non-chemical control methods. This section will try to identify the reasons why so few of these methods have been extensively used.

12.9.1 Technical limitations

Firstly, there are situations where the application of biological control is unnecessary or impossible for a variety of reasons:

- some crops (e.g. lettuce) are grown during too short a period to make IPM an economic investment;
- in other crops, like ornamentals which are grown on about 50% of the greenhouse area, there is a zero tolerance for pest organisms. This makes biological control and IPM difficult to apply. However, experience with IPM in vegetables indicates that the common a priori belief that chemically treated crops contain fewer pest organisms than biologically controlled crops is incorrect;
- pests may occur which cannot (yet) be controlled by natural enemies or selective insecticides, therefore requiring the application of broad-spectrum insecticides. If there is a large probability that such a pest will occur, the control of which will upset biological control, growers are not interested in applying IPM for other pests.

Secondly, factors may hamper use of IPM in crops where it seems feasible:

- the quantity and quality of natural enemies or other resources for IPM at arrival in the greenhouse is insufficient;
- the service that growers obtain from the producer of agents for IPM and/or from extension personnel may be insufficient. For example, growers may lack information on the best moment to introduce a natural enemy or on pesticides suitable for integration. If a grower begins to use IPM, the quality and intensity of the initial guidance determines the success of the programme. Later, guidance can be relaxed. These drawbacks will not necessarily always cause failure, but amateurism in production of natural enemies and guidance has more than once had a negative influence on application of IPM. Governmental institutions demanding certain standards of performance for insecticides should apply similar standards for natural enemies.

Thirdly, IPM may be impeded by a group of various factors:

- new compounds on the pesticide market create difticulties in the application of biological control. Usually the negative effects of new pesticides on natural enemies are not evaluated before these pesticides replace older ones. Such pesticides may totally disrupt a well-balanced IPM programme, e.g. in the mid-1970s when pyrethroids came to the market. This situation is expected to change as governments may soon stipulate the need for data concerning the effects of pesticides on natural enemies as part of applications for registration of new pesticides;
- accidental importation of new pest organisms and difficulties in pests in plant material at quarantine facilities may result in rapid changes in the pest complex. Most pests in greenhouses are imported (van Lenteren *et al.*, 1987); such new imports result in panic actions and initially in extensive chemical control programmes, which upset the IPM programme until proper natural enemies have been found. Better inspection and quarantine facilities would reduce unwanted importations;
- governmental and inter-governmental bodies may provide insufficient support with respect to legislation on pesticides; broad-spectrum pesticides should not be approved for those crops for which reliable and economically attractive IPM programmes have been developed. Furthermore, information on pesticide labels about hazardous effects on natural enemies should help to reduce mistakes in the application of IPM programmes;
- the limited amount of research and the poor education of most extension personnel concerning IPM hampers further introduction. The results obtained in non-chemical pest control are, of course, in the first instance dependent on the amount of research and development work. Funding of this work is limited, especially if one realizes the complications of this type of research, e.g. registration and legislation procedures for natural enemies such as arthropod predators and parasites should be much less demanding than those for microbial natural enemies like pathogens, because the former are inherently less hazardous than the latter. The presently very expensive legislation procedures are prohibitive for registration of biological control agents.

12.9.2 Non-technical limitations

(a) Farmers' attitudes

Until very recently, only few farmers (organizations) asked for, or stimulated, development of non-chemical control methods. The adoption of insecticides was rapid because they allowed the farmer to decide when and where they should be used. Decision criteria were clear, the method was easily understood, it was effective (at least in the short term), reduced labour costs, and it was a practice the farmer could control and decide upon independently of his neighbours, institutions or agencies. Initially it was a straightforward technology. In contrast, integrated control is more complicated because of the requirement for the monitoring of various pests, the integration of different control methods and situation specific prescriptions. The latter systems require a degree of knowledge and sophistication much greater than pesticide technology demands (Perkins, 1982).

Initiatives for development of IPM programmes came and must still come from researchers and policy makers. Being unable to control a pest with chemicals is a stronger reason for farmers to change their ideas on IPM than ideological reasons. As soon as farmers realized that chemical control is no longer sufficient for complete control, their interest in an integrated approach was generated. We should not reproach the farmer for not being interested in IPM, because governments legislate the use of chemicals and often state that when chemicals are used as advised, they do not contaminate food or the environment and do not harm plants, animals or humans. Currently, the attitude of several groups of farmers is changing. European fruit growers and producers of greenhouse vegetables, for example, have experienced the positive aspects of integrated control and seriously worry about the increasing public concern on pesticide usage. Therefore, they now generally prefer to use IPM methods (van Lenteren *et al.*, 1992b).

(b) The viewpoint of the chemical industries

In general, we can state that any complication in a simple, straight chemical pest control programme is appreciated as a negative development by the large industries. Alternatives such as biological and genetic control not only complicate chemical control programmes, but also seem to be unattractive commercially because of a combination of: (i) the impossibility of patenting natural enemies; (ii) complicated mass production; (iii) short shelf-life; (iv) specificity (too small market); and (v) different and more complicated guidance for growers.

Chemical industries will not start the production of other than broad-spectrum pesticides on their own initiative, unless the use of those pesticides is prohibited or when pest organisms substantially develop resistance. We cannot blame the chemical industry for this attitude because their goal is to make a profit. The industry provides pesticides which are allowed for use by a government's legislation and registration policy.

(c) Role of the governments

It is the governmental bodies who are in fact the only ones able to change the pest control picture through measures that make some kinds of chemical control less attractive or impossible (by modification of the pesticide legislation policy, by measures concerning registration, taxation, side-effect labelling or fines for incorrect use of pesticides, etc.), and by stimulating other control methods (through incentives for environmentally friendly techniques, by funding research, but above all by teaching on all levels in order to change ideas about pest control, and improvement of the extension service). It is a rather bizarre situation that public money is used for the development of alternatives for chemical control when, at the same time, their application is often not encouraged due to the overall presence of (too) cheap broad-spectrum pesticides.

How policy decisions can positively influence implementation of IPM was recently illustrated in The Netherlands. Because of the intensive use of pesticides, the Dutch government has decided to opt for a strong reduction in their usage during this decade (Anon., 1991). The Dutch Ministry of Agriculture has developed a long-term task-defining policy – the Multi-Year-Crop Protection Plan – which aims at a 50% reduction in pesticide use by the year 2000, expressed in terms of kilograms of active ingredients. Reports have been published with strategic plans for each sector in agriculture, containing details on how to realize the reduction.

For the glasshouse sector, this reduction in pesticide usage can be achieved as follows:

1. Through legislation, a complete cessation of preventive chemical soil disinfection will be realized. This will result in a ca. 50% reduction of pesticide usage for the whole glasshouse sector. Realistic alternatives for chemical soil disinfection are already in use or are in an advanced stage of development, e.g. soil steaming and production on a substrate other than soil. During the production season, chemical soil treatment will be allowed only when diseases or pests are present. Better detection techniques will be developed and more attention will be given to breeding for disease and pest-resistant plants.
2. Plants are increasingly grown on substrates other than soil. Far smaller quantities of (systemic) pesticides need to be added to plants via the nutrient solutions than with soil-grown crops (van Steekelenburg, 1992). This results in a 90% reduction in the use of these pesticides.
3. Disinfecting methods for above-ground parts of the glasshouse will be developed using less pesticides. Spraying water under high pressure seems a viable option. In this way a reduction in pesticide use of 10% is realistic.
4. Screening of greenhouses with insect-proof material minimizes infection pressure, especially of young plants, delaying plant protection measures.
5. Biological and integrated pest and disease management will be used more widely for control of above-ground pests and diseases. Several options are applied or in development: designing more realistic damage thresholds, development of guided/supervized control programmes, breeding for pest and disease-resistant cultivars, selection and use of new natural enemies, optimization of fertilization, proper disposal of old plant material, more critical choice of new crops (taking sensitivity for pests and

diseases into account), and stricter control on the importing of plant material infected by pests and diseases. A combination of the above listed measures will lead to a further reduction in pesticide use of ca. 5% in glasshouse vegetables (where a drastic reduction has already taken place because of the biological control systems in use) and 10% in glasshouse ornamentals.

For glasshouse crops a total reduction of 65% is anticipated. The reasons behind this option of severely reducing pesticide use are diverse in origin. They originate partly from the agricultural sector, but partly also from environmental concerns. In addition to the actions at the Ministry of Agriculture, the Dutch Ministry of Environment is developing new guidelines for the admission of environmentally alien substances, which will result in an exclusion of 50 to 70% of the pesticides used today.

12.10 FACTORS AFFECTING IPM IMPLEMENTATION AND PESTICIDE USE

Besides removing limitations as mentioned above, a number of other factors may increase acceptance and use of IPM.

12.10.1 Grower-operated and grower-funded IPM

Active participation by growers in development and implementation of IPM often results in a quick decrease of pesticide use and acceptance of IPM by a wider grower community than when IPM is advocated only by scientists or extension personnel. Such a growers' participation approach led to reductions in pesticide use of 80% of active ingredient applied in chrysanthemums and 30% in poinsettias over a 3-year period (1989–1991) in Canada (Murphy and Broadbent, 1993).

12.10.2 Labels for merchandise produced under integrated protection

Several European supermarkets and canning industries are growing and marketing produce under their own responsibility and regulations. Under these regulations, pesticide reductions have been obtained which already surpass the demands of governments set for the year 2000. A special label identifies the products. Sometimes growers themselves take the initiative for such production systems. Following extensive discussions among growers, the organization of vegetable and fruit growers in the Netherlands has created a marketing policy based on 'environmentally conscious production'. Under this production, growers must comply with rules concerning IPM and their products are labelled as such.

12.10.3 Change of legislation and registration procedures

In several countries laws or registration procedures are changed such that chemical pesticides are no longer allowed for use in those crops for which reliable, economically acceptable non-chemical control methods are available.

12.10.4 Provision of information on alternative means of pest control

As a result of the policies to reduce use of chemicals, pesticides have been blacklisted or their use was severely restricted. Information on alternatives was often not easily available and the reliability of alternatives could not always be judged. To improve this situation, the Dutch Ministries of Environment and Agriculture have funded projects to develop databases containing all effective crop-protection measures: hostplant resistance, biological, cultural, other non-chemical and chemical (van der Baan *et al.*, 1991; Oomen, 1992). Such databases can be used not only for improving IPM, but also for strategic goals to decide whether enough alternatives are available for withdrawing the registration of certain pesticides.

12.10.5 Closed production systems

Closed production systems are being developed for vegetables and ornamentals. Not all problems have been solved yet and particularly problems with diseases deserve extra attention, but reductions in total pesticide use in greenhouses are in the order of 50% or more. Further development of closed production systems is therefore strongly stimulated.

12.10.6 Hostplant resistance

In ornamentals, hostplant resistance against pests and diseases does not play an important role in the choice of cultivars to grow, as the grower still expects chemical control to be the solution to all his problems (Fransen, 1992). Hostplant resistance is present in lines of many ornamental species against a whole range of pests in the original wild plants, but many marketable cultivars are very susceptible. Chrysanthemum cultivars have shown large differences in susceptibility to leafminers and thrips. Rose cultivars differ in their susceptibility to powdery mildew (*Sphaerotheca pannosa*), and carnations to *Fusarium oxysporum* and *Phialosphora cinerescens*. It is not necessary to opt for absolute resistance in all cases, because partial resistance may be sufficient in combination with biological control or results in a reduced spray frequency.

12.10.7 Availability of decision-support systems

Development of decision-support systems for crop protection helps the grower to make the best choices from all options available for pest and disease control. As crop protection becomes rather complicated, such systems are essential in guiding the grower. They are being developed for sweet pepper (van Steekelenburg, 1992) and for cucumber (Shipp *et al.*, 1993).

12.10.8 Reporting pesticide usage, monitoring and use of spray thresholds

Variation in pesticide use is not always the result of differences in local pest and disease situations, but also relates to the 'spray-attitude' of the grower. Pesticide use among 15 chrysanthemum growers varied between 23.4 and 71.0 kg active ingredient/ha/year, and among seven gerbera growers it varied between 11.6 and 41.3 kg/ha/year (Fransen, 1992). Awareness of considerable quantitative differences in pesticide use and the financial savings of reduced spraying stimulates a change from preventive sprays to usage of spray thresholds. As a result of such a reporting programme for chrysanthemum and poinsettia in Canada, the pumber of spray applications decreased from 136 to 55 between 1989 and 1991 (Murphy and Broadbent, 1993). A similar Dutch project in chrysanthemum resulted in an overall reduction in pesticide use from 30% to 50% (Fransen, 1992). A next step is to train the growers in recognition of pests, diseases and signs of damage, so that preventive sprays are no longer applied on a calender basis, but on the presence of the pest organisms. The ability to recognize pests, in combination with the use of spray thresholds is crucial if reductions in the use of pesticides are to be achieved.

12.11 SPECIFIC ADVANTAGES OF IPM IN PROTECTED CROPS

Why do greenhouse growers use IPM? There are, of course, the general advantages of biological control such as reduced exposure of producer and applier to toxic pesticides, the lack of residues on the marketed product and the low risk of environmental pollution. More important are the specific reasons that make growers working in greenhouses prefer particularly the biological control aspect of IPM:

- With biological control there are no phytotoxic effects on young plants, and premature abortion of flowers and fruit does not occur.
- Release of natural enemies takes less time and is more pleasant than applying chemicals in humid and warm greenhouses.
- Release of natural enemies usually occurs shortly after the planting period when the grower has plenty of time to check for successful development

of natural enemies; thereafter the system is reliable for months with only occasional checks; chemical control requires continuous attention.
- Chemical control of some of the key pests is difficult or impossible because of pesticide resistance.
- With biological control there is no safety period between application and harvesting fruit; with chemical control one has to wait several days before harvesting is allowed again.
- Biological control is cheaper than chemical control.
- Biological control is permanent: once a good natural enemy – always a good natural enemy.
- Biological control is appreciated by the general public.

12.12 FROM IPM TO INTEGRATED FARMING

Agricultural research has seen a large diversification in this century. In plant breeding the search for better-yielding varieties has been the central theme for years. Plant production concentrated on finding and overcoming limiting factors in plant growth and crop yields. Until very recently plant resistance against pests was not considered to be important, because most pests could easily be controlled with insecticides. This has led to plant varieties becoming less resistant to insects. In addition, over-fertilization resulted in plants which stimulate pest development (e.g. aphids, leafminers, spider mites).

Particularly during the past 50 years, pest control has been seen as an independent area in agriculture (Table 12.9). The viewpoint was that those working in this area could easily present solutions to protect the maximally producing crops. This reductionist approach to agricultural research has resulted in negative feedbacks leading to an increased use of pesticides and inorganic fertilizers. The situation is already changing: pest control is no longer seen as a completely independent subject, now that we know that solutions in chemical control are not unlimited and that pesticides may cause serious side effects. Still, much remains to be done and a more holistic approach, through a strong integration of the research areas of plant breeding/production and crop protection, will prevent developments in one area which may lead to the development of environmentally unattractive activities in the other.

A first step towards the reduction of environmental problems is to test new developments within the framework of a farming systems approach. Entomologists, plant pathologists and weed specialists would work together with agronomists and plant breeders in such an approach. The focal point in the farming systems approach is farm economics in the form of maximizing net income, which is not synonymous with yield maximization. Top yields are obtained currently with excessively high inputs of fertilizers and pesticides. Reducing the inputs may lead to somewhat lower yields, but

Table 12.9 Attitudes in conventional and integrated greenhouse farming

Conventional production: attitude increase production	
• Increase production	• Apply chemical control
• Increase quantity and quality through: breeding, irrigation, fertilization and climate management	• No problem, cheap and effective chemical control was available
↓	↓
Import of products: All plant species welcome, but often with unwanted pests, diseases, resulting in more chemical control	*Export of products:* Blemish and pest free products zero tolerance intensive chemical control needed
Integrated production: attitude minimize negative effects, optimize profits	
• Increase production but not all costs	• Integrated pest management
• Evaluate effects of production increase measures on development of pests and diseases	• Use of non-chemical control strongly reduced use of pesticides
↓	↓
Import of products: Better quarantine and inspection, do not import plants with a rich pest and disease spectrum	*Export of products* Products free of living pest and disease organisms

financial inputs are also lower and the net income may be the same or even better. In farming systems, ecological and environmental effects such as pollution of soil and water by pesticides, can be minimized. In general, integrated farming takes more completely into account the various impacts on ecosystems (preservation of flora and fauna, quality and diversity of landscape, and the conservation of energy and non-renewable resources) as well as sociological considerations (employment, public health and well-being of persons associated with agriculture) than is the case with current farming. An integrated farming system is: 'a coherent farm unit or set of units which relies basically on cultural and biological inputs, with chemicals as integrated supplements'. The main aims are: 'to minimize inputs of non-renewable resources and to provide a better balance between adequate production of yields and farm income on the one hand and ecological, environmental and sociological aims on the other. All these considerations must be compatible with cost effectiveness.' (Wijnands and Vereijken, 1992).

The practices which can be manipulated in such programmes are crop rotation, cultivation, fertilization, pesticide use, cultural control measures, biological control and other alternatives to conventional chemical control. For the greenhouse situation there are specific target areas for integrated farming (Table 12.9): (i) species or cultivars prone to many diseases and pests should not be imported or grown; (ii) the zero-tolerance demand has to be relaxed; (iii) consequences of production-increasing factors on crop protection should be evaluated; (iv) partial disease and pest-resistant plants should be exploited; and (v) quarantine and inspection need improvement to prevent new pest imports.

12.13 THE FUTURE OF IPM

Initially due to resistance problems, we were forced to look for pest control methods other than chemical control. Intensive cooperation between researchers, extension workers, producers of natural enemies and growers has led to considerable success, both in research and application of IPM. In some countries IPM is practised on a large part of the main vegetable crops in greenhouses. Growers have learned to rely on biological control and now ask for new natural enemies before we can provide them with the necessary information. To date, we can safely conclude that IPM in greenhouses has been very successful.

However, a number of conditions have to be met before the technical implementation of IPM becomes a success. Biological control agents should be cheap, easily available, reliable, constant in quality, and as well supervized as chemical control agents. They should fit into the total crop-protection programme well and not be seen as an endeavour separate from other crop-protection measures.

Several current trends will further stimulate the application of IPM in greenhouses. Firstly, fewer new insecticides are becoming available because of escalating costs for development and registration. The few new insecticides that are being developed are not likely to be targeted for greenhouse use because the greenhouse area is small and represents a poor opportunity for chemical companies to recover developmental costs.

Secondly, the recent general use of bumble bees and honey bees for pollination on a large greenhouse acreage, strongly reduces possibilities for chemical control and intensifies demands for IPM.

Thirdly, pests continue to develop resistance to insecticides, a particularly prevalent problem in greenhouses where intensive management and repeated insecticide applications exert strong selective pressure on insects. Therefore we expect a greater demand for new pest control methods.

We should not see IPM as a control strategy that will completely replace chemical control. It is a powerful option and can be applied on a much larger area than at present. For the chemical industry, IPM may result in extended use of products because of slower development of resistance and a more positive perception by laymen of the role of the pesticide industry. In order to serve agriculture as well as the environment and human health, we should deploy the best prevention and reduction strategies to develop effective IPM methods. Designing such environmentally safer IPM programmes is a challenge for our profession.

ACKNOWLEDGEMENTS

The author thanks N. van Steekelenburg and P.J.M. Ramakers (both Research and Experiment Station for Glasshouse Crops), J.J. Fransen (Research and Experiment Station for Floriculture), and Brinkman and Koppert BV (producers of natural enemies) for providing information for this article.

REFERENCES

Anon. (1991) Multi-Year Crop Protection Plan. Government Decision 1991. Ministry of Agriculture, Nature Management and Fisheries.

Beck, N.G., Martin, N.A. and Workman, P.J. (1993) IPM for greenhouse crops in New Zealand: grower acceptance. *Bulletin IOBC/WPRS*, **16**(2), 1–4.

Bennison, J.A. and Corless, S.P. (1993) Biological control of aphids on cucumbers: further development of open rearing units or "banker plants" to aid establishment of aphid natural enemies. *Bulletin IOBC/WPRS* **16**(2), 5–8.

Bravenboer, L. (1963) Experiments with the predator *Phytoseiulus riegeli* Dosse on glasshouse cucumbers. *Mitteilungen der Schweizerischen Entomologischen Gesellschaft*, **36**, 53.

Costello, R.A. and Gillespie, D.R. (1993) The pepper weevil, *Anthonomus eugenii* Cano as a greenhouse pest in Canada. *Bulletin IOBC/WPRS*, **16**(2), 31–4.

de Jager, C.M. and Butöt, R.P.T. (1993) Thrips (*Frankliniella occidentalis* (Pergande)) resistance in chrysanthemum; the importance of pollen as nutrition. *Bulletin IOBC/WPRS*, **16**(5), 109–15.

de Ponti, O.M.B. (1982) Resistance to insects: a challenge to plant breeders and entomologists. *Proceedings 5th International Symposium on Insect–Plant Relationships*, (eds J.H. Visser and A.K. Minks), Pudoc, Wageningen, pp. 337–48.

de Ponti, O.M.B., Romanov, L.F. and Berlinger, M.J. (1990) Whitefly–plant relationships: plant resistance, in *Whiteflies: their Bionomics, Pest Status and Management*, (ed. D. Gerling), Intercept, Andover, pp. 91–106.

Dicke, M., Sabelis, M.W., Takabayashi, J., Bruin, J. and Posthumus, M.A. (1990) Plant strategies of manipulating predator–prey interactions through allelochemicals: prospects for application in pest control. *Journal of Chemical Ecology*, **16**, 3091–118.

Dosse, G. (1959) Über einige neue Raubmilbenarten (Phytoseiidae). *Pflanzenschutzberichte*, **21**, 44–61.

Fokkema, N.J. (1993) Opportunities and problems of control of foliar pathogens with micro-organisms. *Journal of Pesticide Science*, **37**, 411–16.

Fransen, J.J. (1990) Natural enemies of whiteflies: fungi, in *Whiteflies: their Bionomics, Pest Status and Management*, (ed. D. Gerling), Intercept, Andover, pp. 187–210.

Fransen J.J. (1992) Development of integrated crop protection in glasshouse ornamentals. *Journal of Pesticide Science*, **36**, 329–33.

Fransen, J.J. (1993) Integrated pest management in glasshouse ornamentals in the Netherlands: a step by step policy. *Bulletin IOBC/WPRS*, **16**(2), 35–8.

Garibaldi, A. and Gullino, M.L. (eds) (1991) Proceedings IOBC/WPRS Working Group *Integrated Control in Protected Crops under Mediterranean Climate*, 29 September–2 October 1991, Alassio, Italy. *Bulletin IOBC/WPRS*, **14**(5), 215 pp.

Gilkeson, L.A. and Hill, S.B. (1986) Diapause prevention in *Aphidoletes aphidimyza* (Rondani) (Diptera: Cecidomyiidae) by low intensity light. *Environmental Entomology*, **15**, 1067–9.

Gillespie, D.R. and Quiring, D.M.J. (1993) Extending seasonal limits on biological control. *Bulletin IOBC/WPRS*, **16**(2), 43–6.

Gullino, M.L. (1992) Integrated control of diseases in closed systems in the subtropics. *Journal of Pesticide Science*, **36**, 335–40.

Hassan, S.A. (1992) Guidelines for testing the effects of pesticides on beneficial organisms: description of test methods. *Bulletin IOBC/WPRS*, **15**(3), 186 pp.

Helyer, N. (1993) *Verticillium lecanii* for control of aphids and thrips on cucumber. *Bulletin IOBC/WPRS*, **16**(2), 63–6.

Hesselein, C., Robb, K., Newman, J., Evans, R. and Parrella, M. (1993) Demonstration/integrated pest management program for potted Chrysanthemums in California. *Bulletin IOBC/WPRS*, **16**(2), 71–6.

Lahdenperä, M.L., Simon, E. and Uoti, J. (1990) Mycostop – a novel biofungicide based on *Streptomyces* bacteria, in *Biotic Interactions and Soil-borne Diseases*, (ed. A.B.R. Beemster), Elsevier, Amsterdam, pp. 258–63.

Mollema, C., Steenhuis, M.M., Inggamer, H. and Soria, C. (1993) Evaluating the resistance of *Frankliniella occidentalis* in cucumber: methods, genotypic variation and effects upon thrips biology. *Bulletin IOBC/WPRS*, **16**(5), 77–82.

Morewood, W.D. and Gilkeson, L.A. (1991) Diapause induction in the thrips predator *Amblyseius cucumeris* (Acarina: Phytoseiidae) under greenhouse conditions. *Entomophaga*, **36**, 253–63.

Murphy, G.D. and Broadbent, A.B. (1993) Development and implementation of IPM in greenhouse floriculture in Ontario, Canada. *Bulletin IOBC/WPRS*, **16**(2), 113–16.

Ooi, P.A.C., Lim, G.S., Ho, T.H., Manalo, P.L. and Waage, J.K. (eds) (1992) *Integrated Pest Management in the Asia-Pacific Region*, CAB International, Wallingford.

Oomen, P.A. (1992) Chemicals in integrated control. *Journal of Pesticide Science*, **36**, 349–53.

Oskam, A.J., van Zeijts, H., Thijssen, G.J., Wossink, G.A.A. and Vijftigschild, R. (1992) *Pesticide Use and Pesticide Policy in the Netherlands*, Pudoc, Wageningen.

Perkins, J.H. (1982) *Insects, Experts, and the Insecticide Crisis. The Quest for New Management Strategies*, Plenum Press, New York.

Shipp, J.L., Clarke, N.D., Jarvis, W.R. and Papadopoulos, A.P. (1993) Expert system for integrated crop management of greenhouse cucumber. *Bulletin IOBC/WPRS*, **16**(2), 149–52.

Smith, R.F. and Reynolds, H.T. (1966) Principles, definitions and scope of integrated pest control. *Proceedings FAO Symposium on Integrated Pest Control*, Rome, 1965. *FAO*, **1**, 11–17.

Speyer, E.R. (1927) An important parasite of the greenhouse white-fly (*Trialeurodes vaporariorum*, Westwood). *Bulletin of Entomological Research*, **17**, 301–8.

Stengard Hansen, L. (1983) Introduction of *Aphidius aphidimyza* (Rond.) (Diptera: Cecidomyiidae) from an open rearing unit for the control of aphids in glasshouses. *Bulletin IOBC/WPRS*, **6**(3), 146–50.

van de Baan, H.E., Cuijpers, T.A.M.M., van Lenteren, J.C. and Sabelis, M.W. (1991) Pestbase: a relation database for the evaluation of alternatives for environmentally harmful pesticides. *Danish Journal of Plant and Soil Science*, **85**, 57–62.

van den Bos, J. (1983) The isolating effect of greenhouses on arthropod pests: a case study on *Clepsis spectrana* (Lepidoptera: Tortricidae). PhD Thesis, Wageningen Agricultural University.

van Dijk, M.J., de Jong, van der Knaap, J.C.M. and van der Meijden, E. (1993) The interaction between *Liriomyza trifolii* and different chrysanthemum cultivars. *Bulletin IOBC/WPRS*, **16**(5), 101–8.

van Houten, Y.M. and van Stratum, P. (1993) Biological control of Western Flower Thrips in greenhouse sweet peppers using non-diapausing predatory mites. *Bulletin IOBC/WPRS*, **16**(2), 77–80.

van Lenteren, J.C. (ed.) (1993) *Proceedings Working Group 'Integrated Control in Glasshouses', 25–29 April 1993*, Pacific Grove, California, USA. *Bulletin IOBC/WPRS*, **16**(2), 192 pp.

van Lenteren, J.C. and de Ponti, O.M.B. (1990) Plant-leaf morphology, host-plant resistance and biological control. *Proceedings 7th International Symposium on Insect–Plant Relationships, 3–8 June 1989*, Budapest, Hungary, Akadémiai Kiadó, Budapest. *Symposia Biologia Hungarica*, **39**, 365–86.

van Lenteren, J.C. and Woets, J. (1988) Biological and integrated pest control in greenhouses. *Annual Review of Entomology*, **33**, 239–69.

van, Lenteren, J.C., Woets, J., Grijpma, P., Ulenberg, S.A. and Minkenberg, O.P.J.M. (1987) Invasions of pest and beneficial insects in the Netherlands. *Proceedings of the Royal Dutch Academy of Sciences, Series C*, **90**, 51–8.

van Lenteren, J.C., Minks, A.K. and de Ponti, O.M.B. (eds) (1992a) *Biological Control and Integrated Pest Management: Towards Environmentally Safer Agriculture*, Pudoc, Wageningen.

van Lenteren, J.C., Benuzzi, M., Nicoli, G. and Maini, M. (1992b) Biological control in protected crops in Europe, in *Biological Control and Integrated Crop Protection: Towards Environmentally Safer Agriculture*, (eds J.C. van Lenteren, A.K. Minks and O.M.B. de Ponti), Pudoc, Wageningen, pp. 77–89.

van der Linden, A. (1993) Development of an IPM program in leafy and tuberous crops with *Liriomyza huidobrensis* as a key pest. *Bulletin IOBC/WPRS*, **16**(2), 93–5.

van Steekelenburg, N.A.M. (1992) Novel approaches to integrated pest and disease control in glasshouse vegetables in the Netherlands. *Journal of Pesticide Science*, **36**, 359–62.

Wardlow, L.R. (1992) The role of extension services in integrated pest management in glasshouse crops in England and Wales, in *Biological Control and Integrated Crop Protection: Towards Environmentally Safer Agriculture*, (eds J.C. van Lenteren, A.K. Minks and O.M.B. de Ponti), Pudoc, Wageningen, pp. 193–9.

Wardlow, L.R. (1993) Integrated pest management techniques in protected ornamental plants. *Bulletin IOBC/WPRS*, **16**(8), 149–57.

Wardlow, L.R. and van Lenteren, J.C. (eds) (1993) *Proceedings Working Group 'Integrated Control in Glasshouses'*, 8–11 September 1992, Cambridge, UK. *Bulletin IOBC/WPRS*, **16**(2), 192 pp.

Wardlow, L.R. and O'Neill, T.M. (1992) Management strategies for controlling pests and diseases in glasshouse crops. *Journal of Pesticide Science*, **36**, 341–7.

Wearing, C.H. (1988) Evaluating the IPM implementation process. *Annual Review of Entomology*, **33**, 17–38.

Wijnands, F.G. and Vereijken, P. (1992) Region-wise development of prototypes of integrated arable farming and outdoor horticulture, in *Biological Control and Integrated Pest Management: Towards Environmentally Safer Agriculture*, (eds J.C. van Lenteren, A.K. Minks and O.M.B. de Ponti), Pudoc, Wageningen, pp. 125–38.

Wilson, T.F. (1985) Estimating the abundance and impact of arthropod natural enemies in IPM systems, in *Biological Control in Agricultural IPM Systems*, (eds M.A. Hoy and D.C. Herzog), Academic Press, New York, pp. 303–22.

Index